92
Structure and Bonding

Editorial Board:

M. J. Clarke · J. B. Goodenough

C. K. Jørgensen · D. M. P. Mingos · G. A. Palmer

P. J. Sadler · R. Weiss · R. J. P. Williams

Springer-Verlag Berlin Heidelberg GmbH

Less Common Metals in Proteins and Nucleic Acid Probes

Volume Editor: M. J. Clarke

With Contributions by
C. B. Allan, G. Davidson, J. Figlar,
W. R. Harris, J. M. Kelly,
A. Kirsch De Mesmaeker, M. J. Maroney,
C. Moucheron, S. J. Nieter Burgmayer

 Springer

In references Structure and Bonding is abbreviated
Struct.Bond. and is cited as a journal.

Springer WWW home page: HTTP://www.springer.de

ISSN 0081-5993

ISBN 978-3-662-14751-1   ISBN 978-3-540-69667-4 (eBook)
DOI 10.1007/978-3-540-69667-4

CIP Data applied for

This work is subject to copyright. All rights are reserved, whether the whole or
part of the material is concerned, specifically the rights of translation, reprinting,
reuse of illustrations, recitation, broadcasting, reproduction on microfilm or in
any other way, and storage in data banks. Duplication of this publication or parts
thereof is permitted only under the provisions of the German Copyright Law of
September 9, 1965, in its current version, and permission for use must always be
obtained from Springer-Verlag. Violations are liable for prosecution under the
German Copyright Law.

© Springer-Verlag Berlin Heidelberg 1998
Originally published by Springer-Verlag Berlin Heidelberg New York in 1998
Softcover reprint of the hardcover 1st edition 1998

The use of registered names, trademarks, etc. in this publication does not imply,
even in the absence of a specific statement, that such names are exempt from the
relevant protective laws and regulations and therefore free for general use.

Typesetting: Scientific Publishing Services (P) Ltd, Madras
Cover: Medio V. Leins, Berlin
SPIN: 10552944   66/3020 – 5 4 3 2 1 0 – Printed on acid-free paper

Volume Editor

Prof. M. J. Clarke
Merkert Chemistry Center
Boston College
2609 Beacon St. Chestnut Hill
Massachusetts 02167-3860, USA
E-mail: Clarke@bc.edu

Editorial Board

Prof. Michael J. Clarke
Merkert Chemistry Center
Boston College
2609 Beacon St. Chestnut Hill
Massachusetts 02167-3860, USA
E-mail: Clarke@bc.edu

Prof. Christian K. Jørgensen
Départment de Chimie Minerale
de l'Université
Section de Chimie – Sciences II
30 quai Ernest Ansermet
CH-1211 Genéve 4, Switzerland

Prof. Graham A. Palmer
Department of Biochemistry
Wiess School of Natural Sciences
Rice University
P.O. Box 1892
Houston, Texas 77251, USA
E-mail: GRAHAM@TAFFY.RICE.EDU

Prof. Raymond Weiss
Institut Le Bel
Laboratoire de Cristallochimie
et de Chimie Structurale
4, rue Blaise Pascal
F-67070 Strasbourg Cedex, France
E-mail: weiss@chimie.u-strasbg.de

Prof. John B. Goodenough
Center of Materials Science and Engineering
University of Texas at Austin
Austin, Texas 78712, USA
E-mail: jgoodenough@mail.utexas.edu

Prof. David M. P. Mingos
Chemistry Department
Imperial College of Science
Technology and Medicine
South Kensington
London SW7 2AY, Great Britain
E-mail: d.mingos@ic.ac.uk

Prof. Peter J. Sadler
Department of Chemistry
The University of Edinburgh
Joseph Black Chemistry Building
King's Building, West Mains Road
Edinburgh EH9 3JJ, Great Britain
E-mail: P.J.Sadler@ed.ac.uk

Prof. Robert J. P. Williams
Inorganic Chemistry Laboratory
University of Oxford
Oxford OX1 3QR, Great Britain
E-mail: susie.compton@icl.ox.ac.uk

Preface

Bioinorganic chemistry encompasses both naturally occurring metal ions and the use of metal complexes as probes of biological systems. Some of the more interesting elements in the chemistry of life are the less commonly occurring ones such as nickel and molybdenum. This volume elucidates the chemistry of these elements in important enzymes and also explores the chemistry of new probes of biological structure and function.

In proteins, Ni may serve as either a Lewis acid or redox center. Nature has evolved ways of acquiring, transporting and incorporating Ni into a wide variety of proteins, including: ureases, hydrogenases, carbon monoxide dehydrogenases, acetyl coenzyme A synthases, methyl coenzyme M reductases and superoxide dismutases. In his chapter on nickel proteins, Mike Maroney provides a survey of the bioinorganic chemistry of this element beginning with its uptake in bacteria. Drawing on recent crystal structures of Ni proteins, he then affords a better understanding of Ni-active sites at the mechanistic level.

Pterins display multi-electron redox reactivity in biological systems. They are also curiously tethered to Mo in a number of vital enzymes. When non-innocent pterins combine with transition metal ions, which also display multi-electron redox activity, the result challenges the chemist's traditional concepts of oxidation state. Sharon Burgmayer explores the chemistry of the pterin complexes with Mo and other elements (Fe, Cu, Ru) along with the chemistry of the closely related flavins in an effort to determine the possible interactions between these multi-redox active partners in biological systems.

Transferrin normally transports iron in the blood, but in serum is only partially saturated with iron. This enables this protein to also bind and transport significant concentrations other metal ions. Wes Harris's review emphasizes metals for which transferrin appears to play a significant role in serum transport. These include highly toxic metal ions such as Pu^{4+} and others such as Ga^{3+} and In^{3+}, whose biodistribution is important in the chemistry of radiopharmaceutical agents.

Targeting DNA with transition metal probes has become an area of widespread research interest. Much of this has been engendered by the longer-lived excited states displayed by some metal polypyridyl complexes when bound to DNA. Conversely, for other polypyridyl complexes, the excited state may be quenched by electron transfer involving the nucleobases. Such photochemical interactions can result in oxidative damage, photocleavage and the formation of covalent photoadducts between the metal complexes and the nucleobases. John Kelly considers design strategies necessary to generate molecular DNA photoprobes for DNA sequences or structures of particular biological interest.

Taken as a whole, this volume illustrates new directions for traditional protein biochemists, molecular biologists endeavoring to discover new ways to probe DNA, and those engaged in the design of metallopharmaceuticals.

Michael J. Clarke

Contents

Contents of Volume 88

Metal Sites in Proteins and Models
Iron Centres

Volume Editor. P. J. Sadler

Contents of Volume 89

Metal Sites in Proteins and Models

Phosphatases, Lewis Acids and Vanadium

Volume Editors: H.A.O. Hill, P.J. Sadler, A.J. Thomson

Contents of Volume 90

Metal Sites in Proteins and Models
Redox Centres

Volume Editors: H.A.O. Hill, P.J. Sadler, A.J. Thomson

Contents of Volume 91

Bioorganic Chemistry
Trace Element Evolution from Anaerobes to Aerobes

Volume Editor: R.J.P. Williams

The Structure and Function of Nickel Sites in Metalloproteins

Michael J. Maroney*, Gerard Davidson, Christian B. Allan, James Figlar

Department of Chemistry and Program in Molecular and Cellular Biology, University of Massachusetts, Amherst, MA 01003–4510, USA
E-mail: mmaroney@chem.umass.edu

The known Ni enzymes now include ureases, hydrogenases, carbon monoxide dehydrogenases, acetyl coenzyme A synthases, methyl coenzyme M reductases and superoxide dismutases. These enzymes employ Ni in roles varying from Lewis acid sites to involvement in redox catalysis. In addition, mechanisms involving proteins that specifically bind Ni have evolved for the acquisition, transport, and incorporation of Ni. Various biophysical techniques have been used to elucidate aspects of the structure and function of the Ni sites in these enzymes and proteins. This pursuit has been greatly advanced since 1995 by the availability of crystal structures of *Klebsiella aerogenese* urease, various hydrogenases, and two states of methyl coenzyme M reductase from *Methanobacterium thermoautotrophicum*. The knowledge of the structure and function of the various Ni proteins and enzymes is briefly overviewed. Proposals for catalytic mechanisms based on these studies and on model chemistry are presented.

Keywords: Urease; hydrogenase; carbon monoxide dehydrogenase; acetyl coenzyme A synthase; methyl coenzyme M reductase; superoxide dismutase; cofactor F_{430}

*Corresponding author

1
Introduction

The study of specific roles for Ni in biology began with the discovery in 1975 that the active site of urease contained specifically bound Ni atoms [1]. This enzyme was already well-known, for in 1926 it became the first enzyme to be crystallized [2]. The nearly 50-year gap between these discoveries illustrates the problems of detecting Ni(II) via common spectroscopic techniques and likely helped to perpetuate the myth that Ni is not a biologically important metal. Since 1975, a number of enzymes have been shown to involve Ni as part of the active site [3–6]. These enzymes now include urease [7, 8], carbon monoxide dehydrogenases (CODHs) [9, 10], acetyl coenzyme A synthase (ACS) [9, 11], hydrogenases (H_2ases) [12, 13], methylcoenzyme M reductase (MCR) [14–16], and the recently discovered superoxide dismutases (SODs) from *Streptomyces* [17, 18] as well as a number of proteins that are involved in the acquisition, transport, and incorporation of Ni into these enzymes [19, 20]. Thus, a rich biochemistry of Ni has evolved, featuring Ni in enzymes involved with hydrolytic and redox chemistry, and involving specific systems for constructing biological Ni centers [21]. This article will survey the knowledge of the structure and function of biological Ni metallocenters through mid-1997, with an emphasis on structural information obtained from single crystal X-ray diffraction studies and from spectroscopic techniques, and results obtained since these subjects were last reviewed (as indicated by the references above).

2
Ni Proteins with Non-Redox Roles

2.1
Ni Acquisition, Transport, and Metallocenter Assembly

Organisms that express nickel metalloenzymes have developed sophisticated systems of accessory proteins that are involved in the acquisition and transport of Ni, as well as proteins that are involved in the biosynthesis of the Ni active sites and maturation of the proteins [21]. From studies of several bacterial systems, Ni transport and incorporation systems have been shown to contain similar features. There are no crystal structures of Ni transport proteins or Ni proteins involved in metallocenter assembly, and only a few such proteins have been isolated in quantities sufficient for spectroscopic studies.

The acquisition of Ni involves transport of the metal across the cytoplasmic membrane. Nonspecific transport systems include various low-specificity Mg^{2+} transport systems that also take up Ni^{2+} [22, 23]. In addition, several Ni-specific transport systems have been characterized. When *Escherichia coli* is cultured anaerobically, it expresses an ATP-dependent Ni-specific transport system [24, 25]. This system is encoded by the *nik* operon, which is composed of five genes, *nikA – E*. The protein NikA is a 56-kDa periplasmic protein that specifically binds Ni. One atom of Ni is bound per protein with a K_d of less than 0.1 μM [26]. Other metals (e.g., Co, Cu, Fe) bind to apo NikA protein less tightly by at least one order of magnitude. NikB and NikC are integral membrane proteins, and NikD and NikE contain ATP-binding sites. *Nik* mutants that are defective in Ni transport are unable to synthesize any of the three Ni,Fe H_2ases found in *E. coli* [27, 28]. In addition, NikA appears to be involved in the Tar-dependent negative chemotaxis associated with higher concentrations of Ni^{2+}, which are toxic to both eukaryotic and prokaryotic organisms [29].

The structure of the Ni center in periplasmic Ni-binding protein NikA has been examined from the analysis of data from Ni K-edge X-ray absorption spectroscopy (XAS) [30]. The X-ray absorption near edge structure (XANES) (Fig. 1) reveals that the pre-edge peak involving a $1s{\rightarrow}3d$ electronic transition is very small, indicating a centrosymmetric coordination environment (e.g., octahedral or square planar). The absence of a peak assigned to a $1s{\rightarrow}4p_z$ transition (with shakedown contributions) rules out a square planar geometry. The intensity of the white line (the most intense feature of the spectrum) is indicative of a ligand environment composed of hard donor atoms. The analysis of the extended X-ray absorption fine structure (EXAFS) data is consistent with the XANES analysis. The data may be fit with a shell of 6 oxygen atoms at 2.06(2) Å. An additional peak in the Fourier-transformed spectrum can be fit by a heavier scattering atom at a longer distance. For example, an S-donor atom at 2.57(2) Å is accommodated by the data. The analysis of EXAFS data cannot generally distinguish between O- and N-donor ligands. However, in contrast to the UreE Ni binding sites (see below), the data show little evidence of scattering arising from second and third coordination sphere C,N atoms that are expected if the ligand environment were composed largely of histidine imidazole groups (Fig. 1). Although ligation by a single histidine cannot be ruled out, the data are consistent with a ligand environment composed of multiple O-donor ligands. This result is consistent with the amino acid content of the protein, which shows that 12% of the amino acids present are potential carboxylate ligands. The average Ni-O distance determined is typical of a high-spin six-coordinate complex, including those with carboxylate ligation [31].

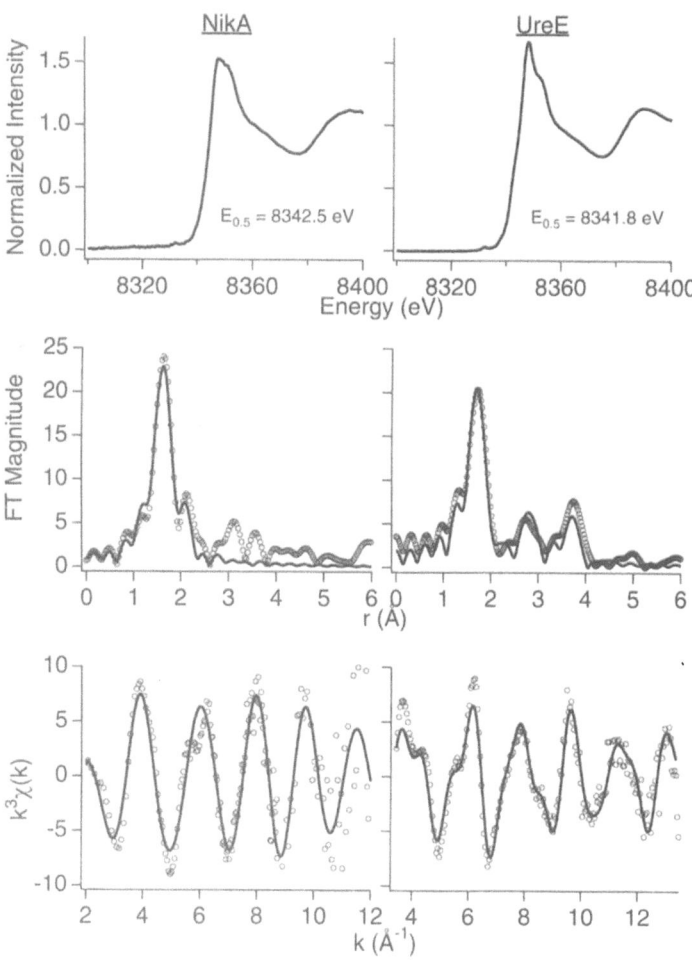

Fig. 1. A comparison of XAS data for Ni-binding proteins NikA (*left*) and UreE (*right*). *Top*: Ni K-edge XANES spectra. *Middle*: Fourier-transformed EXAFS spectra (*circles*) and fits (*lines*). *Bottom*: EXAFS spectra (*circles*) and fits (*lines*). The fit shown for the NikA is: of 6 O(N) @ 2.06(2) Å+1 S @ 2.57(2) Å. The fit shown UreE: 4 N(his) @ 2.08(2) Å+2 N @ 2.11(2) Å+4 C @ 3.03 Å+4 C @ 3.09 Å+4 N @ 4.23 Å+1 C @ 4.29 Å

The EXAFS fits are improved by including a second shell of scattering atoms. The data accommodate a S-donor ligand at 2.57 Å, which is very near the range exhibited by high-spin six-coordinate Ni(II) complexes with S-donor ligands (2.39–2.53 Å) [32–36]. The apparent seven-coordinate Ni center might reflect the inaccuracies of determining the number of scattering atoms from EXAFS analysis, or might indicate the presence of a bidentate carboxylate ligand that would occupy a single vertex of an octahedron but present two donor atoms to the Ni center. It is also possible that this second distance is due to another metal (other than Ni) near the Ni site. It is difficult to discriminate between long S-coordination and a metal at comparable long distances using limited EXAFS alone [37]. The existence of a second metal is not supported by analytical results, but the distance found (ca. 2.5 Å) is fairly typical of M-M distances in dinuclear metal sites, including carboxylate-bridged dinuclear centers such as those found in hemerythrin and other dinuclear Fe proteins. The fact that the mature NikA protein does not contain cysteine [26] indicates that the feature is due to a second metal or to methionine coordination.

Another strategy for Ni acquisition in microorganisms is via Ni permeases. Permeases are single integral membrane proteins that do not require energy to transport Ni. Examples of Ni permeases associated with hydrogenase biosynthesis have been characterized in *Alcaligenes eutrophus* (HoxN) [38] and in *Bradyrhizobium japonicum* (HupN) [39], and are involved with urease biosynthesis in *Helicobacter pylori* (nixA) [40, 41]. All of these proteins share sequence homology; however, no structural information is available regarding the ligands involved in binding to Ni.

Once Ni has crossed the cytoplasmic membrane, it is bound by a second class of Ni-binding proteins whose role is to sequester the Ni and make it available for incorporation into Ni metalloenzymes. Proteins that appear to fit this function include UreE, a protein involved in the synthesis of urease in *Klebsiella aerogenes* [42], HypB, a protein involved in *B. japonicum* H_2ase biosynthesis [43], and CooCJ, proteins involved in *R. rubrum* CODH biosynthesis [44].

The only protein for which structural information regarding the Ni site is known is UreE. In *K. aerogenes*, the urease operon (*ure*) contains seven genes [19]. The first three, *ureA–C*, are the structural genes that encode the three subunits of the enzyme. The remaining four genes, *ureD–G*, encode proteins that are involved with the energy dependent incorporation of Ni and maturation of the protein. UreDFG are not known to bind Ni, but are required for the synthesis of the urease active site. In contrast, UreE binds Ni but is not required for urease biosynthesis. The presence of UreE does, however, increase the level of activation of the enzyme. Homologous urease genes have been characterized in several other bacteria, and it appears that all ureases evolved from a common ancestor [7]. UreD (and its homologs) also has a DNA sequence that shows significant homology to proteins involved in H_2ase expression, including HoxN (23% in the case of the homologous gene from *H. pylori*), the high-affinity Ni permease mentioned above, and HupN in *B. japonicum* [7]. UreG has 25% sequence homology to HypB, and both proteins have nucleotide binding sites that are believed to supply the energy required for Ni incorporation in both urease and H_2ase [45]. In addition, HypB and CooJ have a polyhistidine region that is reminiscent of UreE and implies a Ni-binding function [19, 44].

UreE is a dimeric cytoplasmic protein with a MW of 35 kDa that binds three Ni atoms in each subunit (K_d ~10 μM) [42]. The *average* structure of these six Ni atoms has also been examined by XAS (Fig. 1) and by magnetic circular dichroism spectroscopy [42]. As in NikA, the data are consistent with a high-spin Ni^{2+} ion in a distorted octahedral geometry composed of six O- or N-donor ligands. One group of ligands (3–5) are at a distance of 2.08 Å, and another group (3–1) are at a distance of 2.11 Å. However, in UreE the presence of strong features in the EXAFS data arising from atoms in the second and third coordination sphere of the Ni clearly indicates the presence of multiple (3–5) histidine ligands. The histidine ligation was suggested by the amino acid sequence of the protein, which features a carboxy terminus containing ten His residues, although these regions are not required for Ni binding [46]. Nonetheless, the presence of His-rich regions in the amino acid sequence are a characteristic of this class of Ni-binding protein.

Various proteins are involved in the specific incorporation of Ni into apoproteins, in properly folding the enzymes and in post-translational modifications of the proteins. There is almost no structural detail regarding the Ni sites of any of these proteins, many of which do not bind Ni. Nonetheless, a predilection for post-translational modifications is a distinguishing feature of Ni metallobiochemistry. In the case of urease, complexes between the apoprotein and UreDFG facilitate the formation of active holoenzyme [47]. Several mechanisms are possible, including modification of an active site lysine residue with CO_2 to form a carbamate ligand in the dinuclear Ni active site [48], followed by Ni incorporation from UreE (and other sources), while preventing incorporation of other metals [7]. Alternatively, these proteins might be involved in an energy-dependent removal of undesired metals and/or improperly bound Ni [19]. Unproductive binding of Ni is known to occur if Ni is bound to the protein prior to modification of the active site lysine [48]. Activation of these enzymes requires removal of the improperly bound Ni.

Hydrogenase active site assembly involves the incorporation of Ni and Fe with its CN^- and CO ligands [20]. Several genes, are known to be involved in the maturation of the Ni-containing subunit of hydrogenases (e.g., HypB, HypC, HypD, HypE, HypF, HycI) but only a few of the functions of the gene products have been elucidated [49–53]. Among these are proteases (e.g., HycI) that cleave a carboxy terminal fragment of the Ni-containing subunit following incorporation of Ni [49, 54]. This carboxy terminal extension is not cleaved in the absence of Ni, and Ni is not incorporated into protein that is lacking this polypeptide [55]. Other proteins may serve as chaperones [56], may be involved in the assembly of the Fe center in the active site or the Fe,S clusters that are also present, or may be involved in regulating the expression of H_2ases by serving as H_2, O_2, or Ni sensors [51, 57]. Potential examples of the latter include HupUV (*Rhodobacter capsulatus, B. japonicum*) [51, 58, 59] and HoxBC (*A. eutrophus*) [60]. Although it is not even known if the proteins bind

Ni (it is suspected), the sequence of the genes encoding these proteins resemble those of the Ni-containing subunit of H_2ases (without the carboxy-terminal extensions), and at least one of the proteins has been shown to catalyze H/D exchange – a reaction that is catalyzed by H_2ases [61].

The *Rhodospirillum rubrum* CO redox system is encoded by *coo*-operons. Genes involved in the expression of CODH (*cooFSCTJ*) have been identified [62], including the structural genes for the CODH enzyme (*cooS*) and its Fe,S redox partner (*cooF*). CooCTJ proteins are involved in Ni insertion [44], despite the fact that Ni-deficient *R. rubrum* CODH can be reconstituted by $NiCl_2$ (see below). CooTJ are associated with metal specificity; apo-CODH binds Cd, Zn, Co, and Fe 300 times more tightly than Ni and gives inactive enzyme. Other CO-inducible genes (*cooMKLXU*) are associated with the expression of a CO-inducible H_2ase (*cooH*) and resemble the genes involved in H_2ase expression mentioned above [63]. The CO-inducible genes are regulated by CooA (a CO sensor?). The products of *cooCJ* are reminiscent of proteins involved in urease and H_2ase expression [44]. A run of 15 His residues in the C-terminal coding region of *cooJ* is reminiscent of *ureE* and *hypB*, and a nucleoside triphosphate-binding motif near the N terminal coding region of *cooC* is also found in *hypB* and *ureG*.

A branch of the general pathway for the biosynthesis of tetrapyrroles is responsible for the synthesis of F_{430} [64]. There is no information regarding Ni insertion into F_{430} [21]. The fact that apoMCR can be reconstituted by isolated cofactor suggests that a cofactor assembly mechanism may not be needed. On the other hand, the structure of the enzyme (see below) shows that organization of all six subunits of the holoenzyme is required in order to create the binding site for F_{430} [65]. Such a structure implies that proteins aiding in the organization of this assembly are likely.

Genes encoding proteins involved in the production of active NiSOD have yet to be identified. Like the active Ni centers in *C. thermoaceticum* CODH [66], the Ni in SOD can be removed by chelators [17, 18]. Unlike CODH, the activity of the enzyme is only restored to a small extent (12%) by recombination with Ni. This suggests a role for other factors in the Ni site assembly in SOD.

2.2
Urease

Ureases (urea amidohydrolases) are enzymes found in bacteria, fungi and plants that catalyze the hydrolysis of urea to carbamate and ammonium ions (Eq. 1) [3, 7].

$$\underset{H_2N}{\overset{O}{\|}}\underset{NH_2}{} + H_2O \longrightarrow NH_4^+ + \underset{H_2N}{\overset{O}{\|}}\underset{O^-}{} \quad (1)$$

Carbamate spontaneously hydrolyzes at physiological pH to yield more ammonium ion and bicarbonate. The catalysis afforded by the enzyme is impressive; the uncatalyzed hydrolysis of urea has $t_{1/2} = 3.6$ years at 38 °C and is accelerated by a factor of $\sim 10^{14}$ by the enzyme [3]. The enzyme allows organisms that express it to use urea as a nitrogen source, and is thus important for the utilization of urea fertilizers. Ureases are of great medical interest since they play important roles in bacterial pathogenesis. Urease involvement has been implicated in urolithiasis (infection induced stones due to urea hydrolysis in urine), and is an important virulence factor in *Helicobacter pylori*, which can cause peptic ulcerations and gastritis, and in bacteria that cause urinary tract infections [7, 67].

2.2.1
The Structure of the Urease Active Site

Despite the fact that crystals of jack bean urease have been available since 1926 [2], these crystals have proven unacceptable for crystal structure determinations and the best structural information has been obtained from the bacterial urease from *K. aerogenes*. In general, there is a high degree of amino acid sequence homology between this urease and other ureases [7], particularly in the region containing the active site, suggesting that the active site structure and catalytic mechanism is similar for all ureases. Thus, *K. aerogenes* urease has been adopted as a typical example for this discussion, and data refer to this enzyme unless otherwise noted.

The crystal structure of resting *K. aerogenes* urease has been reported at 2.2 Å resolution [68]. The protein consists of three subunits, α (60.3 kDa), β (11.7 kDa), and γ (11.1 kDa) in a trimer of trimers, $(\alpha\beta\gamma)_3$, that has three-fold symmetry. This quaternary structure differs from other ureases, such as the jack bean enzyme (α_6), but a high degree of sequence homology is maintained ($\geq 50\%$ sequence identity). In the case of the jack bean enzyme, the α subunit (91 kDa) is roughly equivalent to $\alpha\beta\gamma$ in the bacterial enzyme. Each α subunit in *K. aerogenes* urease binds two Ni atoms in a dinuclear active site, separated by 3.5 Å. The structure of the active site is summarized in Fig. 2 and Table 1. One of the Ni centers, Ni-1, is ligated by the ε N atom of His_{272}, the δ N atom of His_{246}, and an O atom from a carbamate ligand derived from Lys_{217}. The carbamate ligand also provides an O-donor to Ni-2, forming a bridge similar to those found in carboxylate-bridged dinuclear Fe centers [69]. In addition, Ni-2 is ligated by the ε-N atoms of His_{134} and His_{136}, and an O atom from Asp_{360}. This refinement has been revised in recent work to include three water molecules in the active site – a bridging water and water molecules bound to Ni-1 and Ni-2 [70]. However, the O-O distances between these water molecules are too short to allow occupancy of all three sites simultaneously. Including the three active site water molecules, the coordination environments of Ni-1 and Ni-2 are best described as involving distorted five-coordinate N_2O_3 complexes. In addition to the metal ligands, the active site contains the strictly conserved residues His_{219}, His_{320}, Arg_{336}, Met_{364}, Ala_{167}, Gly_{277}, and Ala_{363}, which form a pocket that is lined with the polar side chains or with amide O atoms of these residues and filled with water. A cysteine residue (Cys_{319}) is also

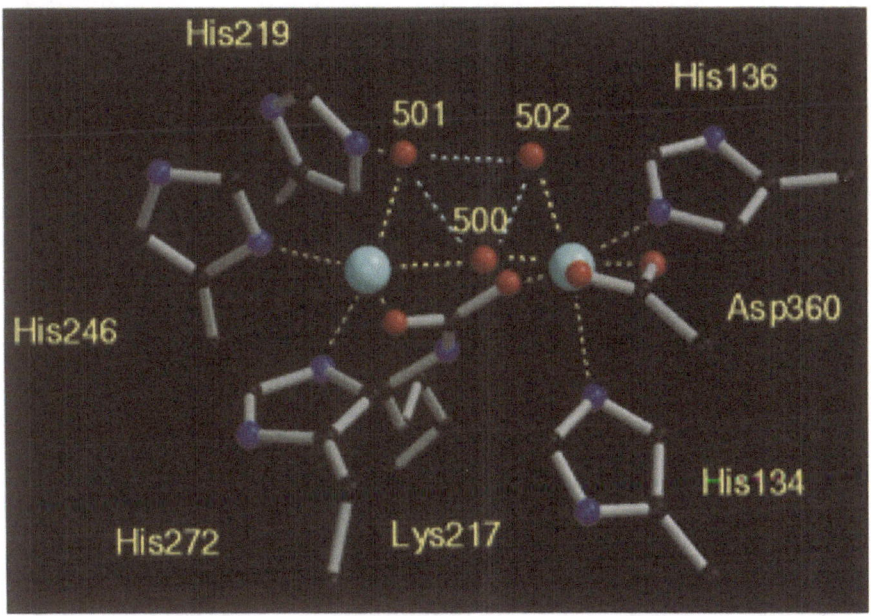

Fig. 2. The structure of the active site of *K. aerogenes* urease, showing the location of three water binding sites (500, 501, 502) and a histidine residue (His$_{219}$) important for substrate binding

found in the active site pocket. This cysteine residue is highly conserved, but is substituted by threonine in *Staphylococcus xylosus* urease [71].

Crystal structures are also available for the apoenzyme [72], for several mutants [70, 72, 73], and for the acetohydroxamate-inhibited complex [70]. The structure of the apoenzyme at 2.3 Å resolution is nearly identical to the holoenzyme, with the exception of the missing carbamate group and Ni atoms. This suggests that the protein structure is not determined by the metals, and is preorganized to bind the Ni specifically and with high affinity. In the H219A and H320A mutants, histidine residues that are strictly conserved and are located in the active site pocket, but are not ligands to the Ni, are altered. These changes affect the catalytic activity of the enzyme (see below) and have only minor perturbations on the active site structure and overall structure of the enzyme [72]. In the H219A mutant structure at 2.5 Å resolution, the active site residues shift by less than 0.3 Å. The changes in H-bonding interactions appear to be the major changes caused by this mutation. In the H320A mutant, the active site is perturbed in that the Ni-2 aquo ligand in the native structure bridges the two Ni sites with Ni$_1$-O and Ni$_2$-O distances of 2.4 and 2.3 Å, respectively. This change is due to the loss of a hydrogen bonding network that involves His$_{320}$ and another water molecule in the native structure. The effect on the structure is to make Ni-1 more nearly tetrahedral and to alter the geometry at Ni-2 so that it is more square pyramidal. The other Ni ligands move by ≤0.25 Å. Again, the overall structure of the protein is not significantly affected.

Table 1. Structural information regarding the urease active site

Sample	Metal–Ligand	Distance (Å)	Reference
	Urease Crystal Structures		
Native *K. aerogenes*	Ni_1-$N(His_{246})$	2.1(2)	68
(2.2 Å res.)	Ni_1-$N(His_{272})$	2.2(2)	
	Ni_1-$O(Lys_{217})$	2.0(2)	
	Ni_2-$N(His_{134})$	2.3(2)	
	Ni_2-$N(His_{136})$	2.2(2)	
	Ni_2-$O(Lys_{217})$	2.1(2)	
	Ni_2-$O(Asp_{360})$	2.1(2)	
	Ni_2-OH_2	2.0(2)	
	Ni_1-Ni_2	3.5(2)	
	ave. Ni-N(His)	2.2(2)	
	ave. Ni-O	2.05(20)	
H134 A Mutant *K. aerogenes*	Ni_1-$N(His_{246})$	2.4(2)	73
(2.0 Å res.)	Ni_1-$N(His_{272})$	2.6(2)	
	Ni_1-$O(Lys_{217})$	2.3(2)	
	Ni_2-OH_2	2.2(2)	
	Ni_2-OH_2	2.0(2)	
	Ni_2-OH_2	2.3(2)	
	ave. Ni-N(His)	2.5(2)	
	ave. Ni-O	2.2(20)	
	Structural Data from EXAFS Analysis		
Native *K. aerogenes*	(2) Ni–N(His)	2.02(2)	74
	(3) Ni–N(O)	2.08(2)	
	Ni– Ni	not detected	
H134 A Mutant K. aerogenes	(2) Ni–N(His)	2.08(2)	73
	(4–5) Ni–N(O)	2.09(2)	
Resting Jack Bean	(2) Ni–N(His)	2.05(2)	74
	(3) Ni–N(O)	2.04(2)	
	Ni– Ni	not detected	
K. aerogenes +$HSCH_2CH_2OH$	(2) Ni–N(His)	2.06(2)	74
	(2) Ni–N(O)	2.03(2)	
	Ni–S	2.23(2)	
	Ni–Ni	3.25(2)	
Jack Bean +$HSCH_2CH_2OH$	(2) Ni–N(His)	2.05(2)	74
	(2) Ni–N(O)	2.05(2)	
	Ni–S	2.23(2)	
	Ni–Ni	3.28(2)	

In the H134A mutant, a histidine residue that is a ligand of Ni-2 in the native structure is altered. This causes the loss of Ni-2 and results in a urease containing only Ni-1. The structure of Ni-1 also changes. In addition to the three native protein ligands (His_{272}, His_{246}, Lys_{217}carbamate), Ni-1 also has three aquo ligands, making it a distorted octahedral N_2O_4 site [73].

The crystal structure of the resting enzyme is largely in agreement with features predicted from various spectroscopic and physical studies. The average structure of the Ni sites in the resting enzymes from *K. aerogenes* and from jack bean, as well as the complexes formed with 2-mercaptoethanol, were examined by XAS [74]. The comparison of the active site structures obtained on the resting *K. aerogenes* enzyme from crystallographic and XAS analysis are in good agreement (Table 1), indicating that the structure of the active site in the crystalline state and in frozen solution are the same.

The structures obtained by XAS analysis do not show any significant differences in the structures of the active sites of the plant and bacterial enzymes. The results of the pre-edge XANES analysis indicated that the areas of the peaks assigned to $1s \rightarrow 3d$ transitions in the native enzymes (0.058 and 0.070 eV) were significantly greater than those expected for an octahedral geometry, and lie in a region consistent with five-coordinate Ni complexes

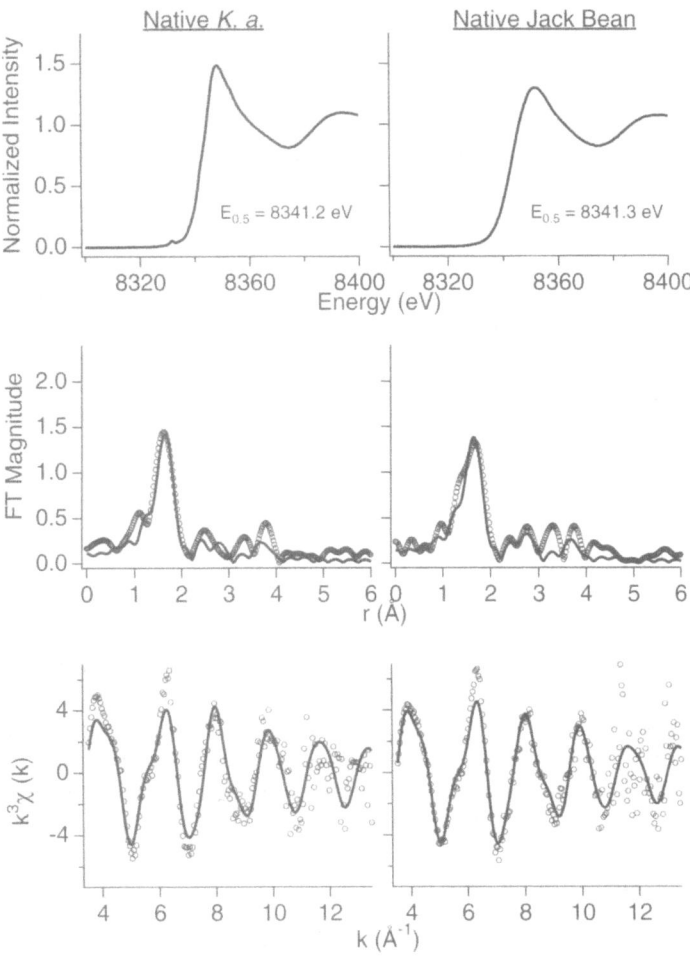

Fig. 3A.

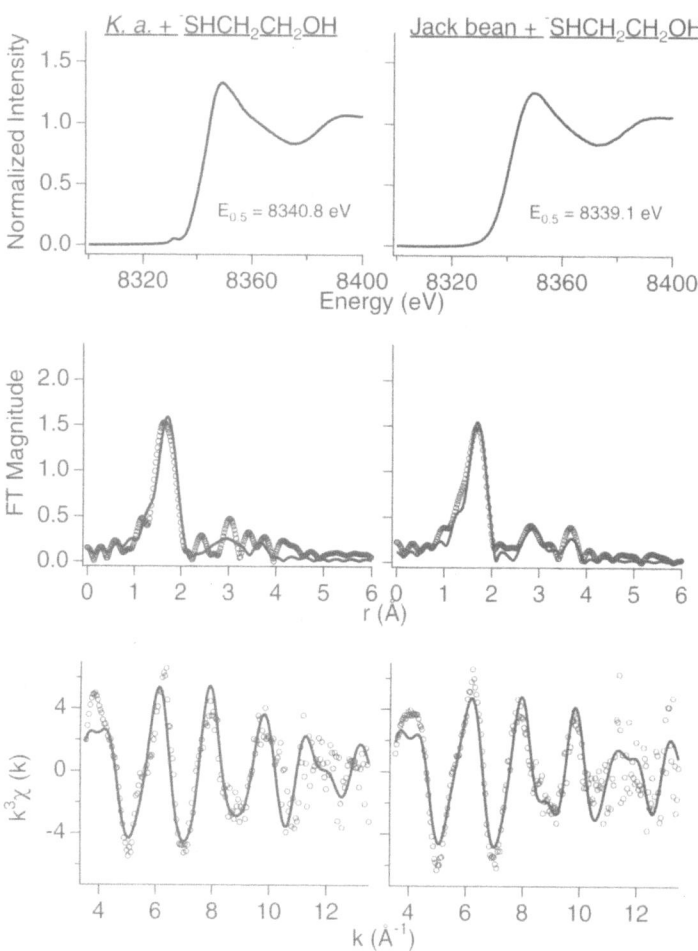

Fig. 3B.

(Fig. 3A). Upon addition of 2-mercaptoethanol, the intensity of the transitions increased to values at the high end of those observed in five-coordinate complexes (0.093, 0.103 eV) and are also appropriate for distorted tetrahedral complexes [75]. The analysis of the EXAFS (Fig. 3A) indicated that the ligands in the native enzymes were all O,N donors. The analysis of features arising from scattering atoms in the second and third coordination sphere of Ni indicated the presence of 2-3 His ligands, with the best fits to the data giving an average structure of $Ni(His)_x(O,N)_{5-x}$ ($x = 2$-3). The addition of 2-mercaptoethanol to the enzyme is accompanied by the appearance of new peaks in the electronic absorption spectrum at 322, 374, and 432 nm that are appropriate for thiolate→Ni LMCT transitions, and indicate that 2-mercaptoethanol (and by analogy other thiolates) bind to Ni in the enzyme [76]. The

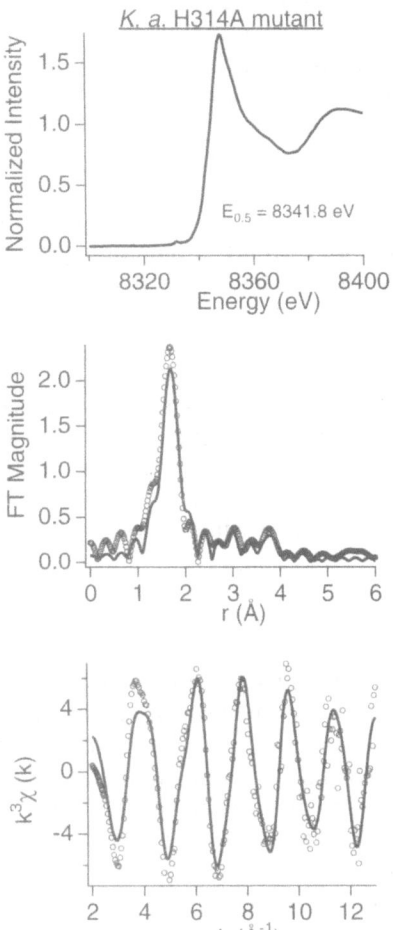

Fig. 3C.

Fig. 3. A Comparison of the XAS data for native urease samples from *K. aerogenes* and Jack bean. *Top*: Ni K-edge XANES spectra. *Middle*: Fourier-transformed EXAFS spectra (*circles*) and fits (*lines*). *Bottom*: EXAFS spectra (*circles*) and fits (*lines*). The fit shown for the bacterial enzyme is: 2 N(his) @ 2.02(2) Å+3 N @ 2.08(2) Å+2 C @ 2.87 Å+2 C @ 2.97 Å+2 C @ 3.03 Å +2 N @ 4.17 Å+2 C @ 4.20 Å. The fit shown for the plant enzyme is: 2 N(his) @ 2.05(2) Å+3 N @ 2.04(2) Å+2 C @ 2.87 Å+2 C @ 2.96 Å+2 C @ 3.03 Å+2 N @ 4.17 Å+2 C @ 4.20 Å.
B Comparison of the XAS data for the 2-mercaptoethanol inhibitor complexes of urease samples from *K. aerogenes* and Jack bean. *Top*: Ni K-edge XANES spectra. *Middle*: Fourier-transformed EXAFS spectra (*circles*) and fits (*lines*). *Bottom*: EXAFS spectra (*circles*) and fits (*lines*). The fit shown for the bacterial enzyme is: 2 N(his) @ 2.06(2) Å+2 N @ 2.03(2) Å+1 S @ 2.23 Å+2 C @ 2.99 Å+2 C @ 3.06 Å+1 Ni @ 3.25 Å+2 N @ 4.19 Å+2 C @ 4.22 Å. The fit shown for the plant enzyme is: 2 N(his) @ 2.05(2) Å+2 N @ 2.05(2) Å+1 S @ 2.23 Å+2 C @ 2.99 Å+1 Ni @ 3.28 Å+2 C @ 3.06 Å+2 N @ 4.19 Å+2 C @ 4.22 Å.
C XAS data obtained for the H314 A mutant of urease from *K. aerogenes*. *Top*: Ni K-edge XANES spectra. *Middle*: Fourier-transformed EXAFS spectra (*circles*) and fits (*lines*). *Bottom*: EXAFS spectra (*circles*) and fits (*lines*). The fit shown is: 2 N(his) @ 2.09(2) Å+4 N @ 2.09(2) Å+2 C @ 3.05 Å+2 C @ 3.14 Å+2 N @ 4.22 Å+2 C @ 4.28 Å

fits obtained from the analysis of the EXAFS for the complexes formed with 2-mercaptoethanol are consistent with an average Ni structure of Ni(His)$_2$ (O,N)$_2$S (Fig. 3B), suggesting the replacement of a ligand other than His by the S-donor ligand. In addition, a feature interpreted as arising from a Ni-Ni vector at 3.25 Å in the bacterial enzyme and 3.28 Å in the plant enzyme was observed. These results, along with changes in the magnetic properties of the dinuclear Ni site (the Ni centers become antiferromagnetically coupled in the 2-mercaptoethanol adduct) [77–79], suggested a bridging role for the thiolate ligand and provided the first details of the dinuclear active site.

XAS data has also been reported from the H134A mutant enzyme (Fig. 3C, Table 1). The $1s{\rightarrow}3d$ transition is much smaller than in the native enzyme (0.03 eV), consistent with a structural change to a more symmetric six-coordinate site. The EXAFS analysis revealed an O,N coordination environment with an average distance of 2.09(2) Å. This distance is more uniform and shorter than obtained from the crystal structure, and is probably due to the lower resolution (2.0 Å) of the crystal structure. Additional features in the second and third coordination sphere of the Ni provided evidence for two His ligands. The data were consistent with a Ni(His)$_2$(N,O)$_{4-5}$ structure.

2.2.2
The Role of Nickel in Urea Hydrolysis

A general mechanism for the catalysis of urea hydrolysis by urease is shown in Fig. 4 [7]. It is based on the mechanism originally proposed by Blakeley and Zerner involving separate but proximal Ni atoms [80]. It consists of the binding of urea to one of the Ni atoms, followed by activation of water (or hydroxide) bound at the second Ni atom via deprotonation with a basic group. The hydroxide ligand then serves as a nucleophile, attacking the carbonyl carbon to generate a tetrahedral transition state. Transfer of the proton from the protonated base, or protonation by an acidic group, produces ammonia and carbamate.

Roles for a base and an acid in the catalytic mechanism were originally suggested by the pH-dependence of the enzyme kinetics. Urease catalysis has simple Michaelis-Menten kinetic behavior [67]. There is no evidence of inhibition by substrate or allosteric behavior. The value of K_m (2.8 mmol/l in *K. aerogenes*) varies with the source of the protein (0.17–130 mmol/l) and shows some correlation with the ecological niche of the organism [7]. The value of V_{max} for the *K. aerogenes* urease (V_{max} = 2800 μmol of urea hydrolyzed/min/mg protein at 37 °C) is typical of the bacterial ureases (1000–5500 U/mg) and displays a pH optimum near pH = 7.75. Studies of the pH dependence of the catalysis indicate that K_m is not very sensitive to pH but that V_{max} is, consistent with the presence of a general base (pK_a = 6.65) and a general acid (pK_a ~8.85).

The fact that the structural characterizations show that the active site contains a dinuclear Ni center is consistent with the original mechanism involving two Ni centers. Analysis of the native crystal structure suggests that the Ni involved in binding urea was Ni-1. Access to the Ni-1 is via a flap

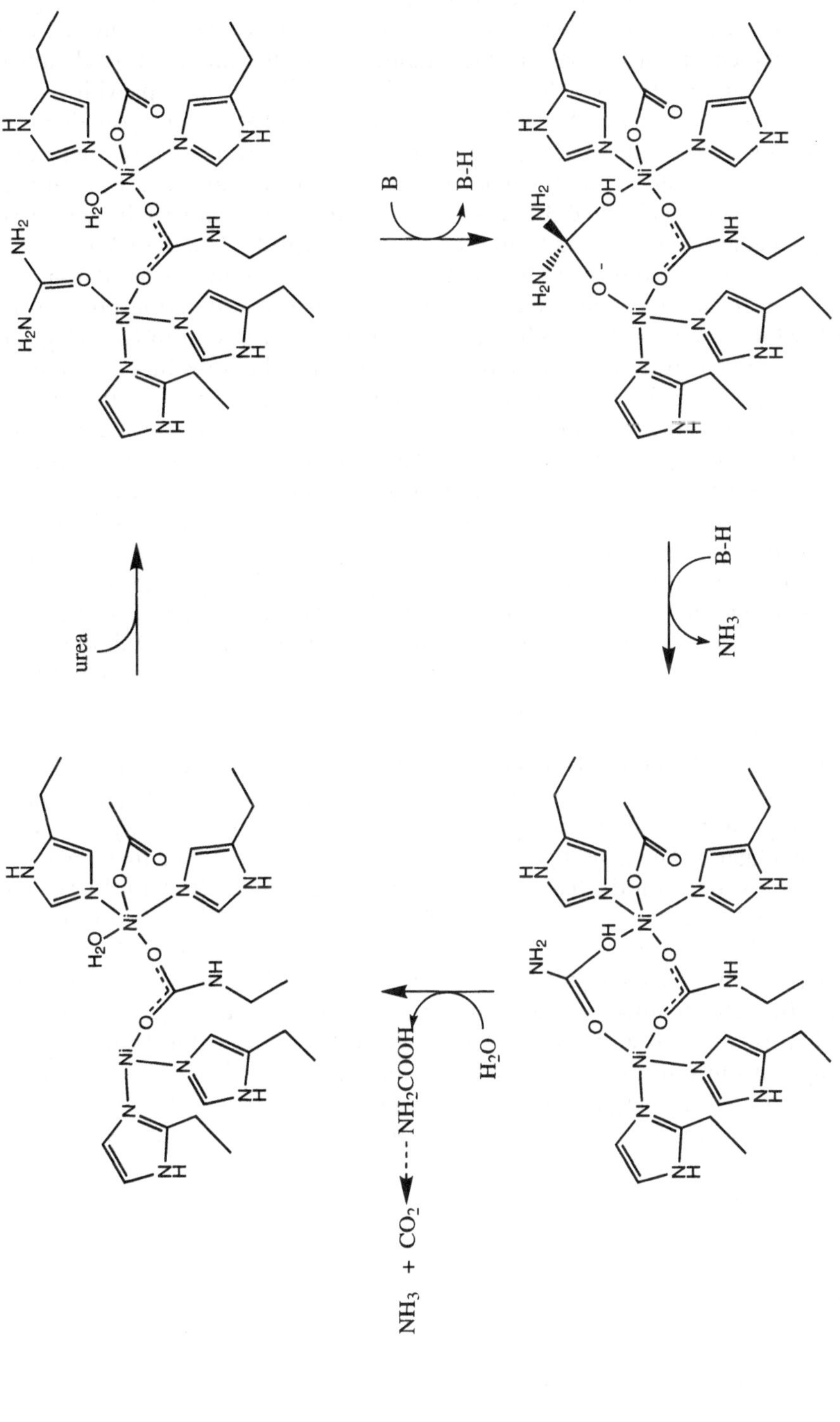

Fig. 4. General mechanism for the hydrolysis of urea by urease

composed of residues 308–336 from the α-subunit [68]. This flap, containing the strictly conserved His$_{320}$ and the highly conserved Cys$_{319}$ residue, is highly mobile and defines one side of a channel leading to the active site. The fact that Cys$_{319}$, which lies at the base of the channel, is readily modified by chemical reagents supports the idea that the channel can readily open to provide access to the dinickel active site. Site-directed mutagenesis of His$_{219}$ suggests that it is important to substrate binding [81]. The H219A mutant has a very high value of K_m (1100 mmol/l), resulting in a large decrease in activity (ca. 2% of native). This His residue is located 3.1 Å from Ni-1 and 3.7 Å from Ni-2 in the active site pocket and is H-bonded from the δ-N to the peptide N atom of Asp$_{221}$. It is therefore believed to be protonated at the ε-N. This potential H-bond donor is 3 Å away from, and is directed towards, the fourth coordination position of Ni-1, which is empty or weakly interacting with water bound to Ni-2 (Fig. 2).

One of the features of the proposed mechanism is that urea binds via the O-atom to polarize the carbonyl carbon for nucleophilic attack by hydroxide. This bonding mode is supported by the structure of a model compound, [Ni$_2$(OAc)$_3$(tmen)$_2$](OTf) (Fig. 5) [82]. The structure of the model consists of two six-coordinate Ni centers in N$_2$O$_4$ ligand donor environments, separated by 3.47 Å. Each Ni is bound to a bidentate tmen ligand (N,N,N',N'-tetramethylethylenediamine) and to three acetate ligands that form bridges between the Ni atoms. Two of these bridges are $\mu_{1,2}$ bridges and one is a μ_1 bridge, where the second acetato O atom is bound only to one Ni. This leaves a vacant site open on the second Ni, where urea binds via the O-donor. The tendency of Ni to bind to urea via the O atom was further demonstrated in a mononuclear complex, [Ni(OAc)(urea)$_2$(tmen)](OTf), where both urea ligands are bound via the O-donor atom [82].

When urea is modeled in the active site of the enzyme with its O-donor bound to Ni-1, the O-donor forms an H-bond with the ε-N of His$_{219}$ that may help to position and further polarize the substrate for attack by the Ni-2 hydroxo ligand. The aquo ligand is activated by the base, a role that has been assigned to His$_{320}$ on the basis of site-directed mutagenesis [81]. In the H320A mutant, K_m is not strongly affected (8.3 mmol/l) and still lies in a normal range for bacterial ureases. However, the enzyme retains only 0.0003% of the activity of the native enzyme. The ε-N atom of this His residue is 4.3 Å away from the water bound to Ni-2 in the structure [68], but this His is located in the flap region and a change in the structure could bring the N atom close enough to the water to deprotonate it. Nonetheless, evidence from chemical modifications also argues that H$_{320}$ is the catalytic base [81]. The native enzyme is inactivated by reaction with diethylpyrocarbonate, a reagent used in the modification of His residues. The residual activity of the H320 A mutant is not eliminated by this reaction.

After deprotonation, the hydroxo group on Ni-2 attacks the C atom of urea to generate a tetrahedral intermediate that bridges the two Ni atoms. This intermediate is supported by a number of studies of inhibition of the enzyme by molecules that might form stable transition state analogs (e.g., hydroxamates and phosphoroamide compounds) [7, 76]. The structure of a model of hydroxamate-inhibited urease predicts how this may happen (Fig. 5) [83]. The

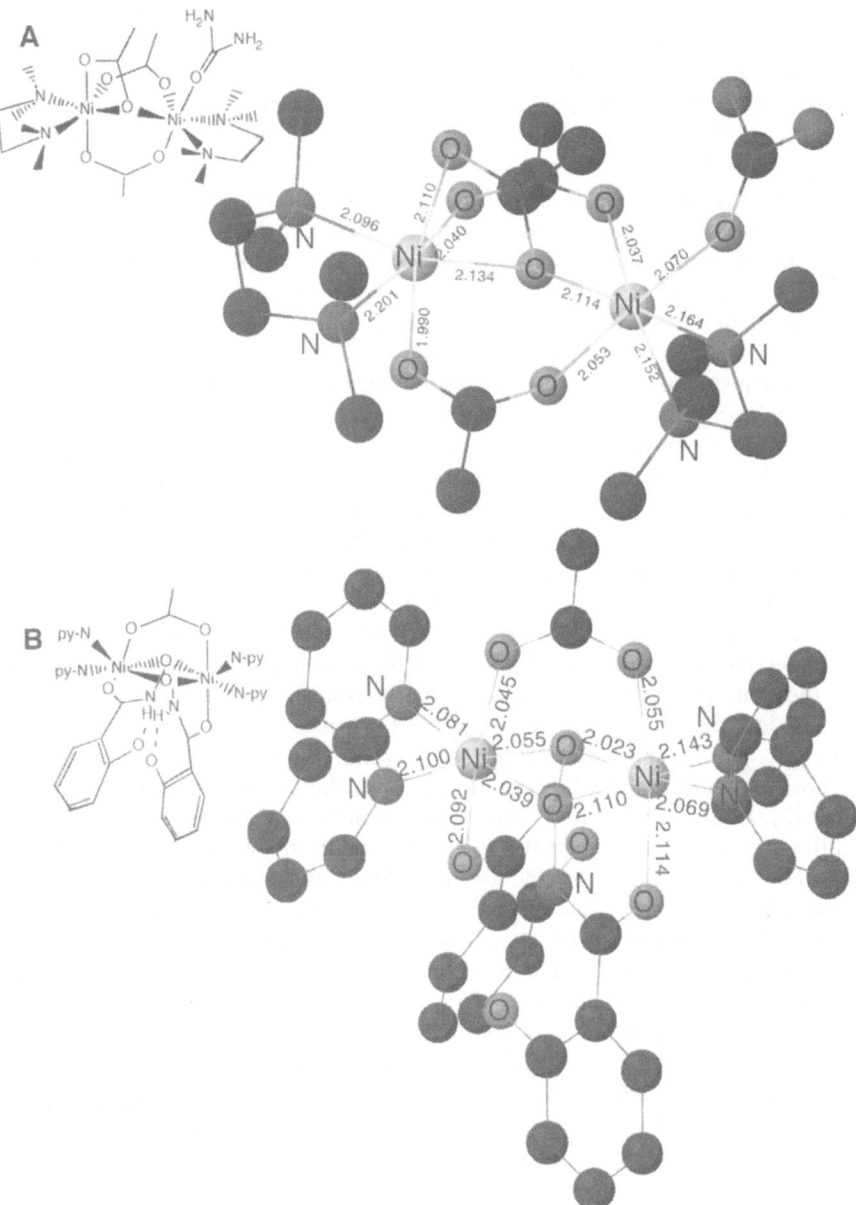

Fig. 5A,B. Structure of urease model compounds: **A** [Ni$_2$(OAc)$_3$(tmen)$_2$](OTf), illustrating a urea O-Ni binding mode [82]; **B** [Ni$_2$(Hshi)(H$_2$shi)(py$_4$)(OAc)], illustrating the binding mode of a hydroxamate group that is found in the crystal structure of the *K. aerogenes* urease acetohydroxamic acid (AHA) enzyme/inhibitor complex [83]

dinuclear model compound, [Ni$_2$(Hshi)(H$_2$shi)(py$_4$)(OAc)], features Ni atoms separated by 3.016 Å in N$_2$O$_4$ ligand donor environments that are reminiscent of those found in [Ni$_2$(OAc)$_3$(tmen)$_2$](OTf). In this case the N-donors are provided by pyridine ligands and the Ni sites are bridged by the acetato group in $\mu_{1,2}$ mode. In addition, there are two molecules of salicylhydroxamate. These bind in a bidentate fashion to each Ni, with one of the two O-donors serving as a bridge between the Ni sites. This predicted bonding mode has recently been confirmed by the crystal structure of the C319A mutant urease-acetohydroxamic acid (AHA) complex. The AHA is found to bind to both Ni centers in the active site via the O atoms, with the carbonyl O atom bound to Ni-1 and the other O atom bridging the two Ni centers. The binding of AHA displaces the active site water molecules.

In the last step of the mechanism, the tetrahedral intermediate decomposes with assistance from the general acid to form carbamate and ammonia. Zerner et al. originally proposed that Cys$_{319}$ was the acid group. However, Cys$_{319}$ is not required for catalysis. Studies involving site-directed mutagenesis of Cys$_{319}$ reveal that this residue can be changed and the enzyme still has activity [84]. In particular, the C319A mutant retains 48% of the wild-type activity and exhibits only a slight increase in K$_m$. Structural studies of C319 mutants show that the effects of replacing Cys$_{319}$ are minor and involve primarily changes in the structure of the water molecules in the active site, and the mobility and position of the active site flap containing Cys$_{319}$ [70]. Thus, one role for Cys$_{319}$ suggested by the structure is in positioning other key residues in the active site.

The identity of the general acid is not known. In fact, it has been suggested that His$_{320}$ may serve as a proton shuttle, removing the proton from water and depositing it on urea and thus acting as both the general base and the general acid [7]. Alternatively, His$_{320}$ may be the general acid involved in protonation of the urea N atom. Such a role is consistent with the observation that His$_{320}$ is located on the opposite side of urea from the proposed catalytic water molecule. However, in order to function as the catalytic acid, His$_{320}$ would have to be protonated. A proposed "reverse protonation" mechanism has been described, consistent with the enzyme kinetics, the structure of the active site, and involving His$_{320}$ as the catalytic acid [8]. In this scheme, the active enzyme is proposed to feature a protonated His$_{320}$ and a deprotonated water molecule. Given the difference in the values of the pK_a of these two groups, only 0.3% of the enzyme would exist in the active state of protonation at the optimal pH of 8. Nonetheless, the low amount of enzyme in the active state could be compensated for by enhanced reactivity of the active state.

3
Ni Proteins with Redox Roles

3.1
Hydrogenase

Hydrogenases (hydrogen:acceptor oxido-reductases) catalyze the reversible, two-electron oxidation of H$_2$ (Eq. 2).

$$H_2 \rightleftharpoons 2H^+ + 2e^- \qquad (2)$$

Nickel is found in the majority of known H_2ases, which have become known as [NiFe]H_2ases because of the presence of Fe in all of the Ni containing enzymes [12]. The presence of Ni in the enzyme was originally associated with a rhombic $S = 1/2$ epr signal [85, 86], an assignment that was confirmed via the observation of ^{61}Ni hyperfine in isotopically labeled enzymes [87]. The Fe is present in the form of Fe,S clusters and in a unique Fe center. The unique Fe atom together with the Ni center comprise the metallic components of the active site. Nickel is not an absolute requirement for H_2ase enzymes; several enzymes containing only Fe as a metal cofactor are known. In fact, one enzyme has been isolated that uses an organic cofactor [88]. In addition, some of the [NiFe]H_2ases contain a selenocysteine residue that is a ligand of the Ni atom [89, 90].

There is a great deal of information regarding the structure of the Ni site in [NiFe]H_2ases stemming from a combination of biochemical, crystallographic and physical studies. This group of H_2ases are typically isolated as heterodimeric proteins with molecular weights near 100 kDa. The amino acid sequences of some 60^+ [NiFe]H_2ases has been determined by sequencing the structural genes for the subunits [12, 91]. These studies reveal that the sequences of the larger of the two subunits (ca. 65 kDa.) are highly conserved, while those of the smaller subunit (35 kDa.) are less so. Among the conserved regions of the large subunit are two Cys-X-X-Cys sequences, one near the N-terminus and one near the C-terminus, that contain the Cys ligands involved in binding the active site metals. In enzymes containing selenocysteine, it is the first of the two C-terminal Cys residues that is substituted (Cys$_{530}$ in the *D. gigas* sequence). The small subunit is ferredoxin-like and has several sequences that contain cysteine residues in arrangements appropriate for constructing Fe,S clusters. The number and types of clusters present are variable among the enzymes.

3.1.1
Structural Studies of the Ni Site in Hydrogenase

Crystallographic studies [92–94] have provided definitive information on the nature of the metal cofactors and their spatial relationship. The crystal structure of a mixed oxidation state crystal (40% Form A, 10% Form B, 50% epr silent, see below) of the H_2ase from *D. gigas* has been the subject of a recent review [95]. The data show that the small subunit contains all three of the Fe,S clusters found in this enzyme, while the Ni,Fe active site is buried in the large subunit (Fig. 6). The Fe,S clusters are arranged in a line, spaced about 12 Å apart, and are of three types. The distal cluster is an Fe$_4$S$_4$ cluster and is unique in that one of the terminal Fe ligands is a histidine imidazole group that is surface exposed, and thus provides a site for electron transfer into or out of the enzyme. The middle cluster is an Fe$_3$S$_4$ cluster, while the proximal cluster is a conventional Fe$_4$S$_4$ cluster and lies at a distance of ca. 13 Å from the Ni center (closest approach is via a terminal cysteine ligand of Ni). The arrangement of the Fe,S clusters and the active site suggest a facile electron transport chain between the active site and the surface of the small subunit.

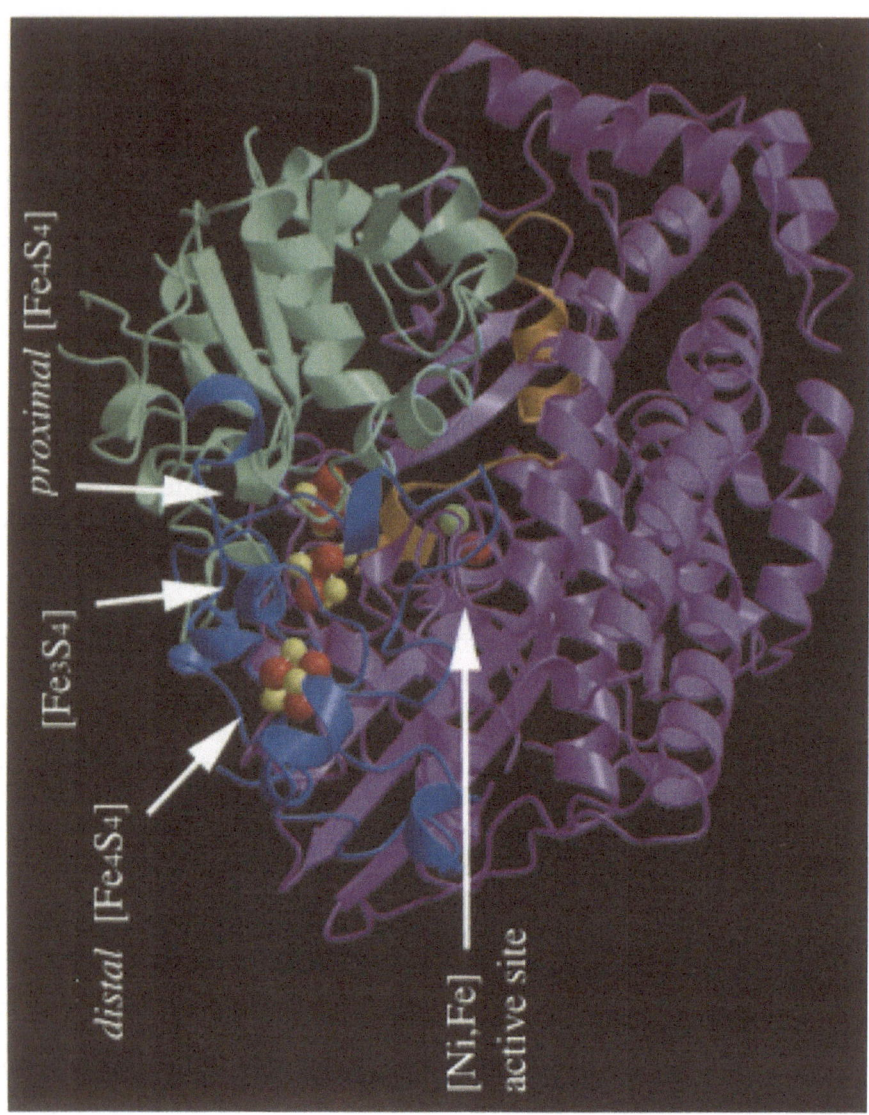

Fig. 6. Structure of *D. gigas* hydrogenase showing the subunit structure and the spatial relationship between the redox centers

Perhaps the most important revelation from the crystal structure of the *D. gigas* enzyme was that the Ni site was part of a cluster that also contained an Fe atom (Fig. 7) [93]. In the most recent refinement, the two metals are separated by 2.9 Å and bridged by two cysteinate thiolates – one each from the N- and C-terminal Cys-X-X-Cys sequences. In addition, there is a third low-Z ligand that may be an oxygen-donor ligand derived from O_2 or from water. This suggestion is consistent with [17]O hyperfine broadening observed in the epr spectra from oxidized (Forms A and B) enzymes [96], although the weakness of the interaction is puzzling. The short Ni-O distance observed (1.7 Å) suggests that this might be an oxo group that bridges to the Fe. In addition to the bridging groups, the Ni atom is bound to the enzyme by the remaining two conserved cysteine residues, making it a five-coordinate S_4O center. The Ni-S distances are not equivalent, with the terminal ligands having Ni-S distances of 2.2–2.3 Å and the bridging ligands having Ni-S distances of 2.6 Å.

Analysis of XAS data collected on the same enzyme (*D. gigas* H_2ase) is consistent with the general structural features of the Ni site, but differs in metric details [97]. Analysis of the pre-edge XANES spectrum of oxidized (Form A) enzyme indicates that the site is five coordinate, in agreement with the crystal structure. Figure 8 compares an EXAFS spectrum calculated from the crystallographic parameters with one following refinement of the EXAFS parameters. The results from the EXAFS analysis (4 S @ 2.18(2) Å+1 O @ 1.95(2) Å) indicate that the Fe atom does not contribute significantly to the

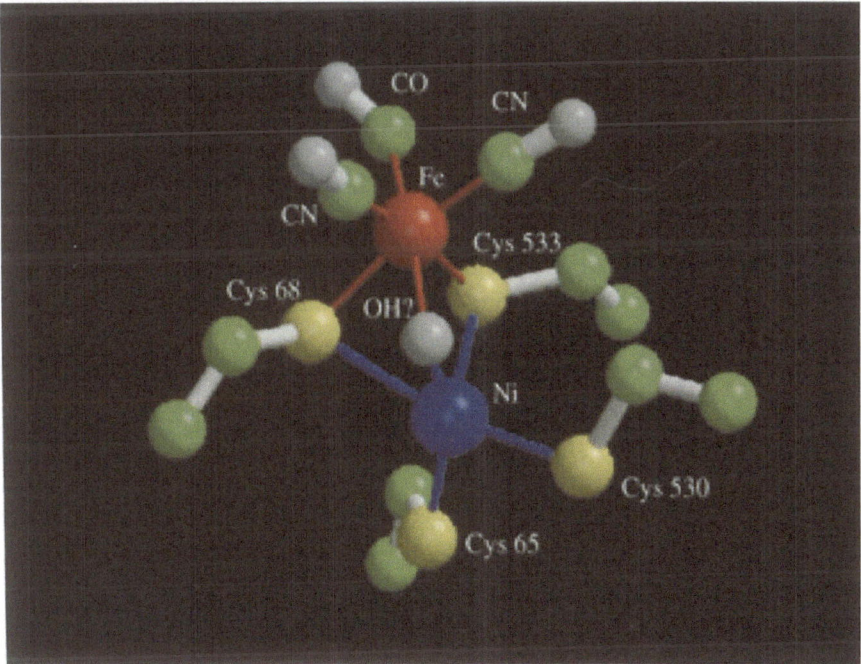

Fig. 7. Structure of the Ni-containing active site in oxidized H_2ase from *D. gigas*

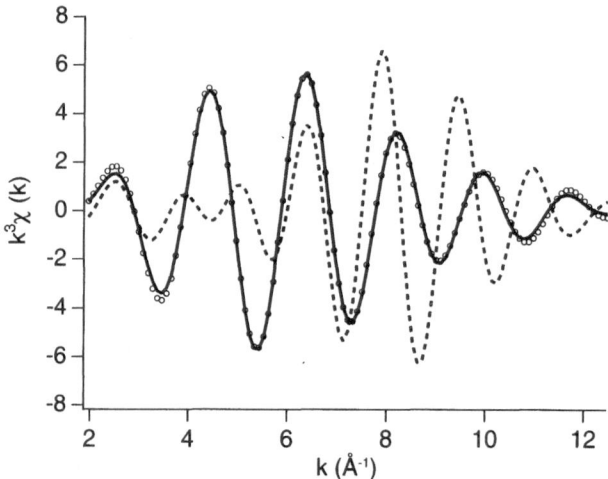

Fig. 8. Comparison of Fourier-filtered EXAFS spectra (Fourier transform window k = 2–12.5 Å^{-1}; Backtransform window r = 1.1–2.6 Å) taken on *D. gigas* H$_2$ase Form A (*circles*) with a spectrum calculated from crystallographic data (*dotted line*, O @ 1.7 Å+2 S @ 2.25 Å+2 S @ 2.6 Å+Fe @ 2.9 Å) and a refined fit (*solid line*, O @ 1.95(2) Å+4 S @ 2.18(2) Å)

EXAFS spectrum of this sample and that the Ni-S distances may not be as different as they appear in the crystal structure. The presence of an O (or N) scattering atom in the primary coordination sphere is confirmed, but the Ni-O distance is not as short as found in the crystal structure, and is more consistent with a hydroxo ligand than with an oxo group.

XAS has also been used to survey the structures of the Ni sites in a number of H$_2$ases from various species of bacteria [97]. The results indicate that the structure of the Ni site is remarkably similar in many enzymes, the obvious exception being the soluble enzyme from *Alcaligenes eutrophus*. This enzyme has a unique Ni site structure in the as-isolated (oxidized) state (3±1 N @ 2.09(2) Å+3±1 S @ 2.35(2) Å), but changes to one that is quite similar to other H$_2$ases upon reduction.

The crystal structure of the *D. gigas* enzyme shows that the active site Fe center is bound to the enzyme by only the bridging Cys residues, with Fe-S distances of 2.2 Å. The remaining ligands in the six-coordinate Fe complex are diatomic ligands originally detected by IR spectroscopy that have been assigned to a combination of 2 CN$^-$ ligands and a CO ligand (see below) [98]. Based on the hydrophobic environment and the lack of H-bonding interactions, the diatomic ligand that lies in a trans position to the O-bridging ligand is believed to be the CO ligand [99]. The assignment of these vibrations and the importance of hydrogen-bonding effects on the vibrations have been examined in detail using a model of the Fe center, CpFe(CO)(CN)$_2^-$, which confirms the assignments made for the enzyme and demonstrates the sensitivity of the CN$^-$ vibrations to interactions with the cation present [100].

3.1.2
Ni as a Redox Center in Hydrogenase

One of the challenges in understanding the mechanism of hydrogenase catalysis revolves around how to couple an inherently two-electron redox process to an enzyme active site that undergoes a series of one-electron redox reactions. The redox states of the enzyme were initially defined by their epr spectral characteristics [101], but it is now recognized that each state also has a characteristic IR spectrum [93, 99, 102]. The observation of ^{61}Ni hyperfine coupling in epr signals that are sensitive to redox poise and the presence of substrate (H_2) and inhibitors (e.g., CO) clearly associates the Ni with the active site. Similarly, the changes in the IR spectrum assigned to CN^- and CO ligands of the Fe center that also respond to changes in redox poise and the presence of H_2 or CO clearly indicate that the Fe atom is also a component of the active site. The typical [NiFe]H_2ase exhibits four redox levels. When the enzyme is isolated in air, a combination of two oxidized and catalytically inactive states are obtained. These two oxidized states are distinguished by their 77 K epr spectra (Form A, g = 2.31, 2.23, 2.02 and Form B, g = 2.33, 2.16, 2.01). Both of the oxidized Forms may be activated by reduction with H_2 or other reducing agents, and are initially reduced to a redox level that is epr-silent at 77 K (SI). Forms A and B are also distinguished by their kinetics of the reduction. Form B is instantaneously reduced upon exposure to H_2 (it is kinetically 'ready'), whereas Form A requires a period of incubation under H_2 (it is kinetically 'unready'). The SI state has recently been shown by IR Spectroscopy to consist of at least two forms to consist of two forms [99]. Fernandez and coworkers showed that two SI redox level intermediates were spectrally distinct. Furthermore, SI_u was shown to be inactive with respect to H/D exchange, while SI_r was active.

Reduction of the enzyme under more reducing conditions produces another epr-active redox state of the enzyme, Form C (g = 2.19, 2.14, 2.02). Form C displays two distinct epr spectra at low temperature. A signal that exhibits coupling due to the interaction of the Ni center with one Fe_4S_4 cluster, which is reduced in the potential range where the Form C 77 K epr signal is observed, and a signal which exhibits no such coupling. Studies of the coupled Form C spectrum using a spin-dipole model have characterized the nature of this interaction and provided a distance estimate between the two spins that is in good agreement with the crytallographic distance between the Ni (or the Cys_{530} S atom) and the proximal Fe_4S_4 cluster [103–106].

Form C may be reduced further to another redox level that is epr-silent at 77 K and contains only fully reduced redox cofactors (R). Redox titrations have been performed in tandem with epr spectroscopy to determine the potentials and the number of electrons involved with the changes in the 77 K epr spectra associated with Ni [12, 107–110]. These titrations indicate that all the redox chemistry associated with the Ni site occurs between ca. –100 and –400 mV and is coupled to the uptake or release of protons. At least in the presence of dyes, the reductions all appear to be one-electron processes. However, for *C. vinosum*, redox titrations conducted using H_2 partial pressure to determine the redox potential indicate that the Form C/R couple is a two-electron process [107]. It is not clear how this is achieved, since the addition of two electrons to a half-integer spin system should give another half-integer spin system, and R is epr-silent at 77 K.

The association of the epr signals with Ni in an oxidized enzyme, along with their general similarity to epr spectra from Ni(III) complexes, led to the assignment of these signals to a *formally* Ni(III) center in the oxidized enzyme. The sequential disappearance (in SI), reappearance (in Form C), and disappearance (in R) of the epr signals associated with Ni during reduction of the enzyme led to a number of mechanistic proposals based on Ni-centered redox chemistry. These proposals, summarized in Table 2, range from invoking oxidation states III-0 (Ni as a three-electron redox center) [12] to proposals utilizing only one-electron redox chemistry for Ni (either the III/II couple or the II/I couple) with the odd oxidation states assigned to epr-active species (Table 1) [111, 112]. With the discovery of the active site Fe atom, additional mechanisms transferring the redox chemistry to Fe have also been proposed (Scheme D) [93]. However, these schemes involve spin coupling of low-spin Fe(I) or Fe(III) centers to an S = 1/2 Ni center in order to produce the epr-silent species and are at odds with the magnetic properties of the active site [106, 113]. The magnetic properties of the active site are consistent with an Fe center that is low-spin Fe(II) and diamagnetic in all redox states of the enzyme. This electronic configuration is consistent with expectations for an Fe center with 2 CN$^-$ and a CO ligand. The scheme involving both oxidation states III and I for Ni requires that the potentials associated with the III/II and II/I redox couples be compressed into a 300 mV potential range between –100 and –400 mV. This is unprecedented redox chemistry for a single Ni complex. Presumably, if metal-centered redox chemistry involving both Ni(III) and Ni(I) were to occur in the enzyme, a major structural change in the Ni site would be required. Similarly, the schemes based on a one-electron redox couple would also require a change in structure at the Ni site. For example, the III→II→III→II scheme requires that the Ni(II) complex derived from reduction of the oxidized enzyme (Forms A or B→SI) be reoxidized (SI→ Form C) at a *lower* potential.

Table 2. Summary of redox chemistry associated with Ni in hydrogenases

Sample	77 K epr signal	Approx. E_m (mV vs NHE, pH 7.7, 40°)	Scheme A	Scheme B	Scheme C	Scheme D
Form A	g = 2.31, 2.23, 2.02	A→SI$_{unready}$–210	Ni(III)	Ni(III)	Ni(I)	
Form B	g = 2.33, 2.16, 2.01	B→SI$_{ready}$–135	Ni(III)	Ni(III)	Ni(I)	
SI	none	SI$_{ready}$→C–365	Ni(II)	Ni(II)	Ni(II)	Ni(I)-Fe(I, III)
Form C	g = 2.19, 2.14, 2.02	C→R –430	Ni(I)	Ni(III)	Ni(I)	Ni(I), Fe(II)
R	none		Ni(0)	Ni(II)	Ni(II)	Ni(I)-Fe(I, III)

The redox activity of the Ni center has been probed by using XAS [37, 114]. Examination of the Ni K-edge energy provides a method for detecting changes in the electron density residing on the Ni center. If the redox reactions observed for H_2ase were Ni-centered, one would expect a ~2 eV shift in the K-edge energy for each oxidation state change. In no case has a shift of ~6 eV been observed for the reduction of Forms A or B to R. The shift observed ranges from <1 eV for the *T. roseopersicina* enzyme to 1.5 eV for the enzymes from *D. gigas* and *C. vinosum*. This result effectively rules out mechanisms involving both Ni(III) and Ni(I) as realistically describing the changes in electron density that occur at Ni during redox catalysis, and at most represents an anemic one-electron metal-centered redox process. The differences in the magnitude of the edge energy shifts observed in different enzymes may be due in part to changes in the coordination number of Ni during reduction [37]. For the *T. roseopersicina* enzyme, XANES analysis shows that the Ni site appears to be more six-coordinate with the additional ligand being an O,N donor. For the *D. gigas* and *C. vinosum* enzymes, the oxidized forms are five-coordinate, while the reduced forms appear to be more six-coordinate. Also, if protonation of cysteinate ligands occurs, this would also affect the electron density at the Ni or Fe centers.

The analysis of EXAFS spectra of redox-poised samples also reveals only small perturbations in the structure of most of the Ni sites [37]. The EXAFS spectra are dominated by scattering from S atoms at ~2.2 Å in every case except for oxidized *A. eutrophus* H_2ase. The Ni-S distance is not very sensitive to the redox state of the enzyme. Compelling evidence for a short Ni-O interaction is evident in samples of Form A, a result that is consistent with recent crystallographic data that indicates the low Z bridging atom is lost upon reduction [115]. When a Ni-Fe vector can be accommodated by the data, it is found between 2.4 and 2.9 Å. For *D. gigas* and *C. vinosum* H_2ases there is a shortening of the apparent Ni-Fe distance between Form A and more reduced states. In the case of *T. roseopersicina* samples, the apparent Ni-Fe distance is relatively constant and near 2.5 Å.

Changes in the electron density at the Fe site are reflected in changes in the v_{CO} and v_{CN} stretching vibrations (Fig. 9) [93, 99, 102]. However, the magnitude of these changes is rather small. The largest difference in v_{CO} values reported is for Form A ($v_{CO} = 1950$ cm^{-1}) vs. the photoproduct of Form C ($v_{CO} = 1898$ cm^{-1}) and is only 52 cm^{-1}. This difference can be compared with the difference in v_{CO} values seen in the IR spectra of the low-spin Fe complexes, $[Fe^{II}(PS_3)(CO)(CN)]^{2-}$ ($v_{CO} = 1904$ cm^{-1}) and $[Fe^{III}(PS_3)(CO)(CN)]^{-}$ ($v_{CO} = 2006$ cm^{-1}) (PS$_3$ = tris(2-phenylthiolato)phosphine), which show a 102 cm^{-1} difference for the redox pair [116]. Thus the electron density changes at Fe are less than expected for a one-electron redox process in a similar ligand environment. Furthermore, the shifts in v_{CO} observed do not reflect a uniform increase in the electron density on Fe during reduction of the enzyme (Fig. 9) [117]. For example, the value of v_{CO} is higher in Form C than in SI. There is an approximately linear relationship between the v_{CO} and v_{CN} stretching frequencies, with the exception of Form C and Form R. Given the fact that the v_{CN} frequencies are expected to be less sensitive to changes in

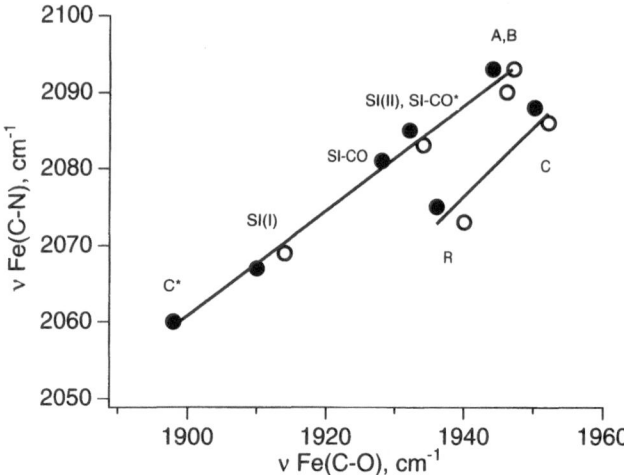

Fig. 9. Plot of ν_{CN} vs. ν_{CO} for the Fe sites of *C. vinosum* (*solid points*) and *D. gigas* (*circles*) hydrogenases

the Fe center [116] and are susceptible to changes in H-bonding interactions [100], the CO frequency is the best indication of the electron density on the Fe. The facts that all of the epr active forms (Forms A, B, C) have similar values of ν_{CO}, and that the epr silent states are similar and have lower frequencies than Forms A, B, and C, are consistent with a two-state redox model. In this model the epr active states of the cluster represent an oxidized active site, while the epr silent states represent a reduced cluster. If one plots the values of ν_{CO} (as a reflection of electron density on Fe) vs. the Ni K-edge energy (as a measure of the electron density on Ni), one observes an approximately linear relationship with the exception of the Form C and R samples (Fig. 10) [117] as was the case for the analysis of the ν_{CO} and ν_{CN} stretching vibrations (Fig. 9). This indicates that the Fe and Ni centers are responding to the same one-electron change in the redox status of the cluster. This conclusion is in agreement with redox titrations carried out by monitoring changes in the IR spectrum. These titrations involved the same number of electrons and gave the same potentials as titrations of the epr signal associated with the Ni center [99].

This observation raises the question of how Forms B and C, and SI and R, differ from each other if they represent the same redox state. One possibility is that they represent different states of protonation of the active site. Alternatively, the differences might indicate the presence of a hydride in the active site of the reduced enzyme.

3.1.3
Ni as a Substrate/Inhibitor Binding Site in Hydrogenase

Another role that has been suggested for the Ni site in H_2ases is as a binding site for the substrate (H_2) or inhibitors, such as CO. Most of the discussion has

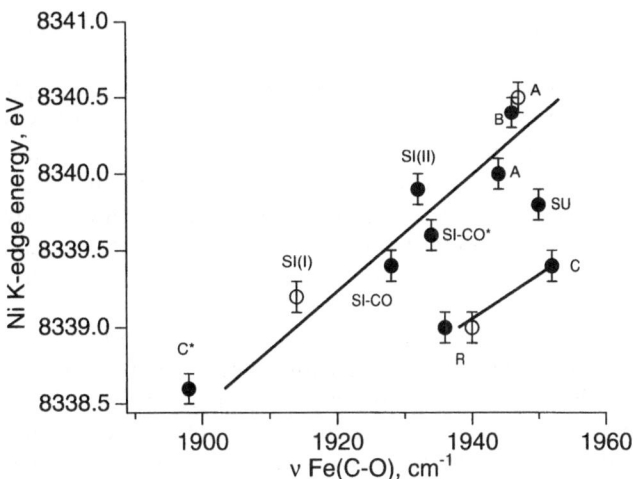

Fig. 10. Plot of Ni K-edge energy vs. ν_{CO} for the hydrogenases from of *C. vinosum* (*solid points*) and *D. gigas* (*circles*)

focused on the reduced epr-active redox state, Form C. The epr signal associated with Form C has been attributed to Ni(III) or Ni(I) complexes of H^- or H_2, based on epr and ENDOR spectroscopic studies of Form C and the changes observed upon exposure to CO [12, 96, 118–120]. In addition, the kinetics of the photolysis of Form C shows a deuterium isotope effect in D_2O that is also consistent with the existence of a M-H in Form C [118]. However, there is no observable hyperfine coupling to a proton in the epr spectrum of Form C, and the kinetics demonstrate only that a bond to a solvent-exchangeable H atom is involved in the photolysis.

^{1}H-ENDOR spectroscopy on *D. gigas* H_2ase reveals three sets of proton resonances: one set corresponding to solvent exchangeable protons with a coupling constant of 17 MHz, one solvent nonexchangeable set with a coupling constant of 12 MHz, and another solvent exchangeable set of protons that are weakly coupled (4 MHz). The nonexchangeable set, attributed to cysteine β-CH_2 protons, is detected in both oxidized (Forms A and B) and reduced (Form C) forms. The weakly coupled ^{1}H-ENDOR resonance was assigned to a protonated Ni ligand, such as a water molecule. Interestingly, this solvent exchangeable proton is not solvent accessible in Form A, suggesting a conformational change that blocks access to the active site (the ready to unready conversion?). The remaining solvent exchangeable ^{1}H-ENDOR resonance (17 MHz) is the best candidate for an H^- or H_2 ligand. Although this is a relatively large ^{1}H-ENDOR coupling constant, it does not resemble Ni-H coupling constants in paramagnetic hydrides, which are in the 300–500 MHz range and clearly detected in epr spectra [121, 122]. To account for the small coupling constant, it was suggested that this proton could be a hydride ligand bound to a Ni(III) center through a d-orbital in the x,y plane that does not contain much spin density. This possibility is supported by epr spectra obtained on $[Ni(CN_4)(H_2O)_2]^-$, where no hyperfine splitting due to equatorial $^{13}CN^-$ ligands is observed [123].

Form C' is light sensitive and is converted to an epr-active photoproduct (Form C*, g = 2.29, 2.13, 2.05) by exposure to visible light at low temperature [12, 118]. The photochemistry is reversible upon annealing the sample at around 200 K, and thus has characteristics of a ligand photodissociation and recombination reaction. It is possible that the ligand involved in the photodissociation is the putative hydride. This possibility was explored using a combination of ^{1}H-ENDOR and XAS measurements [124]. The ^{1}H-ENDOR studies reveal that the photochemistry does indeed lead to the loss of the strongly coupled ^{1}H-ENDOR resonance. However, there is no corresponding change in the Ni K-edge energy, the coordination geometry sensitive XANES features, or in the remaining ligand distances, suggesting that the proton is not bound to Ni. Since the ENDOR experiments do not identify the atom to which the H is bonded, there is no unequivocal evidence of a Ni-H interaction.

However, the presence or absence of a H^-/H_2 ligand cannot be directly detected by analysis of EXAFS. Recent experiments have addressed this issue by examining the Ni ligand environment in another photoactive complex. Carbon monoxide is a competitive inhibitor of H_2ase. One likely mechanism would involve a competition between CO and H_2 for the same binding site. Albracht and coworkers demonstrated that when CO binds to Form C, a new epr signal is generated that shows coupling to both ^{61}Ni and to ^{13}CO, thus implicating Ni as the binding site of CO [96]. Exposure to light led to the observation of the same epr signal observed in Form C*, suggesting that the photoproduct produced from the dissociation of the CO ligand is the same as that produced from Form C. Upon extensive exposure to CO, an epr-silent CO complex is produced. This CO complex has been extensively examined using IR spectroscopy [119]. The extrinsic CO complexed to the active site has a $v_{CO} = 2060$ cm^{-1} and is also light sensitive. This stretching frequency is in the range expected for terminal CO ligation to a metal. Exposure to visible light leads to the reversible photodissociation of the CO ligand. The Ni K-edge XAS studies have probed the role of the Ni in this photochemistry [125]. No evidence for a Ni-CO ligand is found in the EXAFS analysis, and photodissociation of the CO leads to very small changes in the edge energy. Given that it is unlikely that the thiolates bind CO and that the IR spectrum clearly indicates a terminally bound CO ligand, these results support a model wherein the Fe serves as the binding site for this competitive inhibitor.

Recent studies of the accessibility of gases to the Ni,Fe site in H_2ase using a combination of structural models, molecular dynamics calculations, and diffraction data collected on a crystal of *D. fructosovorans* H_2ase equilibrated with an atmosphere of Xe to identify hydrophobic channels connecting the active site and the protein surface, revealed that H_2 came a little closer to the Ni site (1.4 Å) than to the Fe site (1.8 Å) [126]. This suggests that the initial access to the active site may be via the Ni center, but cannot show that a Ni-H bond is present in Form C or in any other stable redox state of the enzyme.

Possible structures for Form C that are consistent with the data include, first, an Fe-H complex. The crystallographic studies of the structure of the active site suggest that if a hydride is bound in the active site, it is likely to occupy the space associated with the low-Z bridging atom. This would place the H atom in a position where it could bind to either Ni or Fe, or form a bridge between the two metals. However, the existence of a hydride in Form C seems to be at odds with the long-term stability of Form C in solution in the absence of H_2 [107, 108] Second, it is possible that Form C corresponds to an enzyme containing a Ni site that features thiol ligands [125]. This model features a solvent-exchangeable proton that is bound to an atom that has spin density in the active site. It is also attractive since it provides a possible explanation for the observation that Form C is stable in the absence of H_2 [108]. This stability might be due to the fact that it represents a form of the enzyme that has retained the H^+ produced from oxidation of H_2. It is also consistent with model studies that demonstrate the ability of metal thiolate ligands to bind protons [125, 127–130].

3.1.4
Hydrogenase Mechanisms: A Role for Thiolate Sulfur?

Two mechanisms outlined in Fig. 11 seem plausible in view of the data presented above. Both mechanisms use the epr silent states (SI and R) to

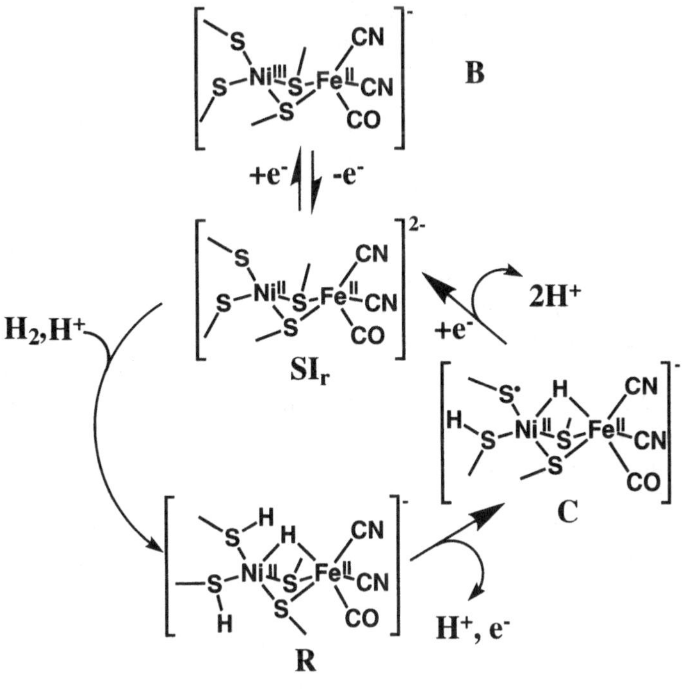

Fig. 11A.

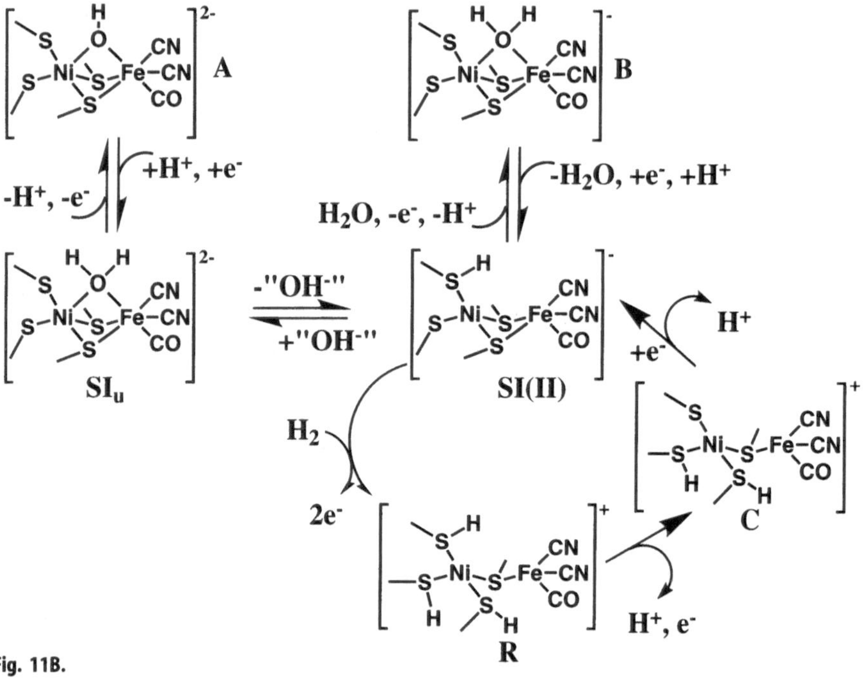

Fig. 11B.

Fig. 11. H$_2$ase mechanisms: **A** catalytic cycle including a stable hydride intermediate; **B** mechanism lacking a stable hydride intermediate and showing hypothetical structures for forms that do not lie on the reaction pathway

catalyze the 2-electron redox chemistry of H$_2$. In the first mechanism [106] the oxidized 'ready' form is ascribed to a covalent Ni(III) complex. This complex is reduced to Ni(II) in the epr-silent intermediate (SI). Operating in the direction of H$_2$ oxidation, SI is reacted with H$_2$ and protonated to produce R, which is then oxidized by one electron to Form C. A second one-electron oxidation regenerates SI with the loss of two protons. It is argued that, given the differences in the magnitude of the [61]Ni hyperfine between Forms B and C, and the observation of large hyperfine from the [77]Se in Form C [131], Form C represents a resonance form that is closer to a Ni(II)-thiyl radical complex [106]. Conversely, the larger [61]Ni hyperfine observed in Form B is more consistent with a covalent Ni(III) complex. The difference between the epr active forms (Forms B and C) or the epr silent states (SI and R) is the presence of a bridging hydride and the state of protonation of the thiolate ligands.

The scheme outlined in Fig. 11B does not assign formal oxidation state of the metals and therefore does not require them to change. The key difference between the two mechanisms is the hydride intermediate. The second mechanism does not involve a stable hydride intermediate. The difference between the epr active forms or the epr silent forms is only the state of protonation. This mechanism suggests that the H$_2$ or H$^-$ complex that may form transiently is converted rapidly into the protonated states, that electron transfer out of the active site is fast, and that the rate-determining step involves proton transport.

3.2
Carbon Monoxide Dehydrogenase/Acetyl Coenzyme A Synthase

Carbon monoxide dehydrogenases (CODHs) are a diverse group of enzymes that catalyze the two-electron redox chemistry of CO (Eq. 3).

$$CO + H_2O \rightleftharpoons CO_2 + 2H^+ + 2e^- \tag{3}$$

As was the case for H$_2$ases, Ni is not an absolute requirement for CODH activity. Aerobic microbes use an enzyme containing Mo, Fe, and flavin to oxidize CO in respiration. Anaerobic microbes utilize enzymes containing Ni and Fe,S clusters [10]. Anaerobic phototrophs oxidize CO to produce H$_2$. CODH is also involved in a primary process for the fixation of CO$_2$ into organic matter, as in acetogenic bacteria that utilize CO, a methyl-donor (a methylated, Co-corrinoid, Fe,S protein, CoFeSP) and coenzyme A (CoA) to synthesize acetyl-CoA (Eq. 4).

$$CH_3\text{-CoFeSP} + CO + CoA \rightleftharpoons acetyl\text{-CoA} + CoFeSP \tag{4}$$

Both the methyl group and the carbonyl group originate from CO$_2$, the former via formate dehydrogenase and tetrahydrofolate (H$_4$folate) methylation, the latter via the reverse of Eq. (3) [9]. Catabolism of acetyl-CoA takes advantage of the reverse of the ACS reaction to obtain energy by cleaving acetyl-CoA to yield methyl and carbonyl groups. The methyl group may be converted to methane (aceticlastic methanogenesis), or oxidized along with CO to form CO$_2$ [10].

In enzymes like that obtained from the acetogen *Clostridium thermoaceticum* the CO redox activity and the acetyl-CoA synthesis (ACS) occur at separate sites [66, 132, 133], both of which involve Ni and Fe,S clusters. The site associated with ACS activity has been designated Cluster-A, while Cluster-C is the site of CO redox catalysis. Cluster-B is an Fe$_4$S$_4$ cluster that is associated with CO redox chemistry and is believed to serve as an electron transport link between external redox partners and Cluster-C [132, 134]. In general, organisms that can catalyze the redox chemistry of CO have enzymes containing analogs to Cluster-C, and those that can synthesize or catabolize acetyl-CoA possess enzymes with active sites that have properties similar to Cluster-A.

3.2.1
The Structure of Cluster-C

At present there is no information regarding the structure of Ni sites in CODHs from crystallographic studies. All of the existing structural informa-tion has been obtained from biophysical probes of the Ni site. Early work on the structure of the Ni site in *C. thermoaceticum* CODH using XAS spectroscopy [135, 136] is complicated by the discovery that both Cluster-A and Cluster-C have Ni centers, and thus XAS data taken on the holoenzyme represent an average of the two distinct sites. Spectral data obtained on the CODH from *Rhodospirillum rubrum* [137] represent the best view of the structure of a C-cluster, since this enzyme has only the CO redox function and does not synthesize acetyl-CoA (i.e., analogs to Clusters B and C are present, but Cluster-A is absent in the *C. thermoaceticum* nomenclature). Thus, it has only the Ni center associated with Cluster-C.

R. rubrum is a phototroph that can grow anaerobically in the dark using CO as a sole source of carbon [10]. It uses CODH to oxidize CO to CO_2 in the production of H_2. The membrane-associated 61,800 Da monomeric CODH forms a 1:1 complex with a 22 kDa Fe,S protein that serves in coupling electron transfer between the CODH and a CO-induced membrane-bound H_2ase [138]. The CODH/ferredoxin/H_2ase system is thus a biological catalyst for the water-gas shift reaction (Eq. 5).

$$CO + H_2O \rightleftharpoons CO_2 + H_2 \tag{5}$$

XAS investigations were carried out on a sample of the *R. rubrum* enzyme oxidized with indigo carmine and on a sample reduced with sodium dithionite (Fig. 12, Table 3) [139]. The Ni K-edge spectra for the oxidized and reduced samples were found to be very nearly superimposable, indicating that, like in H_2ase, the additional electron does not reside in a Ni-centered orbital. Furthermore, there is no significant structural rearrangement in the Ni site upon oxidation or reduction. The geometry-sensitive XANES features do not include a resolved peak near 8338 eV, ruling out a square planar geometry for the Ni site. The intensity of the peak assigned to the $1s \rightarrow 3d$ transition is appropriate for a five-coordinate or distorted four-coordinate Ni center. Analysis of the EXAFS spectrum of the oxidized sample gave a first coordination sphere environment for the Ni site that is composed of 2–3 N/O atoms at 1.87 Å and 2 S atoms at 2.23 Å. No evidence for EXAFS arising from a Ni-Fe vector was observed. This result ruled out the existence of a NiFe$_3$S$_4$ cluster and was consistent with either an isolated Ni center or one that is bridged to an Fe,S cluster by an endogenous ligand.

The Ni center in the *R. rubrum* enzyme has been shown to be essential to the CO redox catalysis. A catalytically inactive Ni-deficient enzyme can be isolated from cells grown photosynthetically in the absence of Ni [140]. Activity is restored upon incorporation of Ni via incubation with aqueous NiCl$_2$ [141], which also restores the epr spectral characteristics of Cluster-C [142]. Line broadening observed in the epr spectra assigned to hyperfine coupling with ^{57}Fe or ^{61}Ni was used to infer the existence of a Ni,Fe,S cluster. A re-examination of the epr data obtained from ^{61}Ni-labeled samples suggest that the small line broadening observed is due to variations in sample preparation [143]. Nonetheless, the existence of a Ni,Fe,S cluster in Cluster-C in the *C. thermoaceticum* enzyme is supported by a number of spectroscopic investigations. Mössbauer studies that show that it contains an Fe,S cluster [144]. Resonance Raman data reveal vibrations that are sensitive to different Ni (365, 333 cm^{-1}) or Fe isotopes (353, 333 cm^{-1}) and hence support the

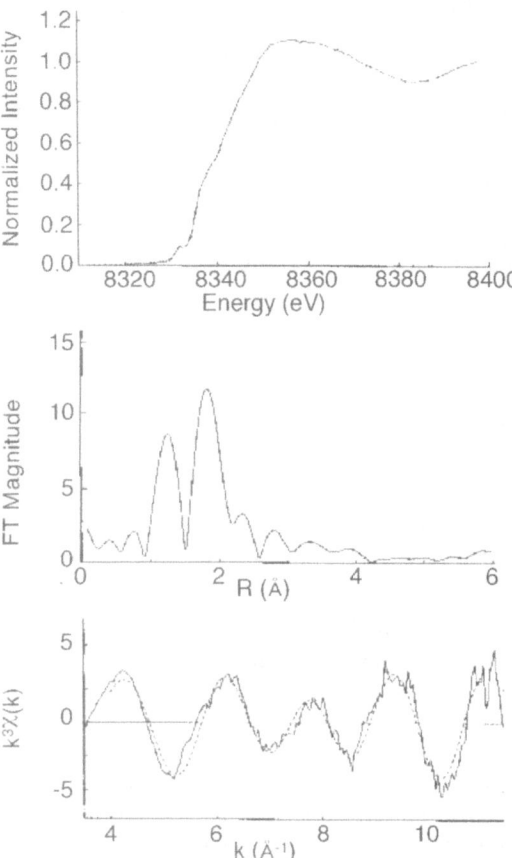

Fig. 12. XASSpectra for the Ni Site in Cluster-C in *R. rubrum* CODH. *Top*: Ni K-edge XANES spectrum. *Middle*: Fourier-transformed EXAFS spectrum. *Bottom*: Unfiltered EXAFS spectrum (*solid line*) and fit (*dashed line*). The fit shown is: 2 N @ 1.89 Å+2 S @ 2.19 Å

inclusion of both Ni and Fe in Cluster-C [145, 146]. Epr and ESEEM spectra taken upon binding azide or thiocyanate to Cluster-C were also interpreted in terms of a Ni,Fe,S cluster [143, 147, 148].

Mössbauer spectroscopic studies show that all of the Fe atoms in the Ni-deficient *R. rubrum* enzyme belong to two Fe_4S_4 clusters [143]. Cluster-B is diamagnetic in the $[Fe_4S_4]^{2+}$ state and exhibits an S = 1/2 epr signal (g = 2.04, 1.93, 1.89) in the reduced, $[Fe_4S_4]^+$ state ($E^{o'}$ = -418 mV) [149]. The Fe_4S_4 cluster that is part of Cluster-C is also diamagnetic in the 2+ state, but has an S = 3/2 ground state when reduced (g = 4–6). Upon incorporation of Ni, the spectral and redox properties of the latter Fe_4S_4 cluster are altered. The resting holoenzyme contains an oxidized Cluster-C (C_{ox}) and an oxidized Cluster-B. It undergoes a one-electron reduction at $E^{o'}$ = -110 mV to a reduced state (C_{red1}) that contains a reduced Cluster C and an oxidized (diamagnetic) Cluster-B. The C_{red1} state exhibits a characteristic S = 1/2 epr signal

Table 3. Comparison of EXAFS data for cluster-C and cluster-A

Sample	Fit #	N and atom	R (Å)	GOF	Reference
Cluster-C, oxidized	1	2.7 N	1.87	0.529	139
R. rubrum CODH		1.8 S	2.23		
(k range fit = 4–11 Å^{-1})					
	2	2 N	1.86	0.487	
		2 S	2.22		
Cluster A, oxidized	1	2 N	1.89	0.058	165
C. thermoaceticum α subunit		2 S	2.19		
(k range = 2–12.5 Å^{-1})					
	2	1 N	1.85	0.059	
		3 S			
Cluster A, reduced	1	2 N	1.89	0.052	165
C. thermoaceticum α subunit		2 S	2.19		
(k range = 2–12.5 Å^{-1})					
	2	1 N	1.86	0.054	
		3 S	2.18		

The values of GOF for analyses carried out in different laboratories was calculated differently and is not directly comparable. It is included here for the purpose of comparing different fits of the same data. Integer values of N indicate that this parameter was not independently refined. Values of the Debye-Waller factors were reported in different manners, and are not directly comparable. However, the values obtained for the N or S scattering atoms in the fits shown were acceptable. The reader is referred to the original reference for details of the analysis

(g = 2.03, 1.88, 1.71) that originates from a reduced, S = 1/2 [Fe$_4$S$_4$]$^+$ cluster, and suggests the presence of a Ni(II) center in both C$_{ox}$ and C$_{red1}$. The C$_{red1}$ cluster has four distinct Fe subsites, two of which were dubbed ferrous components II and III. The Mössbauer parameters of ferrous component II (ΔE_Q = 2.82 mm/s and δ = 0.82 mm/s) suggest a pentacoordinate Fe(II) site similar to one observed in model systems and in aconitase. The Mössbauer studies also showed that only ca. 60% of the cluster-C could attain the C$_{red1}$ state, a result that is consistent with low epr spin quantitation. When R. rubrum CODH was reduced with CO or dithionite, some of the C-clusters developed a distinct epr signal (g = 1.97, 1.87, 1.75), which is assigned to another redox state, C$_{red2}$. The C$_{red2}$ state contains both Cluster-C and Cluster-B in reduced forms and is associated with the presence of an S = 3/2 Cluster-C that may represent a fourth state of Cluster-C. The nature of C$_{red2}$ is not clear. It may represent the same redox state as the C$_{red1}$ cluster, or it may be reduced by two more electrons to generate another S = 1/2 center. It may involve another two-electron redox site, as has been suggested for Cluster-A (see below). The association of the C$_{red1}$→C$_{red2}$ conversion with the inherent two-electron oxidation of CO suggests either of the latter electronic descriptions.

The similarity between the epr and Mössbauer spectral properties of Cluster-C in CODHs from R. rubrum, C. thermoaceticum, and other enzymes, strongly points to a common structure for the site of CO oxidation [143].

Taken together, all of the available data support a structure such as shown in Fig. 13. The presence of S-donor ligands in the coordination environment of Ni

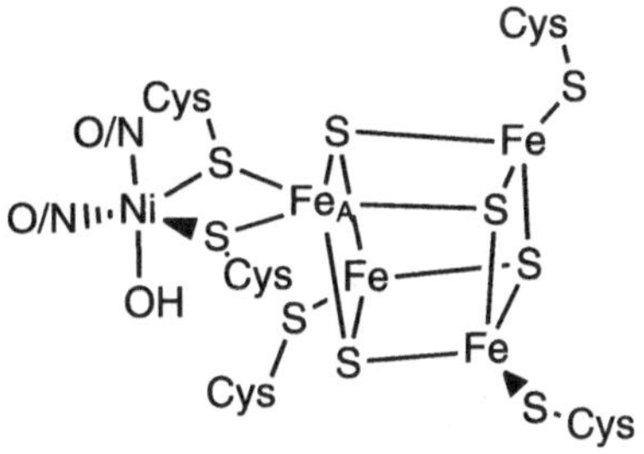

Cluster C

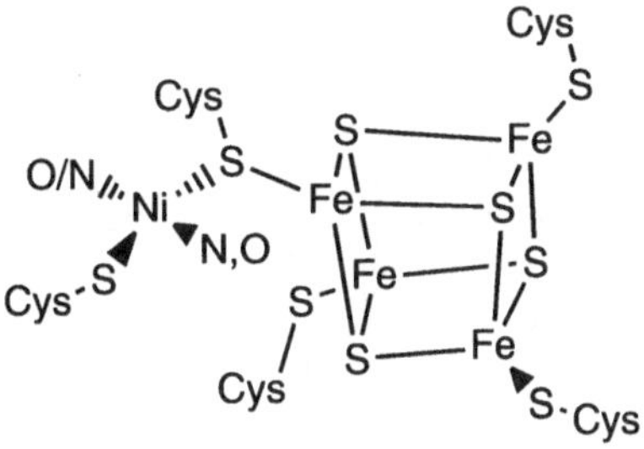

Cluster A

Fig. 13. *Top*: Model for the structure of Cluster-C in CODH. Fe_A is assigned to ferrous component II. *Bottom*: Model for the structure of Cluster A, the ACS site

[139], the precedence of the H_2ase dithiolato bridged Ni,Fe active site, and model chemistry demonstrating the ability of Ni thiolate complexes to form mono- and di-thiolato bridges to Fe [150–152] including Fe,S clusters [153] are consistent with the Ni-Fe bridge being a cysteinate ligand. The role of the Ni thiolate ligands in producing a five-coordinate site at Fe_A is speculative and is not supported by the XAS data, since a Ni-Fe vector was not detected and the data cannot discriminate between *cis-* and *trans-*orientations of the S-donor ligands. However, it cannot be ruled out by the XAS data because of difficulties in analyzing scattering atoms outside the primary coordination sphere. The reason for including it in the structure shown is mechanistic in that it represents an alternative for explaining the higher coordination number of ferrous component II without binding the catalytic OH⁻ ligand to a reduced Fe center [143, 154] or invoking an unknown ligand, L, that departs when CO binds [9].

3.2.2
The Role of Cluster-C in CODH Catalysis

The details regarding how Cluster-C catalyzes CO redox chemistry are not clear. A hypothetical catalytic cycle for CO oxidation is depicted in Fig. 14. The

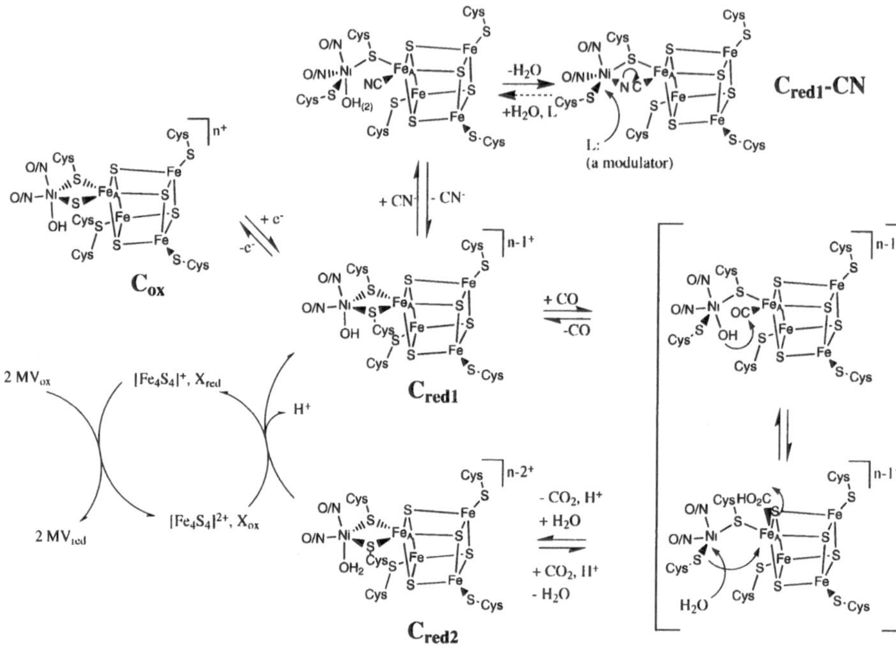

Fig. 14. Bimetallic mechanism for the catalysis of CO redox chemistry at Cluster-C in CODH. The specific roles of Fe and Ni in binding CO/CO$_2$ and OH$^-$ are somewhat ambiguous and it is possible that the CO and OH$^-$ binding sites are reversed

mechanism of CO oxidation involves the binding and deprotonation of water at a metal site to generate a nucleophilic hydroxide group that attacks a metal-bound CO to generate a metal-bound -COOH intermediate [147]. Deprotonation accompanies release of CO$_2$ to generate a Cluster-C that is two electrons more reduced (C$_{red2}$?). Electrons are transferred via Cluster-B to external redox partners (e.g., methyl viologen). In the case of the bifunctional enzymes that also catalyze acetyl-CoA synthesis, the potential redox partners include the Cluster-A ACS site. Kinetic measurements suggest that the reaction proceeds via a ping-pong mechanism, wherein CO$_2$ is released prior to the binding of the oxidant [132, 147] rather than via a ternary complex involving CO$_2$/HCO$_2^+$, CODH, and oxidant.

That Cluster-C binds CO is well established. CO binding induces the reduction of C$_{red1}$ to C$_{red2}$ in a second-order process that is faster than the turnover rate of the enzyme [132, 154]. This process is very sensitive to inhibition by CN$^-$ (much more so than ACS activity), which has been shown to bind to Cluster-C and perturbs the epr signal associated with that site [133]. Mössbauer spectra taken on the Cluster-C-CN complex show perturbations in the spectrum of ferrous component II (ΔE_Q shifts from 2.82 to 2.55 mm/s), suggesting that it is a likely site of CN$^-$ binding [143]. The inhibition by CN$^-$ is of slow and tight binding nature, and is characterized by a CN$^-$ complex that does not readily dissociate [134, 155]. The epr spectrum of C$_{red1}$ is altered upon CN$^-$ binding, and the complex shows coupling to ^{13}CN$^-$ in the ^{13}C-ENDOR spectrum (A_1 = 12.7 MHz, A_2 = 6.5 MHz, attributed to hyperfine anisotropy) [133].

The CN$^-$ inhibition is reversed in the presence of CO, and CO protects Cluster-C from CN$^-$ inhibition, suggesting a competition for a common binding site. However, the situation is not that straightforward. The rapid release of CN$^-$ observed in the presence of CO is inconsistent with the fact that CN$^-$ does not dissociate readily from the C_{red1}-CN complex. Competitive binding would also require that CO bind tightly to C_{red2} in order to explain how CO protects this state from CN$^-$ inhibition, but there is no evidence of CO binding to C_{red2} and the state can be generated using dithionite in the absence of CO. Furthermore, CO_2, CS_2, COS and SO_2 also release CN$^-$ [156], despite the fact that not all of the inhibitors exhibit competitive inhibition (e.g., CS_2) [157]. These observations led to the proposal that CO, CO_2, and CS_2 etc. bind to another site, dubbed the modulator site, that facilitates CN$^-$ release but does not compete for the same site as CN$^-$ in the C_{red1} complex [134, 156]. Furthermore, the binding of the CO, CO_2, and CS_2 affects the redox characteristics of Cluster-A and C_{red1} by facilitating the reduction of both clusters by dithionite [154]. The identity of the modulator is not known. The redox activity associated with the modulator and its existence in R. *rubrum* CODH suggest that it is associated with Cluster-C. One possibility is that the modulator site is the Ni center of Cluster-C [154], although this is difficult to rationalize with the fact that CS_2 appears to bind exclusively to Cluster-A [157].

An alternative model for CN$^-$ inhibition is suggested by vibrational spectroscopic studies of the CN$^-$ complex [146]. The data show that vibrations associated with Cluster-C (M-ligand bands in the 300–400 cm^{-1} region) are affected by CN$^-$ and that new vibrations associated with CN$^-$ ligation (^{13}C and ^{15}N sensitive) are affected by both the Ni and Fe isotopes. The data are consistent with an Fe-CN-Ni bridging mode for CN$^-$ that involves a ~140° C-N-Ni angle. This binding mode is unlikely for CO but is not inconsistent with Mössbauer results for C_{red1}-CN, nor is it inconsistent with CO protection of CN$^-$ inhibition or a model involving a modulator site aiding in the release of CN$^-$. However it suggests another mechanism for the observed slow and tight binding inhibition. The CN$^-$ could rapidly and reversibly bind to one metal and then slowly form the bridged complex, which is more stable. The presence of CO protects the site from initial CN$^-$ binding, and the release of CN$^-$ by CO would involve the loss of the bridge, forming the dissociable monodentate CN$^-$ complex.

Binding of CO to the Fe site is consistent with vibrational spectroscopic data pertaining to the binding of CN$^-$ [146]. However, it does not unambiguously identify the metal that binds CO. A model with CO binding to Fe is also consistent with spectroscopic data obtained on the similar site involved in Cluster-A (see below) and seems reasonable since ferrous component II, which, as part of an $[Fe_4S_4]^+$ cluster, is expected to be more electron rich in C_{red1} and a better π-donor than a Ni(II) site. However, CO binding by Ni cannot be ruled out entirely by the existing data.

That water and its deprotonation at a metal-bound site in Cluster-C are involved in the catalysis is supported by the dependence of the rate of the CO oxidation reaction on pH [147]. The value of k_{cat}/K_M increases with pH in association with a single group with a $pK_a = 7.7\pm0.7$. This is also reflected in changes in the epr spectrum of Cluster-C (C_{red1}) which shift in response to a group with a $pK_a = 7.2\pm0.1$ (low pH: $g = 2.005, 1.815, 1.651$; high pH: $g = 2.015, 1.800, 1.638$). The value of k_{cat} is sensitive to two ionizable groups ($pK_{a1} = 7.05\pm0.13$; $pK_{a2} = 9.46\pm0.32$), neither of which can be associated with the deprotonation of the carboxyl intermediate that is expected below pH = 4. The role of Ni in binding the OH^- nucleophile has a strong parallel in urease chemistry (see above) and is analogous to its role in binding the nucleophile (CH_3^-) involved with acetyl-CoA synthesis at Cluster-A (see below).

A bimetallic mechanism is strongly favored at this point, largely because the two metal sites in the enzyme can be tuned to bind and activate the substrates (CO and OH^-), which have very different preferences for metals (CO prefers soft metals, OH^- prefers hard metals). However, a mechanism where both CO and OH^- are bound to ferrous component II and all of the catalysis occurs at this Fe site, in analogy with aconitase catalysis [158], cannot be ruled out. In such a mechanism, the role of the Ni would involve Fe subsite differentiation and/or serving in the capacity of the modulator site.

The redox chemistry of Cluster-C that is implied by Fig. 14 is consistent with the fact that the conversion of C_{red1} and C_{red2} occurs upon exposure to CO, is kinetically viable, occurs at a potential that is near the CO/CO_2 reduction potential (−520 mV), and is associated with an inherently two-electron redox process (see above). However, the nature of the C_{red1}/C_{red2} interconversion is not clear and might be associated with the reduction of an unknown redox site, or a conformational change rather than a reduction of C_{red1} [154]. That CO_2 is bound to Cluster-C is supported by the ability of CO_2 to affect the epr spectrum of C_{red1} and to facilitate the C_{red1}/C_{red2} conversion [154].

The role of the thiolate bridge in assisting in the displacement of the product (and CN^-) shown in Fig. 14 is an unsupported possibility, and an additional potential role for the Ni thiolate ligands. It may also play a role in serving as a site for deprotonation of the -COOH complex, or in providing the requisite proton upon CO_2 binding to C_{red2}, in analogy with the role of the Ni thiolate ligands as protonation sites in H_2ase (see above). Such roles might also be played by a monothiolato bridge, although this would leave the models of C_{red1} and C_{red2} without the requisite number of Fe_A ligands.

The fact that electrons are removed from Cluster-C one at a time by Cluster-B implies the existence of an intermediate redox state (C_{int}) where Cluster-C has an integer spin state and Cluster-B is reduced. The existence of C_{int} implies a third redox component (X) that can serve to oxidize C_{int} to C_{red1} [154]. However, C_{int} and X have not been characterized and are required only if C_{red2} is two electrons more reduced than C_{red1}, and C_{ox} does not participate in catalysis. Alternatively, C_{ox} could be reduced by two electrons to C_{red2}, in which case C_{red1} becomes the C_{int} analog, but this is inconsistent with the observation that CO does not interact with C_{ox} and does react with C_{cred1} to produce C_{red2}.

The assignment of formal oxidation states to the metal sites is problematic. Mössbauer spectroscopy establishes the existence of an $[Fe_4S_4]^{2+}$ cluster in the

C_{ox} state, and a $[Fe_4S_4]^+$ cluster in both C_{red1} and C_{red2}, consistent with the similar epr signals from these states (see above). Assuming that the epr-silent C_{ox} contains Ni(II), then so does C_{red1}. If C_{red2} contains Cluster-C in a state that is two electrons more reduced than C_{red1}, this implies a two-electron reduction of the Ni center. Although low-valent Ni complexes are associated with CO reduction in model systems [159–161], the reduction of the Ni center by two electrons in the enzyme is inconsistent with the lack of any change in the structure or electron density of the Ni center as measured by XAS [139]. (The reduced sample of *R. rubrum* CODH that was used in the XAS experiments was produced by dithionite reduction and should be analogous to the fully reduced sample produced in this manner and studied by Mössbauer and epr spectroscopy [143], although this was not shown.) This situation is analogous to the situation described for [NiFe] H_2ases and suggests the involvement of unknown redox cofactors or ligand/protein redox chemistry. Alternatively, if C_{red1} and C_{red2} are isoelectronic, then no change in the Ni redox state is required, consistent with the XAS results.

3.2.3
The Structure of Cluster-A

In bifunctional enzymes that are capable of coupling CO oxidation and acetyl-CoA biosynthesis, a second Ni site is associated with the ACS activity – Cluster-A. The CODH/ACS from *C. thermoaceticum* is an $\alpha_2\beta_2$ tetramer with subunit weights of 81 and 71 kDa, respectively [9]. The CODH activity has been associated with the β subunit, while ACS activity is assigned to the α subunit [162]. The β subunit bears 75% sequence homology (46% identity) with the *R. rubrum* CODH. Mild treatment with SDS produces an $\alpha\beta_2$ fragment that has the same level of CODH activity as the holoenzyme, demonstrating that CODH activity is associated with the β subunit [163, 164]. The α subunit isolated by the above procedure does not have ACS activity, but contains Ni and an Fe_4S_4 cluster with spectral properties similar to Cluster-A [165, 166]. The isolated α subunit is competent to restore ACS activity to the $\alpha\beta_2$ fragment [166].

The isolation of an α subunit has provided the first opportunity to examine the structure of the Ni site in Cluster-A by XAS [165]. Ni K-edge XAS data obtained for the as isolated and the dithionite reduced Cluster-A in isolated α subunits is shown in Fig. 15 and summarized in Table 3. The as-isolated sample is epr silent and corresponds to the oxidized state of Cluster-A (A_{ox}), while epr spectra taken on dithionite-reduced samples reveal the epr signatures of $[Fe_4S_4]^+$ clusters with S = 1/2 and 3/2 and corresponds to reduced Cluster-A (A_{red}). The facts that the isolated subunit is not catalytically viable and that epr parameters for the reduced subunit are not identical to those from Cluster-A in the holoenzymes indicates that some structural perturbation of Cluster-A occurs in the isolated subunit. This observation is reminiscent of the situation in *R. rubrum* Cluster-C and its Ni-deficient form.

The XAS spectra display both geometry-sensitive near-edge features (a weak 1s→3d transition near 8332 eV and a transition near 8338 eV) indicative of a square-planar coordination environment for the Ni in Cluster-A. However, the

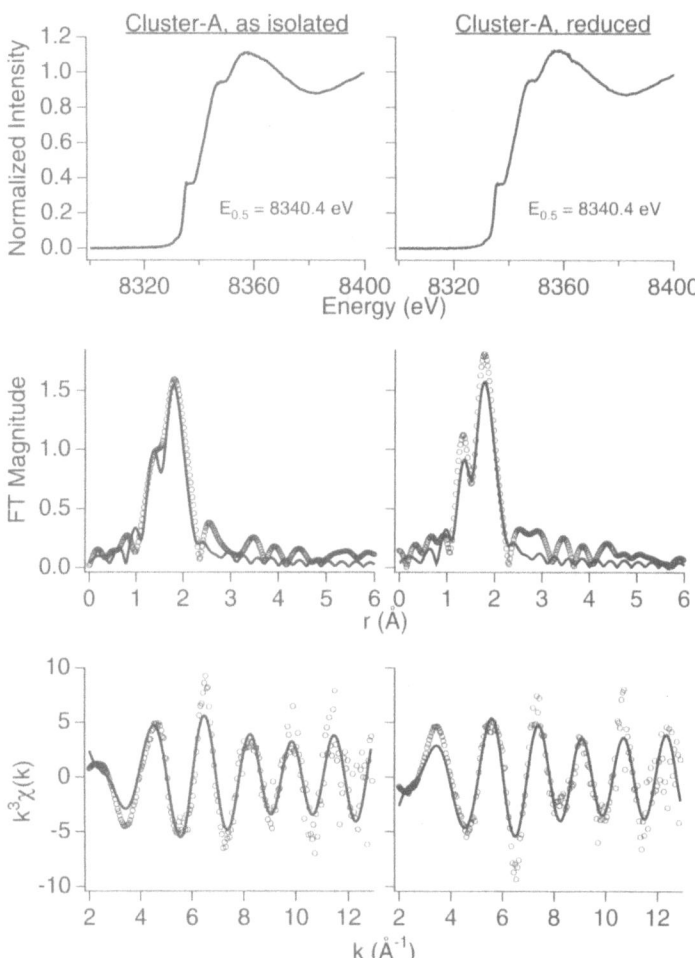

Fig. 15. Comparison of XAS data for the Ni sites in oxidized (as isolated) and reduced Cluster-A, as obtained from Ni K-edge data taken on *C. thermoaceticum* α subunits. *Top*: Ni K-edge XANES spectra. *Middle*: Fourier-transformed EXAFS spectra (*circles*) and fits (*lines*). *Bottom*: EXAFS spectra (*circles*) and fits (*lines*). The fit shown for the oxidized sample is: 1 N @ 1.89(2) Å +2 S @ 2.19(2) Å. The fit shown for the reduced sample is: 2 N @ 1.89(2) Å +2 S @ 2.19(2) Å

transition near 8338 eV is not as intense as those observed in square-planar model compounds, suggesting a distorted environment. Comparison of the intensity of the feature at 8338 eV with model compounds showed that the data was consistent with a geometry that was intermediate between planar and tetrahedral with a dihedral angle of 20–30° (planar=0°, tetrahedral=90°). The edge data obtained from the as-isolated and dithionite-reduced samples is virtually superimposable, indicating that electron density changes associated with reduction of Cluster A over the potential range of ca. +0.1 to −0.6 V are not Ni-centered, and that the Ni environment is not sensitive to the redox state of Cluster-A. Analysis of the EXAFS shows that a model composed of 2 N/O-donor ligands at a distance of 1.89 Å and 2 S ligands at a distance of 2.19 Å fits the data best. No evidence for an Fe scattering near 2.7 Å was found, demonstrating that Ni is not part of an $NiFe_3S_4$ cluster, as was the case for the Cluster-C Ni site (see above).

That Cluster-A is in fact a cluster was first demonstrated by epr and ENDOR data, showing that the unpaired spin associated with the unique S = 1/2 signal of the reduced Cluster-A in the presence of CO (g = 2.08, 2.07, 2.03) had hyperfine interactions with ^{61}Ni, ^{57}Fe, and ^{13}C (from ^{13}CO) [167, 168]. This signal, dubbed the NiFeC signal, is associated with a one-electron reduced form of Cluster-A, which has $E^{\circ\prime}$ = –541 mV. Mössbauer data are consistent with an $[Fe_4S_4]^{2+}$ cluster in A_{ox} and A_{red}, leading to a proposal that the Ni site was Ni(I) in A_{red} [144]. This is seemingly at odds with the epr properties of the α subunit [165] and with assignments based on resonance Raman data [169].

Taken together, the data support a structure for Cluster-A like that depicted in Fig. 13. Again, the inclusion of S-donors in the Ni ligand environment [165], the H_2ase structural precedence, and model chemistry suggest that the bridging group is likely to be a cysteinate ligand (see above). The general similarity in the structures of Cluster-A and Cluster-C has already been noted [9]. The structural models proposed in Fig. 13 differ only in the number of O/N ligands bound to Ni, the geometry of the two thiolate ligands (cis vs. trans), and the bridging mode of the Ni thiolate ligands.

3.2.4
The Role of Cluster-A in Acetyl-CoA Synthesis

The role of Ni in acetyl-CoA synthesis has recently been reviewed [11], and a bimetallic mechanism was proposed (Fig. 16). However, much of the support for this mechanism stems from vibrational spectroscopic investigations [145, 146, 169, 170] that have since been retracted by the authors [171, 172]. The Fe center was proposed to be the CO-binding site. FT-IR measurements revealed a single CO stretch with v_{CO} = 1995 cm^{-1}, clearly indicating the presence of a terminal CO ligand [173]. Subsequently, resonance Raman spectroscopy was used to identify an M-CO stretch that was sensitive to ^{54}Fe substitution but not to ^{64}Ni substitution. This result is not reproducible. Thus, the metal to which CO binds has not been identified. The role of the Ni center in the proposed bimetallic mechanism is as a methyl-binding site. Similarly, the resonance Raman evidence from isotopically labeled samples supporting the formation of a Ni-CH$_3$ intermediate in the methylated enzyme is unreliable, and the site of methylation is unknown. Nonetheless, the possibility of separate metal binding sites for CO and a methyl group remains, and is a feature that might explain the need for a heterometallic cluster in the ACS active site.

The natural methyl-donor for the ACS site in *C. thermoaceticum* is a Me-Co(corrinoid) complex in a protein that also contains an Fe_4S_4 cluster (Me-CoFeSP) [174, 175]. Stopped-flow techniques were used to monitor the electronic absorption spectrum of the Me-Co(corrinoid) complex during

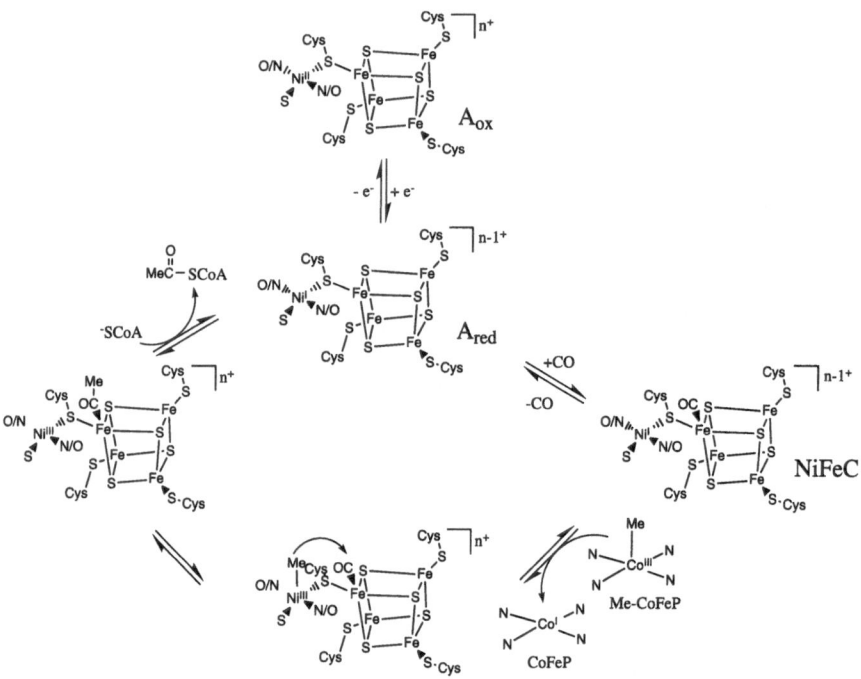

Fig. 16. Bimetallic mechanism for ACS catalysis at Cluster-A. The nature of the Cluster-A-acetyl complex has not been addressed spectroscopically, and it is possible that the acetyl group is ligated to Ni rather than Fe. In addition, the oxidation state of the clusters designated Ni(I) are ambiguous and might also be written Ni(II) with a reduced $[Fe_4S_4]^-$ moiety

methyl transfer to CODH [170]. This study provided a rate of methyl transfer that was kinetically viable and showed that the Co species formed subsequent to methyl transfer was a Co(I)corrinoid.

Two possibilities were proposed for how methyl transfer occurs [11]. Attack by Ni(I) on a CH_3-Co(III) moiety could lead to the formation of a Co(I)corrinoid and a CH_3-Ni(III) species (CH_3^+ transfer, heterolytic CH_3-Co(III) cleavage). Alternatively, prior reduction of CH_3-Co(III) to CH_3-Co(II) could lead to the formation of a Co(I)corrinoid and a CH_3-Ni(II) species (CH_3^+ transfer, homolytic CH_3-Co(III) cleavage). Studies of the stereochemical outcome of the reactions using HTDC methyl groups [176, 177], and the extreme redox potentials required to reduce CH_3-Co(III) support the heterolytic mechanism. A radical mechanism is suggested by model chemistry involving the transfer of methyl group from a CH_3-Co(III) macrocyclic complex, $[CH_3Co(dmgBF_2)_2py]$, to Ni(I)(Me$_4$cyclam). This reaction proceeds via reduction of Co(III) to form Ni(II)(Me$_4$cyclam) and CH_3-Co(II), followed by methyl transfer to another Ni(I)(Me$_4$cyclam) via a radical mechanism [178]. This chemistry has produced the first example of a stable CH_3-Ni(II) complex, $[CH_3Ni(Me_4cyclam)]^+$. The viability of CH_3-Ni(II) species to produce thio-acetyl products has also been demonstrated in model systems [179, 180].

The mechanisms discussed above have been challenged, particularly with respect to bimetallic catalysis and the redox chemistry involved [181]. A catalytic cycle that does not involve Ni-centered redox is consistent with new information and with the XAS data (Fig. 17) [165].

A fraction of the Ni in the enzyme is labile and can be removed by 1,10-phenanthroline (phen) chelation [181, 182]. Enzyme lacking the labile Ni could no longer be methylated by Me-CoFeSP. Reconstituting the enzyme with

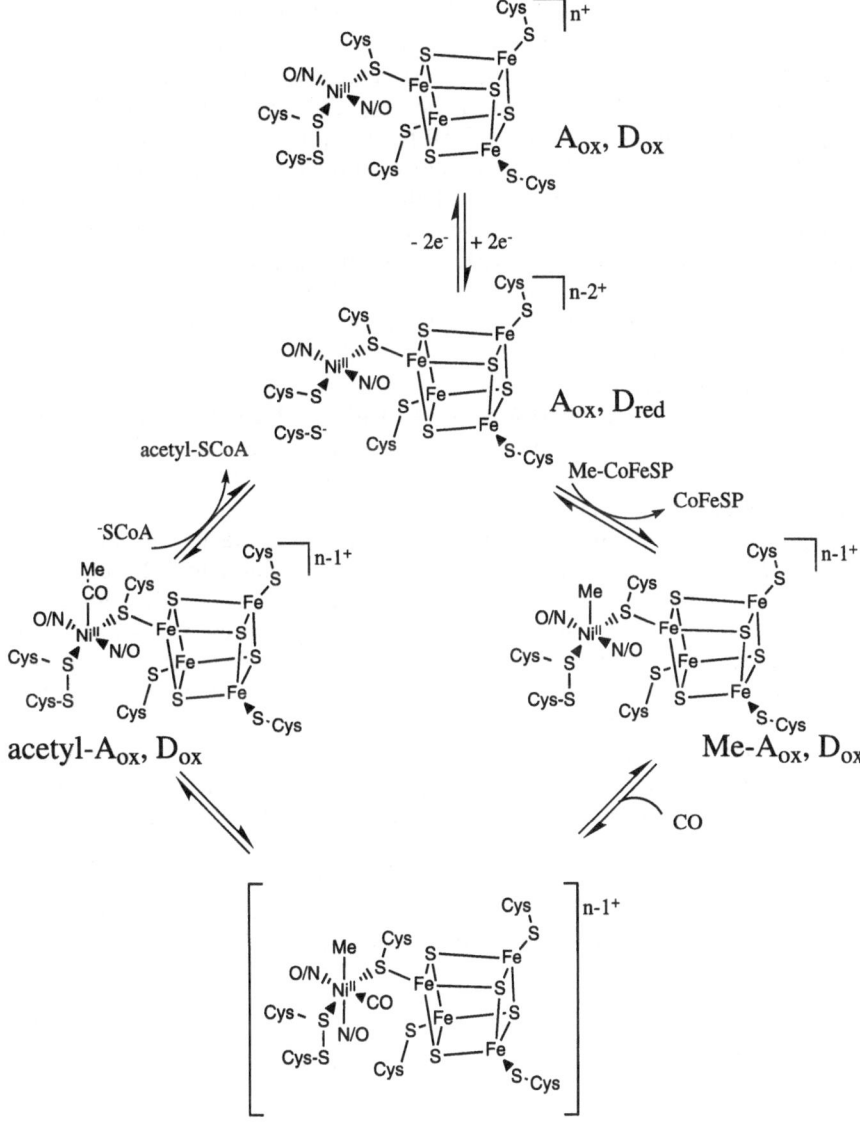

Fig. 17. A mechanism for ACS catalysis involving ligand and protein redox chemistry

$Ni(OH_2)_6^{2+}$ restored its ability to undergo methylation, and methylation inhibited the ability of phen to remove the Ni. The amount of labile Ni that could be extracted correlates with the intensity of the NiFeC signal and with the amount of functional methyl groups bound to the enzyme. This clearly associates the labile Ni with functional Cluster-A [181]. It was argued that the lability of the active site Ni and its inhibition by methylation are properties of the Ni site that are likely related to its role in catalyzing the migratory insertion of CO into a $Ni-CH_3$ bond. Such a process is facilitated by metals with two cis-oriented coordination sites, and a Ni center with open cis sites might be readily removed by a bidentate chelating ligand such as phen.

Since methylated Cluster-A is epr silent, and A_{ox} cannot be reduced by dithionite or redox mediators – yet these same reducing agents are effective in the methylation process, it must be A_{ox} that is methylated. The redox dependence of methylation ($E^{\circ\prime} < {}^-350$ mV) requires another redox site, called D, which is reduced by dithionite before methylation occurs. Since both A_{ox} and CH_3-A_{ox} are epr silent, D is probably a two-electron reducing agent that effectively reduces CH^+_3 to CH_3^-. Since the Ni in A_{ox} is almost certainly Ni(II), the oxidation state of Ni in the CH_3-Ni complex is also Ni(II) [181]. Coupling the methylation to the oxidation of the reduced D site (D_{red}) was proposed as a means of increasing the nucleophilicity of the Ni(II) site in A_{ox} and providing a thermodynamic driving force for what would appear to be an unfavorable methylation process [181].

Reaction of the A_{ox}-Ni(II)-Me species with CO leads to another epr silent species, which is assigned to the Cluster-A-acetyl intermediate [181] that has also been observed in another system [183]. The Cluster-A-acetyl intermediate was assigned to an acetyl-Ni(II) complex based on the site of methylation being Ni and model chemistry that demonstrates the facile formation of Ni-acetyl complexes upon exposure of Me-Ni(II) complexes to CO [180].

In the last step of the proposed mechanism (Fig. 17), $^-$SCoA attacks the Cluster-A-acetyl intermediate and reduces D_{ox} by two electrons to D_{red}. The redox properties of the D site are not consistent with any of the redox properties associated with Cluster-C or Cluster-B, and may correspond to a redox active pair of cysteine residues [181].

The catalytic cycle given in Fig. 17 has no role for the NiFeC intermediate, and it was suggested that, despite properties that indicate that it is possible that it is a catalytic intermediate, it is in fact an artifact of the reduction of Cluster-A under a CO atmosphere [181]. This suggestion is also supported by the observation that the NiFeC epr signal (A_{red}-CO) disappeared upon methylation, suggesting that it was a product of a nonfunctional side reaction [183]. In fact, the proposed mechanism does not involve an A_{red} or the Fe_4S_4 moiety in Cluster-A at all, and leaves one wondering what role is served by the Fe_4S_4 cluster.

At this point, D has not been identified, and given the fact that only six conserved cysteine residues exist among α subunits from various sources, a role for thiolate ligands is suggested [181]. One possibility is that one of the Ni S-donor ligands serves as the bridging group, and that the other is involved in the redox chemistry with a cysteine that is not bound to a metal. This possibility is illustrated in Fig. 17 and is supported by the two electron oxidation of a model Ni thiolate complex [184]. Oxidation of a structurally characterized Ni(II) dimeric system, formed with trans-NS$_2$-donor ligand by two electrons per Ni, leads to the quantitative oxidation of the thiolate ligands to disulfides. The resulting structurally characterized Ni(II) complex contains disulfides that are bound to Ni in a monodentate fashion (Fig. 18) [184].

3.3
Methyl Coenzyme M Reductase

Methyl-coenzyme M reductase (MCR) catalyzes the last step in methane production from CO$_2$ by methanogenic bacteria. This process couples the reaction of a thioether, 2-(methylthio)ethane sulfonate (Me-Coenzyme M, MeS-CoM), and a thiol, 7-thioheptanoylthreoninephosphate (HS-HTP, Coenzyme B, HS-CoB), to form a disulfide with the formation of methane (Eq. 6) [14–16].

$$(6)$$

The disulfide is subsequently reduced to HS-CoM and HS-CoB by hydrogen at a complex containing a heterodisulfide oxidoreductase and H$_2$ase [185]. Methylation of HS-CoM occurs via transfer of a methyl group from N5-methyl-methanopterin [186, 187].

MCR is typically a dimer of trimers, $\alpha_2\beta_2\gamma_2$, with a molecular weight near 300 kDa. It contains two molecules of coenzyme F$_{430}$, which is attached in a non-covalent manner. Coenzyme F$_{430}$ is a Ni tetrapyrrole (hydrocorphin) that is the active site of the enzyme (Fig. 19). The hydrocorphin macrocycle is the most saturated 16-membered tetrapyrrole that has been found in nature and has one ring that bears a negative charge (ring C), making the Ni(II) complex a monocation.

3.3.1
The Structure F$_{430}$

Information regarding the structure of F$_{430}$ was originally obtained on isolated cofactor and is only recently available from crystal structures of intact MCR [65]. When methanogens are grown under conditions that are not limited by the amount of Ni present, there is a large amount of free cofactor present in the cytosol that can be extracted with ethanol. However, the isolated cofactor is

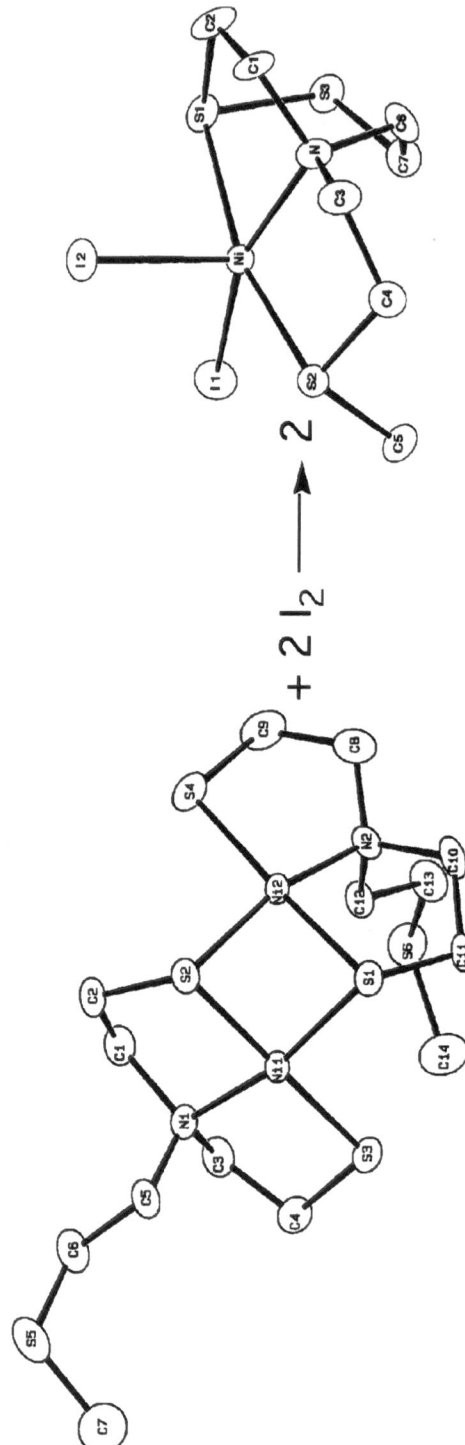

Fig. 18. Two-electron oxidation of a Ni(II) thiolate complex to form a Ni(II) monodentate disulfide provides a model for the hypothetical mechanism of the ACS cluster. The crystal structures shown are from [182]

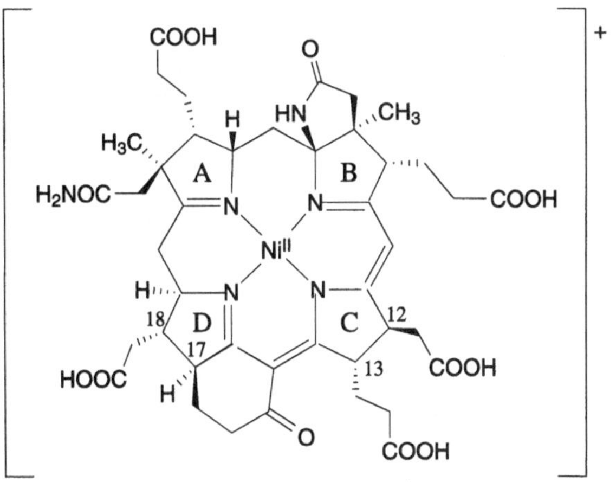

Fig. 19. Structure of isolated cofactor F_{430}

thermally unstable and oxygen-sensitive, leading to modifications of ring C. Thermal degradation leads first to the 13-epimer, 13-epi-F_{430} (C12-S, C13-R) and then to the 12,13-diepimer, 12,13-diepi-F_{430} (C12-R, C13-R). Oxidation leads to the formation of 12,13-didehydro-F_{430} [188, 189]. Reduction of 12,13-didehydro-F_{430} with H_2 and other reducing agents produces native F_{430}. The pentamethylester of F_{430}, $F_{430}Me_5$, may be prepared by methanolysis of the isolated cofactor, and has better solubility properties.

The structure of native F_{430} (Fig. 19) is known from an X-ray diffraction structure of the bromide salt of the pentamethylester, 12,13-diepi-$F_{430}Me_5$ [190] and from NMR studies of native F_{430} and $F_{430}Me_5$ [190–193]. The elegant NMR studies of the cofactor yielded most of the structure and showed that native F_{430} and $F_{430}Me_5$ had no significant structural differences. However, the NMR studies were unable to assign the configuration of C17, C18, and C19 in ring D (Fig. 19). This ambiguity was removed by the crystal structure, which shows the configurations to be S, S, and R, respectively. The crystal structure also reveals that the tetrapyrrole distorts from planarity in order to accommodate the low-spin Ni(II) center. The Ni-N bonds in 12,13-diepi-$F_{430}Me_5$ average 1.86(2) Å and reflect the short Ni-N distances typical of low-spin Ni(II) tetraazamacrocycles (for examples see the review by Telser [16]). In order to accommodate the short Ni-N distances, the macrocycle undergoes an S_4 ruffling that was predicted by the structures of model Ni tetrapyrroles [190, 194]. The degree of ruffling observed in 12,13-diepi-$F_{430}Me_5$ is the largest observed in Ni(II)tetrapyrroles [190] and reflects the short Ni-N bonds and the flexibility of the highly saturated hydrocorphin ligand [195].

XAS investigations have played an important role in understanding the structural differences between the various isolated forms of F_{430}, the MCR-F_{430} complex, and the structural changes associated with reduced forms of F_{430}. Figure 20 and Table 4 summarize the results of XAS studies of F_{430}. The coordination number/geometry of the various samples is clearly indicated by

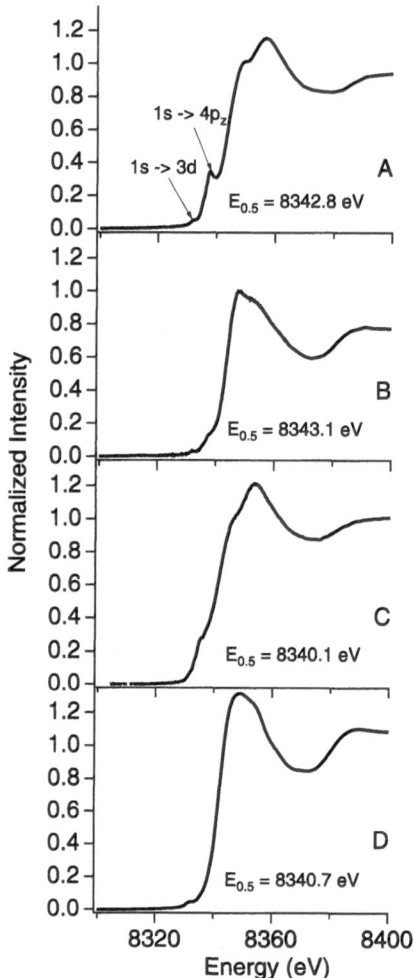

Fig. 20A–D. Ni K-edge XAS spectra of F_{430} samples illustrating the effect of coordination number/geometry on XANES features: **A** 12,13-diepi-F_{430} (4-coordinate planar NiN_4); **B** native, isolated F_{430} (extracted with LiCl, mostly 6-coordinate NiN_4O_2); **C** isolated Ni(I)F_{430} ($NiN_2N'_2$); **D** MCR complex (6-coordinate, N_4O_2)

the pre-edge XANES spectra (Fig. 20). The feature near 8338 eV that is associated with a $1s \rightarrow 4p_z$ electronic transition is present as a resolved maximum in four-coordinate planar versions of the cofactor containing low-spin (S=0) Ni(II). In six-coordinate complexes, where the Ni ion has additional axial ligands and a high-spin (S=1) Ni(II) electronic configuration, this feature is absent. This data shows that the Ni is six-coordinate in both the enzyme and in the native isolated cofactor [196–198]. Other techniques have shown that, at room temperature, the native cofactor is an equilibrium mixture of four-coordinate planar and six-coordinate species [197]. This equilibrium is

Table 4. Structural information regarding F_{430}

Sample	Metal–Ligand	Distance (Å)	Reference
	Structural Data from Crystallography		
M. thermoautotrophicum MCR$_{ox/silent}$ (1.45 Å res.)	Ni–N	2.14	65
	Ni–N	2.11	
	Ni–N	2.10	
	Ni–N	1.99	
	Ni–O(Gln$_{\alpha 147}$)	2.3	
	Ni–S-CoM	2.4	
	ave. Ni–N	2.09	
M. thermoautotrophicum MCR$_{silent}$ (2.0 Å res.)	Ni–OSO$_2$-CoM	2.1	65
12,13-diepi-F_{430}Me$_5$	Ni–N	1.87	188
	Ni–N	1.87	
	Ni–N	1.85	
	Ni–N	1.84	
	ave. Ni–N	1.86	
	Structural Data from EXAFS Analysis		
M. thermoautotrophicum MCR	(5–6) Ni–N	2.09(2)	194
Native F_{430}	(6) Ni–N	2.10(2)	195
	(8) Ni–C	3.03	
F_{430}Me$_5$	(4) Ni–N	1.90(2)	197
13-epi-F_{430}	(4) Ni–N	1.89(2)	198
12,13-diepi-F_{430}	(4) Ni–N	1.89(2)	198
	(8) Ni–C	2.90(2)	
12,13-diepi-F_{430}(1-MeIm)$_2$	(4) Ni–N	2.05(2)	198
	(2) Ni–N	2.15(2)	
	(8) Ni–C(F_{430})	3.02	
	(4) Ni–C(1-MeIm)	3.15	
	ave. Ni–N	2.08(2)	
12,13-diepi-F_{430}(CN)$_2$	(4) Ni–N	2.11	198
	(2) Ni–C	2.01	
	(8) Ni–C(F_{430})	3.13	
	(2) Ni–N(CN)	3.23	
12,13-didehydro-F_{430}	(4) Ni–N	1.88(2)	195
Reduced F_{430}Me$_5$	(2) Ni–N	1.88(3)	197
	(2) Ni–N	2.03(3)	
	ave. Ni–N	1.96	

sensitive to the nature of the solvent, to temperature, and to the presence of other ligands. Other forms of the cofactor, including F_{430}Me$_5$ [199], 13-epi-F_{430} [200], 12,13-diepi-F_{430} [200], 12,13-didehydro-F_{430} [200], are four-coordinate planar Ni(II) complexes under the conditions employed in obtaining the spectra. In no case is a five-coordinate geometry indicated by XANES analysis. The structure of the NiN$_4$ site is not perturbed by the chemistry occurring at the C12 and C13 positions of the cofactor. Using the reversible redox chemistry associated with F_{430} (6-coordinate) and 12,13-didehydro-F_{430} (4-coordinate), it was shown that the axial ligands to the isolated native cofactor are water molecules [197].

The difference in the electronic configuration of the Ni(II) center is also apparent in the EXAFS spectra of various forms of F_{430} (Table 4). The addition of electrons to the $d_{x^2-y^2}$ orbital in the high-spin electronic configuration causes a lengthening of around 0.2 Å in the Ni-N bonds to the cofactor. These distances are ~1.9 Å in the four-coordinate planar forms and ~2.1 Å in the six-coordinate forms. Analysis of features arising from C atoms in the cofactor show larger Ni-C distances in the six-coordinate forms and confirm that the macrocycle expands to accommodate the larger bond lengths.

The six-coordinate nature of F_{430} bound to MCR is also apparent in the resonance Raman and electronic absorption spectra of the cofactor [197, 200, 201]. The magnetic properties of F_{430} bound to MCR were explored via MCD spectroscopy, which clearly indicated a high-spin, six-coordinate Ni(II) center in the enzyme [202]. EXAFS analyses were consistent with O/N-donor axial ligands to MCR bound cofactor F_{430} (Table 4). Measurement of the zero-field splitting parameter, D, via MCD spectroscopy in isolated cofactor complexes showed that it was sensitive to the nature of the axial ligands. The value of D observed in the enzyme (D = +10±1 cm^{-1}) was similar to the bis-aquo ligated isolated native F_{430} complex (D = +9±1 cm^{-1}) and distinct from a bis-imidazole complex (D = −8±1 cm^{-1}). On this basis, O-donor ligands were favored and coordination by His in the enzyme was ruled out [202].

Reduction of planar $F_{430}Me_5$ produces Ni(I)$F_{430}Me_5$ [199]. The reduction is accompanied by a shift in the Ni K-edge to lower energy by 2-3 eV and by distortion of the macrocycle, such that the Ni-N bonds are no longer of equal length (2 @1.88 Å+2 @ 2.03 Å). These results are supported by a number of model studies [194, 203–206] that confirm the structural effects of metal-centered vs. ligand-centered reduction. Ni-centered vs. ligand centered reductions were easily discerned by the presence or absence of a ~3 eV Ni K-edge energy shift, and are also reflected in the epr and electronic absorption spectra. Furthermore, in the Ni(I) complexes, the geometry sensitive features remained apparent on the edge, suggesting that Ni(I) exists in a distorted planar geometry in the tetrapyrrole macrocyles. These studies point to the flexibility of the saturated F_{430} macrocycle as being a key feature in its ability to accommodate low-spin Ni(II), high-spin Ni(II) and Ni(I) centers.

The reduced form of the cofactor bound to MCR has been examined by epr and ENDOR techniques. Reduced isolated native F_{430} gives an axial spectrum (g=2.224, 2.061) [207] that is similar to the spectra obtained for a reduced form of the enzyme (MCR$_{red1}$: g = 2.260, 2.088) [208], 12,13-diepi-F_{430} (g = 2.238, 2.057) [207], $F_{430}Me_5$ (g = 2.250, 2.074, 2.065) [209], and model systems [194, 203]. The g-values observed are consistent with a Ni(I) system, since ligand radical anions would be expected to exhibit isotropic spectra with g-values near 2.00 [194, 203]. No evidence for the axial coordination of the cofactor by N-donors or by solvent (D_2O) was seen in MCR$_{red1}$ by ESEEM [207] or ENDOR spectroscopy [210]. The epr spectra often exhibit ^{14}N-hyperfine coupling to the pyrrole N atoms of around 30 MHz, which was seen in the ENDOR study of

MCR_{red1}. This study also provided evidence for the existence of the two distinct types of N atoms, in agreement with EXAFS analysis (see above).

The recent crystallographic characterization of two forms of MCR have provided great insight into the structure of the intact cofactor and its relationship to bound CoM and the disulfide reaction product (Fig. 21) [65]. The crystallographic data regarding the structure of the Ni site is summarized and compared with the structural data from XAS in Table 4.

The first structure was determined for a sample derived from enzyme isolated in the epr-active M_{ox} state. However, M_{ox} is slowly converted to an epr silent (Ni(II)) state that is inactive ($MCR_{ox/silent}$). The structure of $MCR_{ox/silent}$ was refined to a resolution of 1.45 Å and provides an excellent look at the holoenzyme with bound CoM and CoB [65]. The overall enzyme structure is composed of a series of α-helices arranged in a compact ellipsoidal shape, and shows that the dimer of trimers quaternary structure is functionally important. The two F_{430} cofactors present are bound at identical sites that are roughly 50 Å apart and at the bottom of a 30 Å long narrow channel that is formed by residues from subunits α, α', β, and γ, with the equivalent site formed by residues from α', α, β', and γ'. The Ni center is six-coordinate and lies in the N_4 plane of the nearly planar cofactor, which is shown to have the same conformation determined for isolated cofactor (Fig. 19). The Ni-N bonds to the pyrrole N atoms have distances of 2.14 Å (ring A), 2.11 Å (ring B), 2.10 Å (ring C), and 1.99 Å (ring D). These distances contrast with distances of 1.87, 1.87, 1.85, and 1.84 Å, respectively, found in the structure of isolated, four-coordinate, 12,13-diepi-$F_{430}Me_5$ [190]. The average Ni-N distance in the MCR bound cofactor is 2.09 Å, in excellent agreement with EXAFS data, and consistent with the high-spin six-coordinate geometry found from XAS analysis (see below). The fifth ligand is an O-donor from $Gln_{\alpha147}$ at a rather long distance (Ni-O=2.3 Å). The sixth ligand is the S atom from the bound CoM (Ni-S=2.4 Å). The Ni-S distance is quite similar to the axial Ni-S distances found in model Ni macrocyclic complexes (2.452(4) Å) [211] and 2.369(2) Å [206]. The S-donor ligand was not detected by EXAFS and may not have been present in the form of the enzyme that was examined by XAS.

Cofactor F_{430} is rigidly, but non-covalently, bound to the protein via several interactions. These include 21 H-bonding interactions (15 to peptide amide N atoms) and hydrophobic and aromatic interactions. The cofactor is oriented with the front face pointing toward the channel exit. The channel, which extends from the surface of the protein deep into the interior, also contains CoM and CoB.

CoM is positioned nearly parallel to the front face of the F_{430} plane, in analogous manner to thiolate in the model system cited above, with the S-atom in an axial position of the Ni. The S atom has H-bonding interactions with two Tyr residues and a water molecule that bridges between CoM and CoB. The sulfonate group forms a salt bridge to Arg 120 and has additional H-bonding interactions. CoB resides in the channel in an elongated conformation with the thiol end directed toward the Ni center with a Ni-S distance of 8.7 Å.

The second structure was obtained on a sample that was isolated under reducing conditions. This treatment leads to the MCR_{red} epr-active state (see

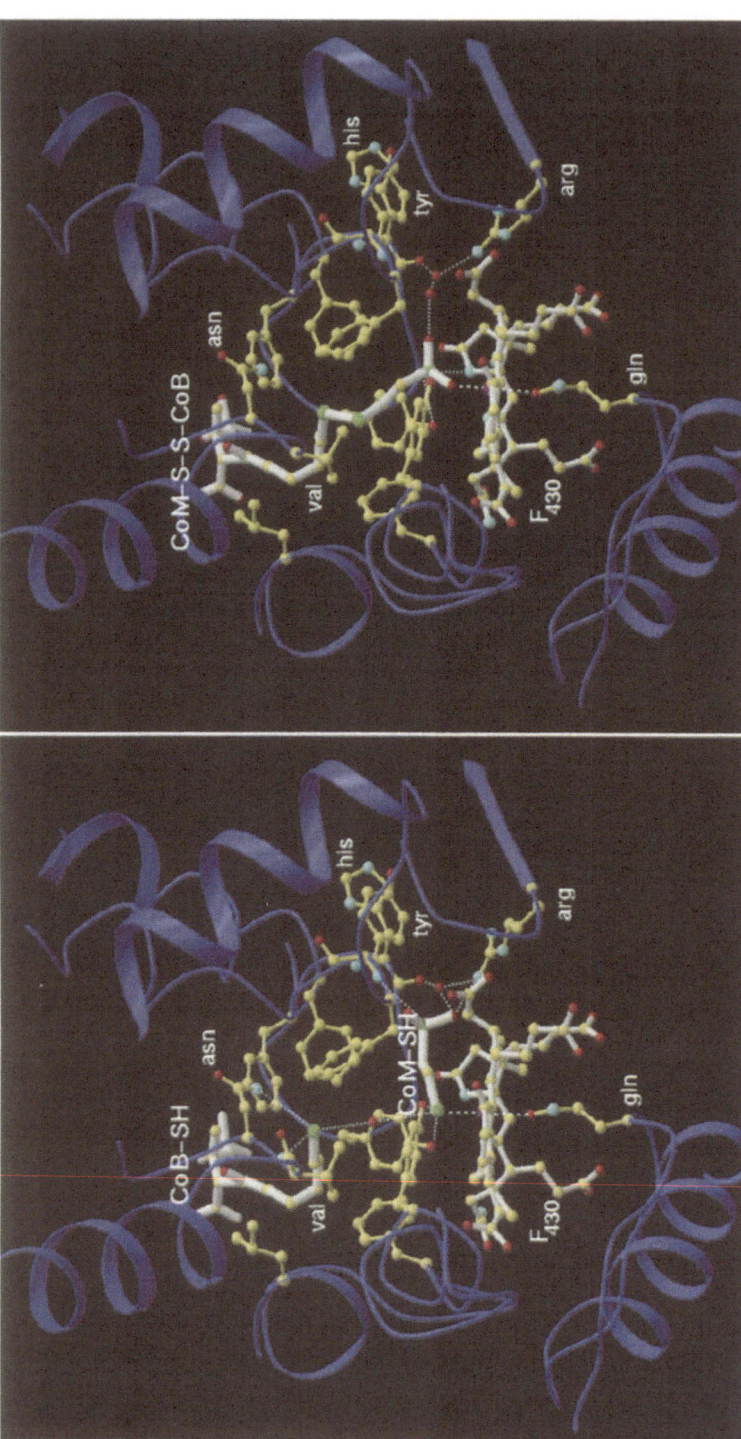

Fig. 21. Structures of F_{430} in methyl coenzyme M reductase from the crystal structures of $MCR_{ox/silent}$ and MCR_{silent}

below). This state is also unstable and decays under strictly anaerobic conditions to an epr-silent form MCR_{silent}, from which the structure was obtained. In the MCR_{silent} (Ni(II)) structure, CoM and CoB are linked as the heterodisulfide product. The CoB is little changed in the structure and nearly superimposable with its position in the structure of $MCR_{ox/silent}$. CoM has moved by more than 4 Å from its position in $MCR_{ox/silent}$ so that its S atom is lifted into the channel to form the disulfide and the sulfonate group is bound by an O-donor atom (Ni-O = 2.1 Å). This bonding mode is the favored one in a model complex that lacks the H-bonding and salt bridges that involve the sulfonate moiety in the $MCR_{ox/silent}$ structure [206].

3.3.2
The Mechanism of Methane Production at F_{430}

Several mechanisms for the catalysis of methane formation have been proposed on the basis of studies of MCR and model systems, and these have been reviewed [5, 6, 14, 16]. Mechanisms involving Ni(I), Ni(III), or both have been postulated [14, 212]. Several epr signals attributable to MCR species have been characterized and support a redox role for the Ni cofactor [213]. Aerobically purified MCR exhibits an epr signal assigned to an inactive species, MCR_{ox1} (g = 2.227, 2.159), that is lost upon removal of the cofactor from the holoenzyme. Isolation of highly active MCR can be achieved if the cells are treated with H_2 prior to cell disruption. These cells were found to exhibit two epr signals corresponding to reduced forms of the enzyme, MCR_{red1} (g = 2.260, 2.088) and MCR_{red2} (g = 2.285, 2.235, 2.184). The intensity of these signals correlates with enzyme activity. Enzyme purified from these extracts in the presence of Me-S-CoM retains the MCR_{red1} epr signal, and MCR_{red2} is converted to MCR_{red1} upon addition of Me-S-CoM [208]. If MCR_{red2} is exposed to CO_2 it becomes epr silent, whereas exposure of MCR_{red2} to O_2 generates an epr active oxidized form, MCR_{ox2} (2.24, 2.112). The association of active enzyme with reduced and epr active forms of F_{430} suggests that reduced cofactor is involved in catalysis.

Studies of alternate substrates identified Et-S-CoM and Me-Se-CoM, among others. Substitution of S by O, gave a weak inhibitor, and substitution of Me with groups larger than Et were not substrates [3, 214, 215]. These results suggested that a Ni-chalcogenide bond forms and that the binding pocket imposes steric constraints. Reaction of MCR with R- and S-[1-^{2}H,^{3}H]-Et-S-CoM determined the sterochemical course of the reaction. The reaction was shown to proceed via inversion of configuration, a result consistent with displacement of S by Ni(I) during catalysis, followed by cleavage of the Ni-Me bond by a proton [216]. However, isolated reduced $F_{430}Me_5$(Ni(I)$F_{430}Me_5$, $E^{o\prime}$ = –0.89 V vs SCE, g = 2.250, 2.074, 2.065) will not react with thioethers [14]. The reaction of Ni(I)$F_{430}Me_5$ with more reactive methyl-donors does occur, and the reaction proceeds through a Me-Ni(II)$F_{430}Me_5$ intermediate that can be prepared by reaction of Ni(II)$F_{430}Me_5$ with Me_2Mg [14]. In addition, the isolated cofactor can be reversibly oxidized to Ni(III)$F_{430}Me_5$ ($E^{o\prime}$ = +1.25 V vs SCE, g = 2.211, 2.020).

The enzyme is highly specific for HS-CoB, and changes in the length of the methylene chain or methylation of the thiol S atom results in inhibition [3].

The reaction mechanism shown in Fig. 22 is supported by the crystal structures of $MCR_{ox/silent}$ and M_{silent} [65]. This mechanism suggests that the reduced enzyme containing $Ni(I)F_{430}$ serves as a nucleophile to the methyl group of Me-CoM, forming a Ni(III)-Me complex. This complex is reduced via oxidation of HS-CoM to a thiyl radical. Protonation of the Ni-Me complex leads to release of methane. The thiyl radical oxidatively couples with the thiolate of S-CoB to form the disulfide product and reduce Ni(II) to Ni(I).

The structures place a number of restrictions on possible mechanisms and suggest certain features that support specific reaction intermediates [65]. The catalytic site is identified as the front side of F_{430}. This site is accessed via only one channel whose dimensions would exclude molecules with a diameter >6 Å, and is completely blocked when CoB is bound. Thus, Me-S-CoM must bind first, which is consistent with a mechanism that involved the formation of a ternary complex in this way, that was proposed on the basis of kinetic studies [217]. The distance between the S atom of CoB and the Ni center (8.7 Å), coupled with the fact that longer analogs are inhibitors, makes it unlikely that CoB binds to the Ni. The enzyme provides a good binding site for the sulfonate group of Me-S-CoM, but it was regarded as unlikely that a Ni-thioether bond forms in the Me-S-CoM complex because this would place the two S atoms involved in forming the disulfide 6.2 Å apart. Models show that the S atoms are in van der Waals contact when the methyl group of Me-S-CoM is placed in van der Waals contact with the Ni. Thus, a Ni-Me intermediate is supported by the structure. The active site is lined with a number of aromatic residues and lies in a generally hydrophobic environment. Assuming that the water molecule in the active site would be displaced by Me-S-CoM, this hydrophobic environment is compatible with radical intermediates in the reaction mechanism. Several potential proton donors to the putative Ni-Me intermediate were identified; transfer of the proton from HS-CoB via CoM is an attractive possibility.

3.4
Superoxide Dismutase

Superoxide dismutases (SODs) catalyze the dismutation of O_2^- as shown in Eq. (9) by exploiting the one-electron redox chemistry of a metal center (Eqs. (7) and (8)) [218].

$$M^{n+} + O_2^- \rightleftharpoons M^{(n-1)+} + O_2 \tag{7}$$

$$M^{(n-1)+} + O_2^- \rightleftharpoons M^{n+} + O_2^{2-} \tag{8}$$

$$2O_2^- \rightleftharpoons O_2 + O_2^{2-} \tag{9}$$

The enzyme appears to play a role in protecting cells from the deleterious effects of reactive oxygen species, including superoxide radical and hydroxyl

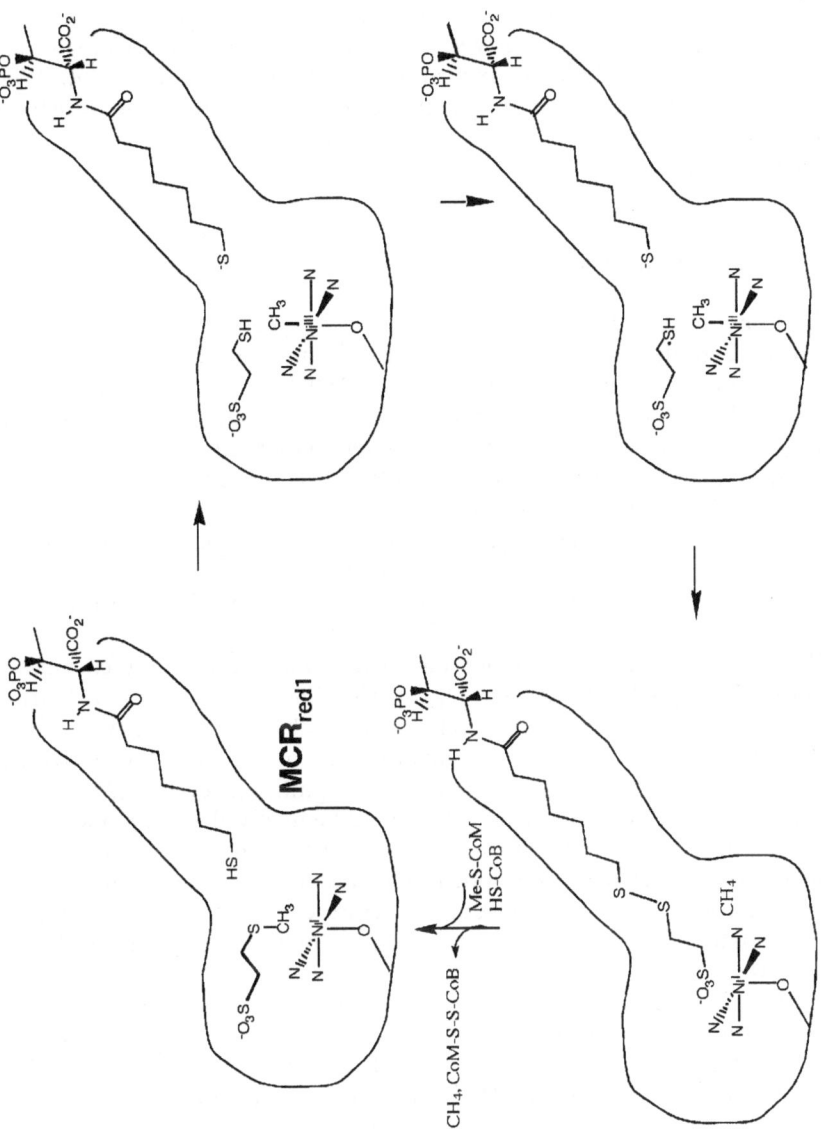

Fig. 22. Reaction mechanism for MCR catalysis (adapted from [65])

radical, and has been associated with a number of pathological conditions [218, 219]. The activity of the enzyme is enhanced by the presence of HS⁻, and a possible role in sulfide detoxification has been suggested [220]. Well-characterized examples that utilize the Cu(II/I), Fe(III/II), and Mn(III/II) redox couples are known [221, 222]. In 1996, Ni joined the list of metals found in SODs [17, 18].

3.4.1
The Structure of the Ni Site in NiSOD

Nickel-containing SODs were isolated from various *Streptomyces* species, which are aerobic soil bacteria. These cytoplasmic enzymes have a subunit mass determined to be around 13 kDa by SDS/PAGE. Gel filtration chromatography showed that the holoenzyme had an apparent molecular mass of around 60 kDa, suggesting a tetrameric quaternary structure that is also found in a few Fe and Mn SODs [17, 18]. With the exception of an unusually large amount of glycine, the amino acid content of NiSODs is unremarkable, and particularly so with respect to cysteine (1–2 residues/subunit). The *N*-terminal amino acid sequences of the NiSODs are highly homologous and distinct from other SODs. The NiSODs were also shown to be immunologically distinct from representative Fe, Mn, and Cu/Zn SODs [17, 18].

That Ni is involved in the redox chemistry catalyzed by these enzymes is supported by a number of studies [17, 18]. The NiSOD has a specific activity of around 50 units/ng-atom of Ni, which is comparable to the activity of bovine erythrocyte Cu/Zn SOD (55.0 units/ng-atom of Cu). Analysis for Fe, Ni, Cu, and Zn showed that only Ni was present at significant levels (0.74–0.89 Ni/ subunit), and thus no other metal is known to be involved and a mononuclear site is implied. Most of the Ni can be removed from the enzyme by dialysis against various chelators, and apoenzyme was found to have only residual activity. Reconstitution with aqueous Ni^{2+} restored only 12.5% of the activity, but information regarding the amount of tightly bound Ni in the reconstituted enzyme was not reported. An electronic absorption at 378 nm was associated with Ni by comparison of spectra obtained on apo- and holo-enzyme.

The resting (oxidized) enzyme displays a rhombic epr signal (g=2.304, 2.248, 2.012) attributed to the presence of a Ni(III) center [17, 18]. This assignment was based on the presence of Ni in the enzyme and the similarity of the spectrum to those seen in Ni(III) oligopeptides [223–227]. Peptide complexes containing Ni(III) are associated with deprotonated amide ligation, suggesting that Ni-SOD may be a rare example of this type of ligation in a protein. A structurally characterized model compound featuring this type of ligation to a Ni(II) center has been reported [228]. This structure is of an intermediate in the oxidative decomposition of a Ni(II) complex of glycyl-glycyl-histidine by O_2 that proceeds through a Ni(III) intermediate (see below).

The ligand and protein environment in NiSOD that stabilizes Ni(III) and prevents the chemistry seen in model systems is largely unknown and remains an interesting structural goal in Ni biochemistry. The high field epr feature in the enzyme is split into a triplet with a hyperfine coupling constant of 20G,

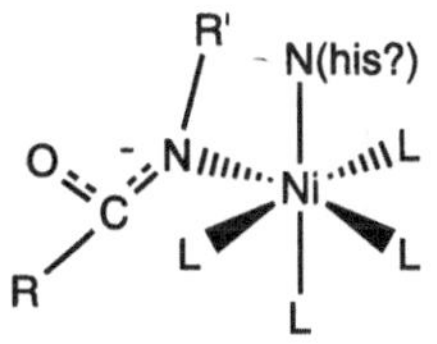

Fig. 23. Structural features associated with the Ni site in SOD

indicating axial ligation of a tetragonally elongated (d_{z^2}), low-spin Ni(III) species by one N-donor ligand [224, 226, 227]. The data are consistent with a hypothetical structure shown in Fig. 23. The details of the ligand environment of the Ni are unknown, except that it likely has an axial N-donor ligand (His?) and probably involves amide coordination in analogy to model systems.

3.4.2
Mechanistic Considerations

If the role of the Ni in SOD catalysis is analogous to the roles played by Cu, Fe, and Mn, it serves as a one-electron redox center in the "ping-pong" mechanism proposed by Fridovich (Eqs. (8) and (9)). The epr data obtained on the oxidized enzyme suggests that it contains a Ni(III) center, and thus the catalytic cycle would involve Ni(III) reduction by O_2^- to generate O_2, and oxidation of Ni(II) by O_2^- to generate O_2^{2-}. Despite the fact that Cu/Zn SOD substituted by Ni in the $Cu(His)_4(OH_2)$ site is not an active SOD [229], complexes of Ni(II) with peptides are known to display SOD-like chemistry in terms of O_2^- consumption, with peptides containing histidine residues being particularly efficient [228, 230]. These systems also disproportionate H_2O_2 and display a range of peptide-based oxidations. In the case of a Ni(II) complex of glycyl-glycyl-histidine (Fig. 24), it was proposed that decarboxylation of an Ni(III)-O_2^- adduct yielding a His-C_α radical, followed by peroxide formation, occurred. Peroxide could undergo O-atom abstraction forming the crystallographically characterized product, [Ni(II)(glycyl-glycyl-α-hydroxy-D,L-histamine)], or undergo H_2O_2 elimination forming a His-C_α-C_β double bond as seen by NMR and in the autoxidation of an analogous Cu peptide complex [231]. However, activation of O_2 by Ni(II) appears to require a "sacrificial reductant" which can be sulfite, or a ligand such as a peptide [232, 233].

Kinetic studies of superoxide dismutation by NiSODs are yet to be carried out. The existence of a Ni-O_2^- intermediate in the mechanism is doubtful, if the kinetics of catalysis by NiSOD is anything like the other enzymes. SODs catalyze the dismutation of superoxide at rates that approach the rate of diffusion. It has been argued that, because this rate is fast compared to ligand exchange rates at Cu^{2+}, the mechanism likely proceeds via outersphere electron transfer. For the case of Ni, such a mechanism would be even more favored because of the even more sluggish rate of ligand exchange, at least at Ni(II) centers.

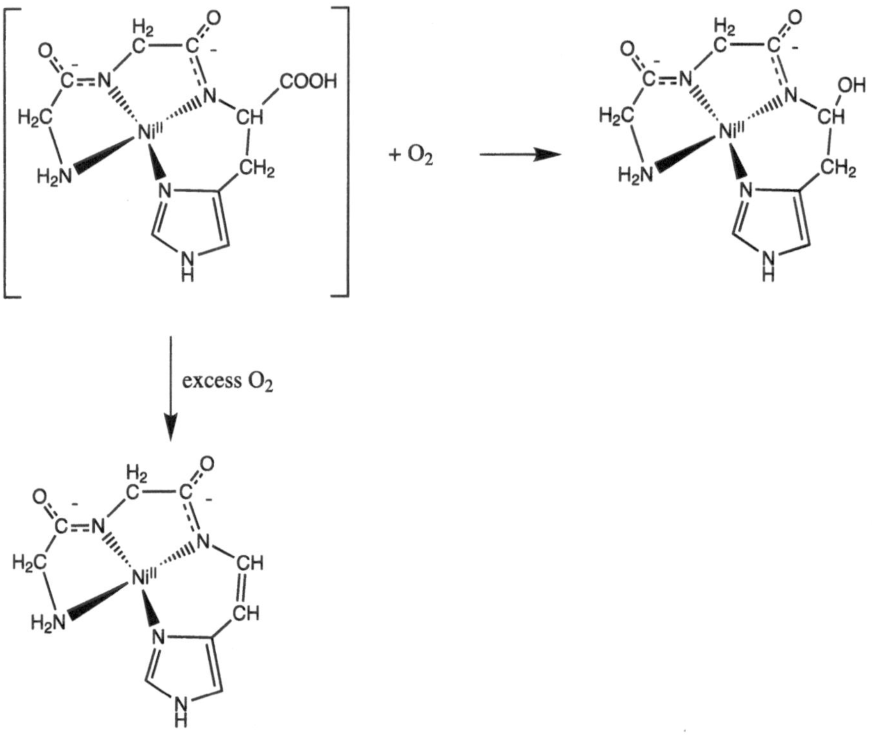

Fig. 24. Reactions of a [Ni(gly-gly-his)] peptide complex with O_2

4
Trends in the Structures and Functions of Biological Nickel Sites

Nature's choice of Ni for the active sites of enzymes is a curious one, given the modern distribution of soluble metals on Earth. There are better Lewis acid centers used in biological hydrolytic chemistry (e.g, Zn), and one-electron redox centers involving Fe and Cu are widespread in biology. One possibility is that the choice of Ni reflects a selection that was made under a different set of conditions. Before O_2 became abundant in the atmosphere, many transition metals would have been present as sulfides, and nickel sulfides are among the more soluble transition metal sulfides [234].

The fact that the biosynthesis of Ni enzymes involves an elaborate system for the acquisition, transport, and specific incorporation of Ni, suggests that the choice of Ni has been made for important specific reasons. In the case of urease, Ni appears to favor the binding of urea by the O atom, which favors the tetrahedral intermediate formed during hydrolysis. The involvement of Ni in catalyzing redox chemistry in H_2ase, CODH, ACS, MCR, and SOD active sites is even more puzzling. Complexes of Ni with N- and O-donor ligands are not noted for undergoing facile redox processes at biologically accessible potentials. Furthermore, with the exception of SOD, the reactions that are catalyzed involve two-electron redox processes that would ordinarily involve more than one metal capable of facile one-electron redox chemistry. The more common biological two-electron redox center, Mo, is typically involved in atom-transfer reactions, not two-electron redox reactions of the types illustrated by Ni metalloenzymes. The association of Ni with cysteinate ligands (e.g., H_2ases, CODH, ACS) and thiolate redox chemistry (e.g., F_{430}) points to the involvement of S either in directly catalyzing the redox chemistry, or in stabilizing *formally* Ni(III) centers where the charge created is distributed over several atomic centers. The modification of the thiolate S-donors by protonation (or alkylation) opens the possibility of stabilizing Ni(I) in a sulfur-rich ligand environment. Alternatively, Ni may serve as a center to coordinate the redox chemistry that is occurring at thiolate ligands, and thus its presence in enzymes with redox roles is favored by its inability to become involved in metal-centered redox chemistry. The association of Ni with other redox cofactors (Fe, Fe,S clusters, tetrapyrroles) further complicates the view of the role of Ni in redox catalysis. It is possible that all of these alternatives are represented in Ni enzymes to one degree or another.

The recent elucidation of several structures of Ni sites in enzymes places the field on a firm structural base. This knowledge coupled with the characterization of the molecular biology of Ni metalloenzymes is opening the field to studies employing site-directed mutagenesis to address the roles of specific ligands in catalysis. It remains for studies employing a range of biophysical techniques to ferret out the specific roles played by Ni in catalysis.

Acknowledgments. The authors are indebted to Prof. Robert A. Scott and Dr. Christina M. V. Stålhandske for sharing the XAS data used to construct several of the figures. The authors also wish to thank Profs. Robert P. Hausinger, P. Andrew Karplus, Robert J. Maier, Joshua Telser, and Rudolf K. Thauer for sharing information prior to publication, and Profs. Juan Fontecilla-Camps, Karplus, and Thauer for providing the illustrations of crystal structures. This work was supported by a grant from the NIH (MJM, GM-38829).

5
References

1. Dixon NE, Gazzola C, Blakeley RL, Zerner B (1975) J Am Chem Soc 97: 4131
2. Sumner JB (1926) J Biol Chem 69: 435
3. Hausinger RP (1993) Biochemistry of nickel. Plenum, New York
4. Kolodziej AF (1994) Prog Inorg Chem 41: 493
5. Hausinger RP (1994) Sci Total Environ 148: 157
6. Maroney MJ (1994) In: King RB (ed) Encyclopedia of inorganic chemistry. Wiley, Sussex, p 2412
7. Mobley HLT, Island MD, Hausinger RP (1995) Microbiological Reviews 59: 451
8. Karplus PA, Pearson M, Hausinger RP (1997) Accts Chem Res (30: 330)
9. Ragsdale SW, Kumar M (1996) Chem Rev 96(7): 2515
10. Ferry JG (1995) Annu Rev Microbiol 49: 305
11. Ragsdale SW, Riordan CG (1996) J Biol Inorg Chem 1: 489
12. Albracht SPJ (1994) Biochim Biophys Acta 1188: 167
13. Maroney MJ (1995) Comments Inorg Chem 17: 347
14. Jaun B (1993) Met Ions Biol Syst 29: 287
15. Jaun B (1994) Chimia 48: 50

16. Telser J (1998) Structure and Bonding (91: 31)
17. Youn H-D, Youn H, Lee J-W, Yim Y-I, Lee JK, Hah YC, Kang S-O (1996) Arch Biochem Biophys 334: 341
18. Youn H-D, Kim E-J, Roe J-H, Hah YC, Kang S-O (1996) Biochem J 318: 889
19. Moncrief MBC, Hausinger RP (1996) Adv Inorg Biochem 11: 151
20. Maier T, Boeck A (1996) Adv Inorg Biochem 11: 173
21. Hausinger RP (1997) J Biol Inorg Chem 2: 279
22. Lohmeyer M, Friedrich B (1987) Arch Microbiol 149: 130
23. Snavely MD, Gravina SA, Cheung TT, Miller CG, Maguire ME (1991) J Biol Chem 266: 824
24. Wu LF, Navarro C, Mandrand-Berthelot MA (1991) Gene 107: 37
25. Navarro C, Wu L-F, Mandrand-Berthelot M-A (1993) Mol Microbiol 9: 1181
26. De Pina K, Navarro C, McWalter L, Boxer DH, Price NC, Kelly SM, Mandrand-Berthelot M-A, Wu L-F (1995) Eur J Biochem 227: 857
27. Wu LF, Mandrand MA (1993) Fems Microbiol Rev 104: 243
28. Sawers RG, Ballantine SP, Boxer DH (1985) J Bacteriol 164: 1324
29. Charon M-H, Wu L-F, Piras C, de Pina K, Mandrand-Berthelot M-A, Fontecilla-Camps JC (1994) J Mol Biol 243: 353
30. Allan CB, Wu L-F, Gu Z, Choudhury SB, Mandrand-Berthelot M-A, Maroney MJ (1997) unpublished results
31. Prout CK, Walker C, Rossotti FJC (1971) J Chem Soc (A): 556
32. Setzer WN, Ogle CA, Wilson GS, Glass RS (1983) Inorg Chem 22: 266
33. Osakada K, Yamamoto T, Yamamoto A, Takenaka A, Sasada Y (1984) Acta Crystallogr C40: 85
34. Yamamoto T, Sekine Y (1984) Inorg Chim Acta 83: 47
35. Rosenfield SG, Berends HP, Gelmini L, Stephan DW, Mascharak PK (1987) Inorg Chem 26: 2792
36. Krüger H-J, Holm RH (1990) J Am Chem Soc 112: 2955
37. Gu Z, Dong J, Allan CB, Choudhury SB, Franco R, Moura JJG, Moura I, LeGall J, Przybyla AE, Roseboom WR, Albracht SPJ, Axley MJ, Scott RA, Maroney MJ (1996) J Am Chem Soc 118: 11,155
38. Wolfram L, Friedrich B, Eitinger T (1995) J Bacteriol 177: 1840
39. Fu C, Javedan S, Moshiri F, Maier RJ (1994) Proc Natl Acad Sci USA 91: 5099
40. Bauerfeind P, Garner RM, Mobley HLT (1996) Infect Immun 64: 2877
41. Mobley HL, Garner RM, Bauerfeind P (1995) Mol Microbiol 16: 97
42. Lee MH, Pankratz HS, Wang S, Scott RA, Finnegan MG, Johnson MK, Ippolito JA, Christianson DW, Hausinger RP (1993) Protein Sci 2: 1042
43. Maier T, Jacobi A, Sauter M, Boeck A (1993) J Bacteriol 175: 630
44. Kerby RL, Ludden PW, Roberts GP (1997) J Bacteriol 179: 2259
45. Maier T, Lottspeich F, Boeck A (1995) Eur J Biochem 230: 133
46. Brayman TG, Hausinger RP (1996) J Bacteriol 178: 5410
47. Moncrief MBC, Hausinger RP (1996) J Bacteriol 178: 5417
48. Park IS, Hausinger RP (1995) Science 267: 1156
49. Maier T, Boeck A (1996) Biochemistry 35: 10,089
50. Bernhard M, Schwartz E, Rietdorf J, Friedrich B (1996) J Bacteriol 178: 4522
51. Maier RJ, Triplett EW (1996) Critical Reviews in Plant Sciences 15: 191
52. Vignais PM, Toussaint B (1994) Arch Microbiol 161: 1
53. Friedrich B, Schwartz E (1993) Annu Rev Microbiol 47: 351
54. Thiemermann S, Dernedde J, Bernhard M, Schroeder W, Massanz C, Friedrich B (1996) J Bacteriol 178: 2368
55. Binder U, Maier T, Boeck A (1996) Arch Microbiol 165: 69
56. Rodrigue A, Batia N, Mueller M, Fayet O, Boehm R, Mandrand-Berthelot M-A, Wu L-F (1996) J Bacteriol 178: 4453
57. Vignais PM, Toussaint B, Colbeau A (1995) Adv Photosynth 2: 1175
58. Elsen S, Colbeau A, Chabert J, Vignais PM (1996) J Bacteriol 178: 5174

59. Black LK, Fu C, Maier RJ (1994) J Bacteriol 176: 7102
60. Friedrich B (1990) Fems Microbiol Rev 87: 425
61. Vignais PM, Dimon B, Zorin NA, Colbeau A, Elsen S (1997) J Bacteriol 179: 290
62. Kerby RL, Hong SS, Ensign SA, Coppoc LJ, Ludden PW, Roberts GP (1992) J Bacteriol 174: 5284
63. Fox JD, He Y, Shelver D, Roberts GP, Ludden PW (1996) J Bacteriol 178: 6200
64. Warren MJ, Scott AI (1990) Trends Biochem Sci 15: 486
65. Ermler U, Grabarse W, Shima S, Goubeaud M, Thauer RK (1997) (Science 278: 1457)
66. Shin W, Lindahl PA (1992) J Am Chem Soc 114: 9718
67. Mobley HLT, Hausinger RP (1989) Microbiol Rev 53: 85
68. Gabri E, Carr MB, Hausinger RP, Karplus PA (1995) Science 268: 998
69. Kurtz DM Jr (1997) J Biol Inorg Chem 2: 159
70. Pearson MA, Michel LO, Hausinger RP, Karplus PA (1997) Biochemistry (36: 8164)
71. Jose J, Schäfer UK, Kaltwasser H (1994) Arch Microbiol 161: 384
72. Jabri E, Karplus PA (1996) Biochemistry 35: 10,616
73. Park I-S, Michel LO, Pearson MA, Jabri E, Karplus PA, Wang S, Dong J, Scott RA, Koehler BP, Johnson MK, Hausinger RP (1996) J Biol Chem 271: 18,632
74. Wang S, Lee MH, Hausinger RP, Clark PA, Wilcox DE, Scott RA (1994) Inorg Chem 33: 1589
75. Colpas GJ, Maroney MJ, Bagyinka C, Kumar M, Willis WS, Suib SL, Mascharak PK, Baidya N (1991) Inorg Chem 30: 920
76. Todd MJ, Hausinger RP (1989) J Biol Chem 264: 15,835
77. Clark PA, Wilcox DE (1989) Inorg Chem 28: 1326
78. Day EP, Peterson J, Sendova MS, Todd MJ, Hausinger RP (1993) Inorg Chem 32: 634
79. Finnegan MG, Kowal AT, Werth MT, Clark PA, Wilcox DE, Johnson MK (1991) J Am Chem Soc 113: 4030
80. Blakeley RL, Zerner B (1984) J Mol Catal 23: 263
81. Park IS, Hausinger RP (1993) Protein Sci 2: 1034
82. Wages HE, Taft KL, Lippard SJ (1993) Inorg Chem 32: 4985
83. Stemmler AJ, Kampf JW, Kirk ML, Pecoraro VL (1995) J Am Chem Soc 117: 6368
84. Martin PR, Hausinger RP (1992) J Biol Chem 267: 20,024
85. Graf EG, Thauer RK (1981) FEBS Lett 136: 165
86. Lancaster JR Jr (1980) FEBS Lett 115: 285
87. Albracht SP, Graf EG, Thauer RK (1982) FEBS Lett 140: 311
88. Thauer RK, Klein AR, Hartmann GC (1996) Chem Rev 96(7): 3031
89. Eidsness MK, Scott RA, Prickril BC, DerVartanian DV, Legall J, Moura I, Moura JJG, Peck HJ (1989) Proc Natl Acad Sci USA 86: 147
90. He SH, Teixeira M, LeGall J, Patil DS, Moura I, Moura JJG, DerVartanian DV, Huynh BH, Peck HD Jr (1989) J Biol Chem 264: 2678
91. Voordouw G (1992) Adv Inorg Chem 38: 397
92. Volbeda A, Charon MH, Piras C, Hatchikian EC, Frey M, Fontecilla-Camps JC (1995) Nature 373: 580
93. Volbeda A, Garcin E, Piras C, de Lacey AL, Fernandez VM, Hatchikian EC, Frey M, Fontecilla-Camps JC (1996) J Am Chem Soc 118: 12,989
94. Montet Y, Amara P, Volbeda A, Vernede X, Hatchikian EC, Field MJ, Frey M, Fontecilla-Camps JC (1997) Nature Structural Biology 4: 523
95. Frey M (1998) Structure and Bonding (90: 97)
96. van der Zwaan JW, Coremans JMCC, Bouwens ECM, Albracht SPJ (1990) Biochim Biophys Acta 1041: 101
97. Gu Z, Dong J, Allan CB, Choudhury SB, Franco R, Moura JJG, Moura I, LeGall J, Przybyla AE, Roseboom W, Albracht SPJ, Axley MJ, Scott RA, Maroney MJ (1996) J Am Chem Soc 118: 11,155
98. Happe RP, Roseboom W, Pierik A, Albracht SPJ, Bagley KA (1997) Nature 385: 126
99. deLacey AL, Hatchikian EC, Volbeda A, Frey M, Fontecilla-Camps JC, Fernandez VM (1997) J Am Chem Soc 119: 7181

100. Darensbourg DJ, Reibenspies JH, Lai C-H, Lee W-Z, Darensbourg MY (1997) J Am Chem Soc 119: 7903
101. Cammack R, Fernandez VM, Hatchikian EC (1994) Methods Enzymol 243: 43
102. Bagley KA, Duin EC, Roseboom W, Albracht SPJ, Woodruff WH (1995) Biochemistry 34: 5527
103. Guigliarelli B, More C, Fournel A, Asso M, Hatchikian EC, Williams R, Cammack R, Bertrand P (1995) Biochemistry 34: 4781
104. Bertrand P, Camensuli P, More C, Guigliarelli B (1996) J Am Chem Soc 118: 1426
105. Dole F, Medina M, More C, Cammack R, Bertrand P, Guigliarelli B (1996) Biochemistry 35: 16,399
106. Dole F, Fournel A, Magro V, Hatchikian EC, Bertrand P, Guigliarelli B (1997) Biochemistry 36: 7847
107. Coremans JMCC, Van GCJ, Albracht SPJ (1992) Biochim Biophys Acta 1119: 148
108. Barondeau DP, Roberts LM, Lindahl PA (1994) J Am Chem Soc 116: 3442
109. Roberts LM, Lindahl PA (1994) Biochemistry 33: 14,339
110. Roberts LM, Lindahl PA (1995) J Am Chem Soc 117: 2565
111. Halcrow MA, Christou G (1994) Chem Rev 94: 2421
112. Moura JJG, Teixeira M, Moura I, LeGall J (1988) In: Lancaster JR Jr (ed) The bioinorganic chemistry of nickel. VCH, New York, p 191
113. Huyett JE, Carepo M, Pamplona A, Franco R, Moura I, Moura JJG, Hoffman BM (1997) J Am Chem Soc (119: 9291)
114. Bagyinka C, Whitehead JP, Maroney MJ (1993) J Am Chem Soc 115: 3576
115. Garcin E, Vernede X, Volbeda A, Hatchikian C, Frey M, Fontecilla-Camps JC (1997) Vth International Conference on the Molecular Biology of Hydrogenases, Albertville, France, p 27
116. Hsu H-F, Koch SA, Popescu CV, Munck E (1997) J Am Chem Soc 119: 8371
117. Woodruff WH, Maroney MJ (1997) (unpublished results)
118. van der Zwaan JW, Albracht SPJ, Fontijn RD, Slater EC (1985) Febs Lett 179: 271
119. Bagley KA, Van Garderen CJ, Chen M, Woodruff WH, Duin EC, Albracht SPJ (1994) Biochemistry 33: 9229
120. Fan C, Teixeira M, Moura J, Moura I, Huynh BH, Le Gall J, Peck HD Jr, Hoffman BM (1991) J Am Chem Soc 113: 20
121. Symons MCR, Aly MM, West DX (1979) Chem Commun: 51
122. Morton JR, Preston KF (1984) J Chem Phys 81: 5775
123. Pappenhagen TL, Margerum DW (1985) J Am Chem Soc 107: 4576
124. Whitehead JP, Colpas GJ, Bagyinka C, Maroney MJ (1991) J Am Chem Soc 113: 6288
125. Maroney MJ, Allan CB, Chohan BS, Choudhury SB, Gu Z (1996) ACS Symp Ser 653: 74
126. Montet Y, Amara P, Volbeda A, Vernede X, Hatchikian EC, Field MJ, Frey M, Fontecilla-Camps JC (1997) Nature Stuct Biol 4: 523
127. Sellmann D, Sutter J (1996) ACS Symp Ser 653: 101
128. Sellmann D, Mahr G, Knoch F, Moll M (1994) Inorg Chim Acta 224: 45
129. Sellmann D, Kaeppler J, Moll M (1993) J Am Chem Soc 115: 1830
130. Sellmann D, Becker T, Knoch F (1996) Chem Eur J 2: 1092
131. Sorgenfrei O, Duin EC, Klein A, Albracht SPJ (1996) J Biol Chem 271: 23,799
132. Kumar M, Lu WP, Liu L, Ragsdale SW (1993) J Am Chem Soc 115: 11,646
133. Anderson ME, DeRose VJ, Hoffman BM, Lindahl PA (1993) J Am Chem Soc 115: 12,204
134. Anderson ME, Lindahl PA (1994) Biochemistry 33: 8702
135. Cramer SP, Eidsness MK, Pan W-H, Morton TA, Ragsdale SW, DerVartanian DV, Ljungdahl LG, Scott RA (1987) Inorg Chem 26: 2477
136. Bastian NR, Diekert G, Niederhoffer EC, Teo B, Walsh CT, Orme-Johnson WH (1988) J Am Chem Soc 110: 5581
137. Bonam D, Ludden PW (1987) J Biol Chem 262: 2980
138. Ensign SA, Ludden PW (1991) J Biol Chem 266: 18,395
139. Tan GO, Ensign SA, Ciurli S, Scott MJ, Hedman B, Holm RH, Ludden PW, Korszun ZR, Stephens PJ, Hodgson KO (1992) Proc Natl Acad Sci USA 89: 4427

140. Bonam D, McKenna MC, Stephens PJ, Ludden P (1988) Proc Natl Acad Sci USA 85: 31
141. Ensign SA, Campbell MJ, Ludden PW (1990) Biochemistry 29: 2162
142. Stephens PJ, McKenna MC, Ensign SA, Bonam D, Ludden PW (1989) J Biol Chem 264: 16,347
143. Hu Z, Spangler NJ, Anderson ME, Xia J, Ludden PW, Lindahl PA, Munck E (1996) J Am Chem Soc 118: 830
144. Lindahl PA, Ragsdale SW, Munck E (1990) J Biol Chem 265: 3880
145. Qiu D, Kumar M, Ragsdale SW, Spiro TG (1995) J Am Chem Soc 117: 2653
146. Qiu D, Kumar M, Ragsdale SW, Spiro TG (1996) J Am Chem Soc 118: 10,429
147. Seravalli J, Kumar M, Lu WP, Ragsdale SW (1995) Biochemistry 34: 7879
148. Kumar M, Lu WP, Smith A, Ragsdale SW, McCracken J (1995) J Am Chem Soc 117: 2939
149. Spangler NJ, Lindahl PA, Bandarian V, Ludden PW (1996) J Biol Chem 271: 7973
150. Colpas GJ, Day RO, Maroney MJ (1992) Inorg Chem 31: 5053
151. Mills DK, Hsiao YM, Farmer PJ, Atnip EV, Reibenspies JH, Darensbourg MY (1991) J Am Chem Soc 113: 1421
152. Osterloh F, Saak W, Haase D, Pohl S (1997) Chem Commun 1997: 979
153. Osterloh F, Saak W, Pohl S (1997) JACS 119: 5648
154. Anderson ME, Lindahl PA (1996) Biochemistry 35: 8371
155. Ensign SA, Hyman MR, Ludden PW (1989) Biochemistry 28: 4973
156. Hyman MR, Ensign SA, Arp DJ, Ludden PW (1989) Biochemistry 28: 6821
157. Kumar M, Lu WP, Ragsdale SW (1994) Biochemistry 33: 9769
158. Flint DH, Allen RM (1996) Chem Rev 96: 2315
159. Sakaki S (1992) J Am Chem Soc 114: 2055
160. Lu Z, Crabtree RH (1995) J Am Chem Soc 117: 3994
161. Beley M, Collin J-P, Ruppert R, Sauvage J-P (1986) J Am Chem Soc 100: 7461
162. Grahame DA, DeMoll E (1996) J Biol Chem 271: 8352
163. Xia J, Sinclair JF, Baldwin TO, Lindahl PA (1996) Biochemistry 35: 1965
164. Xia J, Lindahl PA (1995) Biochemistry 34: 6037
165. Xia J, Dong J, Wang S, Scott RA, Lindahl PA (1995) J Am Chem Soc 117: 7065
166. Xia J, Lindahl PA (1996) J Am Chem Soc 118: 483
167. Gorst CM, Ragsdale SW (1991) J Biol Chem 266: 20,687
168. Fan C, Gorst CM, Ragsdale SW, Hoffman BM (1991) Biochemistry 30: 431
169. Qiu D, Kumar M, Ragsdale SW, Spiro TG (1994) Science 264: 817
170. Kumar M, Qiu D, Spiro TG, Ragsdale SW (1995) Science 270(5236): 628
171. Qiu D, Kumar M, Ragsdale SW, Spiro TG (1997) J Am Chem Soc 119: 11, 134
172. Qiu D, Kumar M, Ragsdale SW, Spiro TG (1997) Science 278: 21
173. Kumar M, Ragsdale SW (1992) J Am Chem Soc 114: 8713
174. Ragsdale SW, Lindahl PA, Munck E (1987) J Biol Chem 262: 14,289
175. Lu W-P, Schiau I, Cunningham JR, Ragsdale SW (1993) J Biol Chem 268: 5605
176. Raybuck SA, Bastian NR, Zydowsky LD, Kobayashi K, Floss HG, Orme-Johnson WH, Walsh CT (1987) J Am Chem Soc 109: 3171
177. Lebertz H, Simon H, Courtney LF, Benkovic SJ, Zydowsky LD, Lee K, Floss HG (1987) J Am Chem Soc 109: 3173
178. Riordan CG, Ram MS, Yap GPA, Liable-Sands L, Rheingold AL, Marchaj A, Norton JR (1997) J Am Chem Soc 119: 1648
179. Stavropoulos P, Muetterties MC, Carrie M, Holm RH (1991) J Am Chem Soc 113: 8485
180. Tucci GC, Holm RH (1995) J Am Chem Soc 117: 6489
181. Barondeau DP, Lindahl PA (1997) J Am Chem Soc 119: 3959
182. Shin W, Anderson ME, Lindahl PA (1993) J Am Chem Soc 115: 5522
183. Grahame DA, Khangulov S, DeMoll E (1996) Biochemistry 35: 593
184. Kumar M, Day RO, Colpas GJ, Maroney MJ (1989) J Am Chem Soc 111: 5974
185. Setzke E, Hedderich R, Heiden S, Thauer RK (1994) Eur J Biochem 220: 139
186. DiMarco AA, Bobik TA, Wolfe RS (1990) Ann Rev Biochem 59: 355
187. Won H, Olson KD, Summers MF, Wolfe RS (1993) Comments Inorg Chem 15: 1

188. Pfaltz A, Livingston DA, Jaun B, Diekert G, Thauer RK, Eschenmoser A (1985) Helv Chim Acta 68: 1338
189. Keltjens JT, Hermans JMH, Rijsdijk GJFA, Van dDC, Vogels GD (1988) Antonie van Leeuwenhoek 54: 207
190. Färber G, Keller W, Kratky C, Jaun B, Pfaltz A, Spinner C, Kobelt A, Eschenmoser A (1991) Helv Chim Acta 74: 697
191. Won H, Summers MF, Olson KD, Wolfe RS (1990) J Am Chem Soc 112: 2178
192. Won H, Olson KD, Park J, Wolfe RS, Hare DR, Summers MF (1995) Bull Korean Chem Soc 16: 649
193. Olson KD, Won H, Wolfe RS, Hare DR, Summers MF (1990) J Am Chem Soc 112: 5884
194. Renner MW, Furenlid LR, Barkigia KM, Forman A, Shim HK, Simpson DJ, Smith KM, Fajer J (1991) J Am Chem Soc 113: 6891
195. Kaplan WA, Suslick KS, Scott RA (1991) J Am Chem Soc 113: 9824
196. Eidsness MK, Sullivan RJ, Schwartz JR, Hartzell PL, Wolfe RS, Flank AM, Cramer SP, Scott RA (1986) J Am Chem Soc 108: 3120
197. Shiemke AK, Shelnutt JA, Scott RA (1989) J Biol Chem 264: 11,236
198. Shiemke AK, Hamilton CL, Scott RA (1988) J Biol Chem 263: 5611
199. Furenlid LR, Renner MW, Fajer J (1990) J Am Chem Soc 112: 8987
200. Shiemke AK, Kaplan WA, Hamilton CL, Shelnutt JA, Scott RA (1989) J Biol Chem 264: 7276
201. Shiemke AK, Scott RA, Shelnutt JA (1988) J Am Chem Soc 110: 1645
202. Hamilton CL, Scott RA, Johnson MK (1989) J Biol Chem 264: 11,605
203. Renner MW, Furenlid LR, Stolzenberg AM (1995) J Am Chem Soc 117: 293
204. Furenlid LR, Renner MW, Smith KM, Fajer J (1990) J Am Chem Soc 112: 1634
205. Furenlid LR, Renner MW, Szalda DJ, Fujita E (1991) J Am Chem Soc 113: 883
206. Ram MS, Riordan CG, Ostrander R, Rheingold AL (1995) Inorg Chem 34: 5884
207. Holliger C, Pierik AJ, Reijerse EJ, Hagen WR (1993) J Am Chem Soc 115: 5651
208. Rospert S, Voges M, Berkessel A, Albracht SPJ, Thauer RK (1992) Eur J Biochem 210: 101
209. Jaun B, Pfaltz A (1986) J Chem Soc, Chem Commun 1327
210. Telser J, Fann Y-C, Renner MW, Fajer J, Wang S, Zhang H, Scott RA, Hoffman BM (1997) J Am Chem Soc 119: 733
211. Wilker JJ, Gelasco A, Pressler MA, Day RO, Maroney MJ (1991) J Am Chem Soc 113: 6342
212. Berkessel A (1991) Bioorg Chem 19: 101
213. Albracht SPJ, Ankel-Fuchs D, Boecher R, Ellermann J, Moll J, Van der Zwaan JW, Thauer RK (1988) Biochim Biophys Acta 955: 86
214. Wacket LP, Honek JF, Begley TP, Wallace V, Orme-Johnson WH, Walsh CT (1987) Biochemistry 26: 6012
215. Wackett LP, Honek JF, Begley TP, Shames SL, Niederhoffer EC, Hausinger RP, Orme-Johnson WH, Walsh CT (1988) In: Lancaster JL Jr (ed) The bioinorganic chemistry of nickel. VCH, New York, p 249
216. Ahn Y, Krzycki JA, Floss HG (1991) J Am Chem Soc 113: 4700
217. Bonacker LG, Baudner S, Mörschel E, Böcher R, Thauer RK (1993) Eur J Biochem 217: 587
218. Fridovich I (1986) Arch Biochem Biophys 247: 1
219. Yim MB, Kang J-H, Yim H-S, Kwak H-S, Chock PB, Stadtman ER (1996) Proc Natl Acad Sci USA 93: 5472
220. Searcy DG, Whitehead JP, Maroney MJ (1995) Arch Biochem Biophys 318: 251
221. Bertini I, Banci L, Piccioli M, Luchinat C (1990) Coord Chem Rev 100: 67
222. Stallings WC, Metzger AL, Pattridge KA, Fee JA, Ludwig ML (1991) Free Rad Res Commun 12/13: 259
223. Sugiura Y, Kuwahara J, Suzuki T (1983) Biochem Biophys Res Commun 115: 878
224. Margerum DW, Anliker SL (1988) In: JR Lancaster J (ed) The bioinorganic chemistry of nickel. VCH, New York, p 29

225. Wang J-F, Kumar K, Margerum DW (1989) Inorg Chem 28: 3481
226. Lappin AG, Murray CK, Margerum DW (1978) Inorg Chem 17: 1630
227. Sugiura Y, Mino Y (1979) Inorg Chem 18: 1336
228. Bal W, Djuran MI, Margerum DW, ET Gray J, Mazid MA, Tom RT, Nieboer E, Sadtler PJ (1994) Chem Commun 1994: 1889
229. Valentine JS, Pantoliano MW (1981) In: Spiro TG (ed) Copper proteins, vol 3, Wiley, New York, p 291
230. Ueda J, Ozawa T, Miyazaki M, Fujiwara Y (1993) Inorg Chim Acta 214: 29
231. Meester Pd, Hodgson DJ (1978) Inorg Chem 17: 440
232. Cheng C-C, Gulia J, Rokita SE, Burrows CJ (1996) J Mol Cat 113: 379
233. Muller JG, Hickerson RP, Perez RJ, Burrows CJ (1997) J Am Chem Soc 119: 1501
234. Frausto da Silva JJR, Williams RJP (1991) The biological chemistry of the elements. Clarendon Press, Oxford

Electron Transfer in Transition Metal-Pteridine Systems

Sharon J. Nieter Burgmayer

Department of Chemistry, Bryn Mawr College, Bryn Mawr, Pennsylvania 19010, USA
E-mail: sburgmay@brynmawr.edu

The combination of a pterin and a transition metal in many enzymes is the motivation for exploring the chemistry of pteridine complexes in detail. Unlike other biological ligands for essential transition metals, pterin is unique in displaying multi-electron redox reactivity, an ability that resembles the redox capabilities of transition metals. It is perhaps because these two partners, metal and pterin, have this chemical similarity that their compounds defy traditional categorization by formal oxidation number. The result challenges the chemist to formulate fresh interpretations of these deceptively ordinary complexes. This review concerns reports of metal-pterin complexes that appeared from the early 1980s through 1996. In a few cases older literature is briefly mentioned to build a context for the newer work. The review comprises four sections. Section 1 introduces the pteridine family and its important contributions to biochemistry. Section 2 is devoted to studies of molybdenum(6+) complexes reacted with reduced pterins. Section 3 describes redox interactions between reduced pterins and the first row metals copper and iron. Finally, Section 4 turns to a discussion of the electronic interactions in flavin complexes of various metals. An Epilogue closes the review.

Keywords: Metalloenzyme; molybdenum; pteridine; tetrahydropterin

Structure and Bonding, Vol. 92
© Springer Verlag Berlin Heidelberg 1998

List of Symbols and Abbreviations

acac	acetylacetonate
bipy	2,2′-bipyridine
BVS	bond valence sum
detc	diethyldithiocarbamate
CDP	cytosine dinucleotide
cyt	cytochrome b
DCIP	dichlorophenolindophenol
DHPR	dihydropteridine reductase
DMF	N,N-dimethylformamide
DMS	dimethylsulfide
DMSO	dimethylsulfoxide
DMSO Red'ase	dimethylsulfoxide reductase
EHMO	extended Huckel molecular orbital
EPR	electron paramagnetic resonance
ESEEM	electron spin echo envelope, to read "electron spin echo envelope modulation" modulation
ESI-MS	electrospray-ionization mass spectrometry
EXAFS	extended X-ray absorption fine structure
FAD	flavin adenine dinucleotide
FMN	flavin mononucleotide
GDP	guanosine dinucleotide
gly	glycine
7,8-H_2pterin	7,8-dihydropterin
H_4B	5,6,7,8-tetrahydrobiopterin
H_4P, H_4pterin	5,6,7,8-tetrahydropterin
H_4DMP	6,7-dimethyl-5,6,7,8-tetrahydropterin
H_4HMP	6-hydroxymethyl-5,6,7,8-tetrahydropterin

HOO-H$_2$B	4a-hydroperoxy-dihydrobiopterin
IR	infrared spectroscopy
MeH$_4$P	N5-methyl-5,6,7,8-tetrahydropterin
MeOH	methanol
Moco	molybdenum cofactor
NMR	nuclear magnetic resonance
PAH	phenylalanine hydroxylase
Phe	phenylalanine
phen	1,10-phenanthroline
piv-H$_4$DMP	2-pivaloyl-6,7-dimethyl-5,6,7,8-tetrahydropterin
PUP	2-pivaloyl-pterin
q-H$_2$P, quin-H$_2$pterin	quinonoid-6,7-dihydropterin
q-H$_2$B	quinonoid-6,7-dihydrobiopterin
quin	quinonoid
TEA	tetraethylammonium
TMAZ	1,3,7,8-tetramethylalloxazine
Tp*	tris(pyrazolyl)hydroborate
tbpb	tris(3-*tert*-butylpyrazolyl)hydroborate
tippb	tris(3-*iso*-propylpyrazolyl)hydroborate
tppb	tris(3-phenylpyrazolyl)hydroborate
TrpH	tryptophan hydroxylase
Tyr	tyrosine
TyrH	tyrosine hydroxylase
UV/vis	ultraviolet/visible electronic spectroscopy

1
Introduction

1.1
Pterins and Pteridines

1.1.1
Structure

Fusion of pyrimidine and pyrazine rings forms the pteridine structure. As such, pteridines may be viewed as heterocyclic cousins of naphthalene. The polar -C=N- bonds make the pteridine structure inherently reactive. To stabilize the electron deficient rings, pteridine is commonly embellished with electron-rich functional groups, such as amine, carbonyl, and chloride [1, 2].

pyrimidine   pyrazine   pteridine

One substitution motif that frequently appears in bioorganic molecules has been given a special name. *Pterins* are molecules having amino and keto substituents at the 2- and 4-positions, respectively, that constitute a sub-class of pteridines. Pterins were the first members of the pteridine family to be structurally characterized. Their function as pigments for butterfly wings [3] led to their unusual name (pterin) that has its root in the Greek word for 'wing', *pteron*. This common name was later modified to pteridine [4] to represent the parent family. As an aid to remembering the pterin structure, note that pterin is structurally related to guanine.

pterin   guanine

The pteridine core appears in a second ubiquitous biomolecule. Fusion of a phenyl ring across the C6,C7 bond and addition of two keto substituents at the 2- and 4-positions generates the isoalloxazine structure found in flavin.

isoalloxazine   flavin

R = -CH$_2$(CHOH)$_3$CH$_2$OPO$_3$$^{2-}$ in FMN
R = -CH$_2$(CHOH)$_3$CH$_2$-ADP in FAD

1.1.2
Pteridine Redox Relatives

The polar -C=N- bonds in pteridines make these molecules susceptible to Michael addition [1, 5] and reduction reactions [1, 2]. It is in the realm of redox reactions that pteridines exhibit a remarkably rich repertoire. Bicyclic pteridines accept a total of four electrons and four protons during the transformation to their most reduced state, tetrahydropteridine, and this reaction is illustrated for an unsubstituted pterin in Scheme 1. The planar structure of oxidized pterin becomes ruffled when the saturated region in the pyrazine ring spanning atoms N5, C6, C7 and N8 adopts a half chair conformation [6–13].

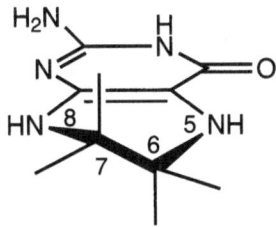

half-chair conformation
within pyrazine ring of
tetrahydropterin

Redox states between the fully oxidized pteridine and the fully reduced tetrahydropteridine are also accessible. The majority of these are at the dihydro-level of reduction and are inter-related by tautomeric isomerism or proton rearrangement. Scheme 1 summarizes the formation of known dihydropterin species and their inter conversion as studied by both electrochemical and chemical methods [14–16].

Reduction of fully oxidized pterin electrochemically or with chemical reductants proceeds predictably with the addition of two electrons at the most electronegative sites, the ring nitrogen atoms; protonation yields the unstable 5,8-dihydropterin isomer, a species detected electrochemically but as yet not isolated [14, 15]. In the absence of two additional electron/proton equivalents to transform 5,8-dihydropterin to the fully reduced 5,6,7,8-tetrahydropterin, the 5,8-dihydropterin isomerizes to 7,8-dihydropterin. Oxidation of tetrahydropterin by 2 equivalents produces 6,7-dihydropterin or the so-called quinonoid-dihydropterin [1, 17]. Of the three tautomers (A, B, C), structure A is favored by most data [18, 19]. Like the 5,8-dihydropterin isomer, the 6,7-dihydropterin is also unstable and rearranges to 7,8-dihydropterin [20]. Since this rearrangement involves H-transfer from C6, the quinonoid isomer can be stabilized and isolated if C6 bears two methyl substituents to block H6 loss [21]. A radical trihydropterin species can be produced from reactions of radical initiators (not shown in Scheme 1) [22].

Two other pteridines at the dihydro-level of reduction will be mentioned separately [1]. Chemical reduction of pteridines not bearing a keto or amino substituent at C4 produce 3,4-dihydropteridines (Eq. 1).

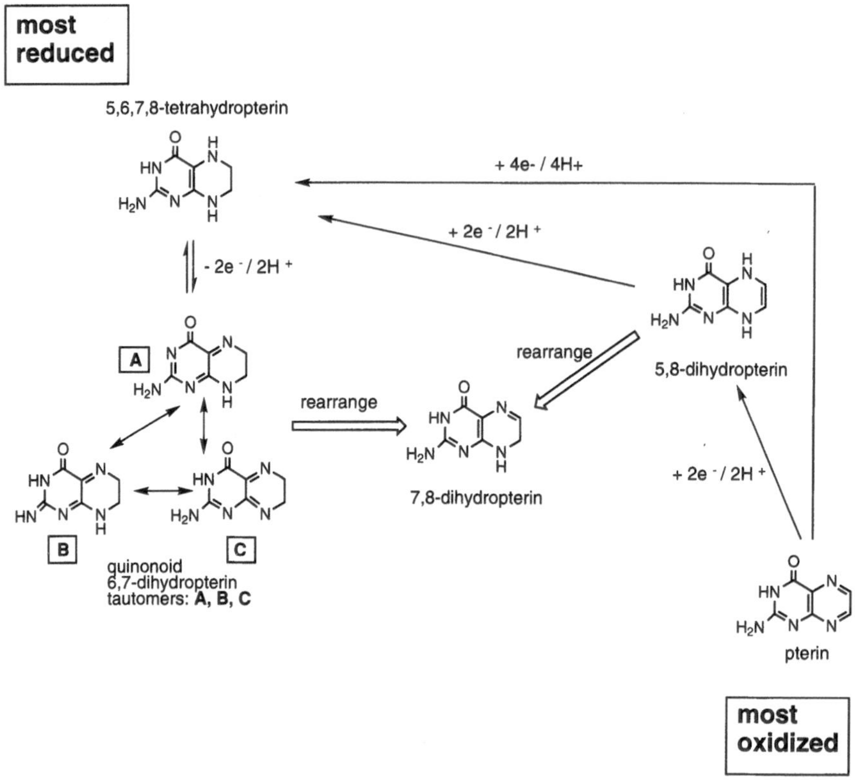

most reduced

5,6,7,8-tetrahydropterin

+ 4e- / 4H+

+ 2e⁻ / 2H⁺

- 2e⁻ / 2H⁺

A

rearrange

rearrange

5,8-dihydropterin

7,8-dihydropterin

+ 2e⁻ / 2H⁺

B   **C**

quinonoid
6,7-dihydropterin
tautomers: **A, B, C**

pterin

most oxidized

Scheme 1. Redox reactions that inter convert tetrahydro-, dihydro- and oxidized pterins

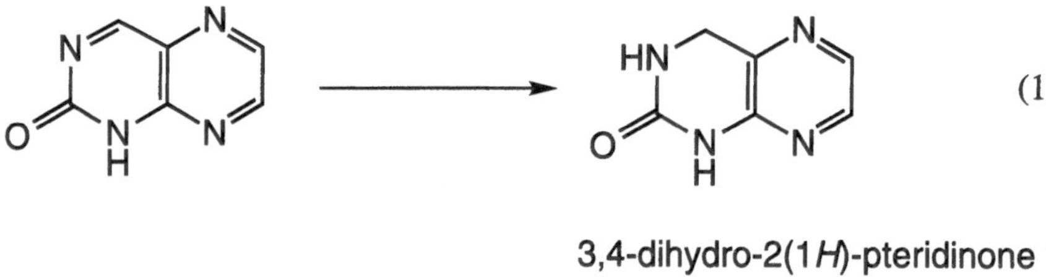

3,4-dihydro-2(1*H*)-pteridinone   (1)

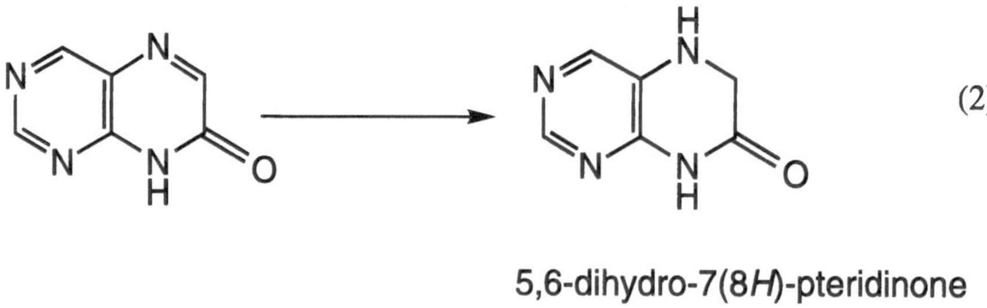

5,6-dihydro-7(8*H*)-pteridinone   (2)

Pteridines and pterins having keto, carboxyl, or amino substituents at C7 that participate in ring conjugation are observed to yield 5,6-dihydropteridines on reduction (Eq. 2). For the purposes of this review whose focus is on metal interactions of pteridinones bearing a carbonyl at C4, neither of these reductions is dealt with further.

1.2
Pteridine Redox in Biological Systems

Members of the pteridine family are key players in important biochemical reactions ranging from metabolism to catabolism. Except for those pteridines whose role is solely pigmentation [1, 3], all other systems make use of pteridine redox or their propensity for nucleophilic addition reactions. An encyclopedic review of these functions is not a goal of this review. Rather, the most widely studied or well-known examples will be briefly described. The first two profiles illustrate enzymes requiring both a pterin and a transition metal. The third and fourth examples have no known metal requirement for their normal function, but studies of metal binding to these pteridines have been reported.

1.2.1
Aromatic Amino Acid Hydroxylases

The most famous member of this trio of enzymes is phenylalanine hydroxylase (PAH), a reputation based on its connection to the genetic disease phenylketonuria (PKU). The recent appearance of an excellent and thorough review of this area permits a deliberately cursory description here [23].

PAH and the related tyrosine and tryptophan hydroxylases (TyrH and TrpH, respectively) are responsible for phenylalanine catabolism (PAH) and neurotransmitter synthesis (TyrH and TrpH).

During the substrate reactions shown in Scheme 2, the catalyses performed by PAH, TyrH, and TrpH each require one iron atom and one molecule of tetrahydrobiopterin per catalytic site. One molecule of O_2 is consumed in the reaction. One oxygen atom is inserted as an hydroxyl into the substrate while the second oxygen atom is reduced to water. The intimate details of the catalytic reaction are not yet fully known. Despite this, multiple functions have been identified for the tetrahydrobiopterin cofactor and these are schematically described below in Eqs. (3) and (4).

Tetrahydrobiopterin is first used by the catalytic system to reduce the iron center to the correct oxidation state. Although there is no net redox reaction required of the iron center, the catalytically active ferrous state can be adventitiously oxidized to the inactive ferric "resting" state, a reaction easily reversed when tetrahydrobiopterin or artificial cofactor [24] is allowed to react with the iron.

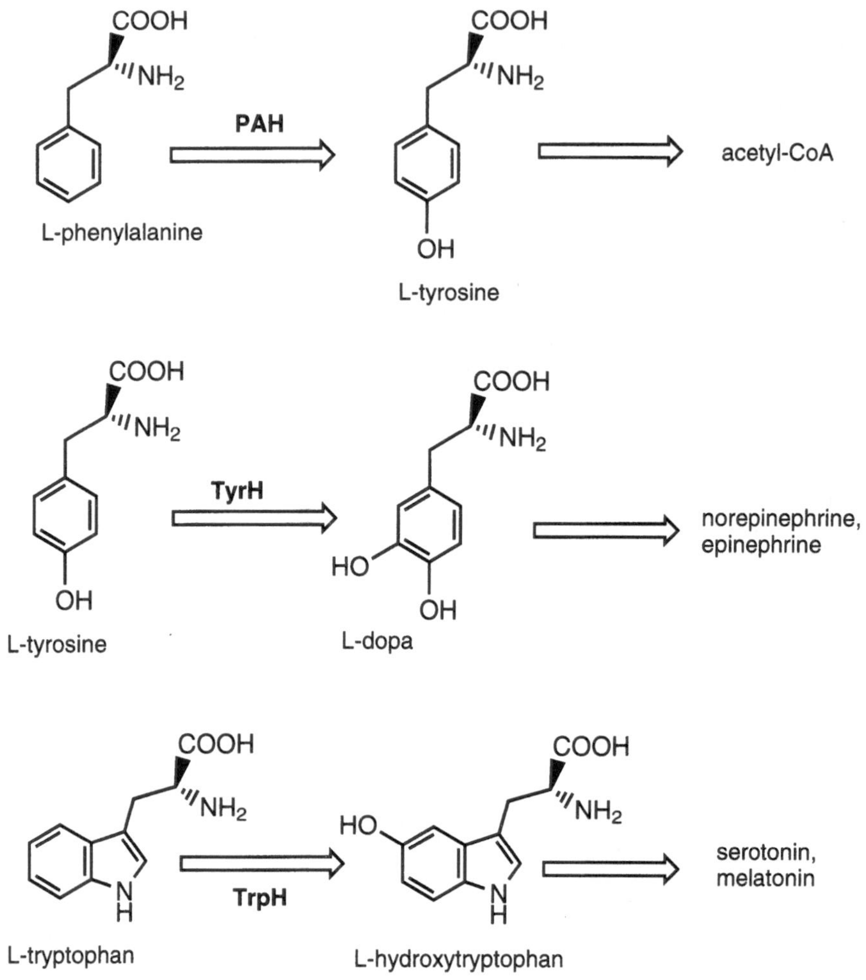

Scheme 2. Substrate reactions of the aromatic amino acid hydroxylases PAH, TyrH and TrypH

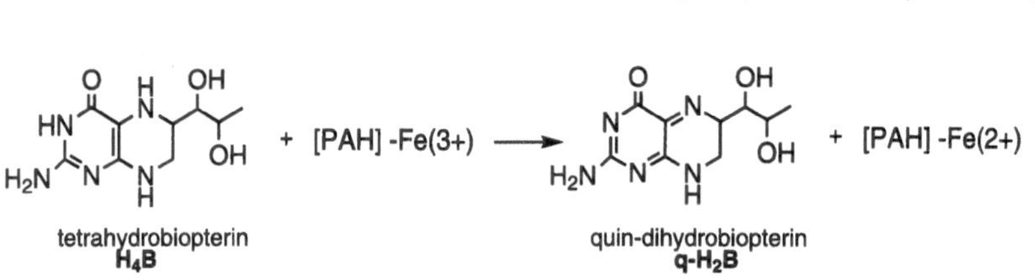

$$\text{tetrahydrobiopterin} \quad + \quad [PAH]\text{-Fe(3+)} \quad \longrightarrow \quad \text{quin-dihydrobiopterin} \quad + \quad [PAH]\text{-Fe(2+)} \qquad (3)$$

tetrahydrobiopterin
H$_4$B

quin-dihydrobiopterin
q-H$_2$B

The intimate mechanism for iron reduction is still surrounded by controversy due to the stoichiometric incompatibility between the two-electron reductant H$_4$pterin and the one-electron oxidant ferric ion. One rationalization is that the second electron of H$_4$pterin is consumed by oxygen under aerobic conditions, while under anaerobic conditions a trihydropterin radical is the postulated product which reduces a second iron site.

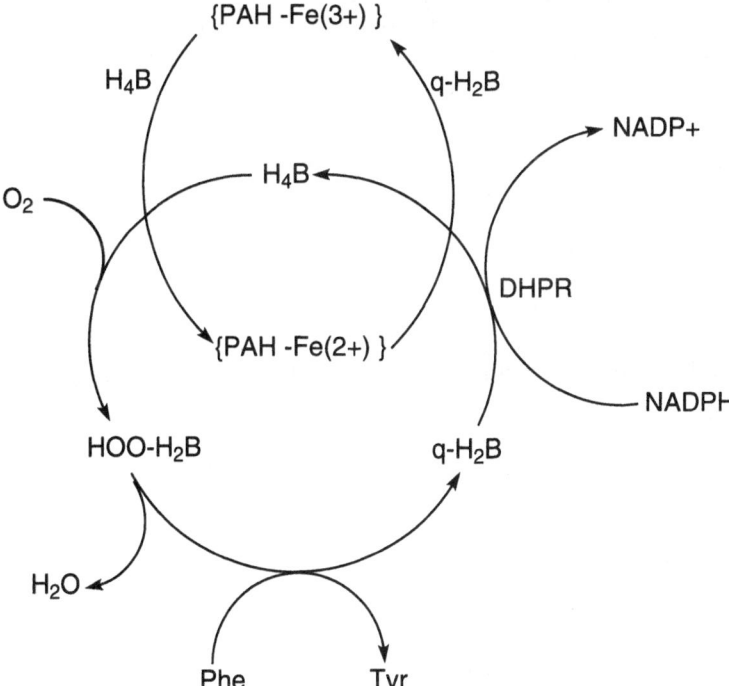

$$\text{(4)}$$

H$_4$B   4a-hydroperoxydihdyrobiopterin   **q-H$_2$B**
HOO-H$_2$B

The next identified process involving the pterin is the addition of O_2 at the C4a bridgehead site (Eq. 4), a reaction whose precedence is well-established in flavin chemistry [25–26]. It seems likely that O_2 activated in this manner generates a reactive species capable of hydroxylating the unactivated aromatic rings. Whether the ferrous center is simultaneously interacting with the pterin and oxygen is unknown.

The reactions in Eqs. (3) and (4) are incorporated into the catalytic system as shown in Scheme 3. Here the notation {PAH-Fe(2+)} represents the iron-bound enzyme, DHPR represents dihydropterin reductase and all other abbreviated notations are defined in Eqs. (3) and (4).

While the above discussion is general for PAH from mammalian sources, there are several examples of so-called 'atypical' PAH. Of these, the most pertinent to this review is the enzyme isolated from the bacterium *chromo-*

Scheme 3. A catalytic scheme for phenylalanine hydroxylase. All abbreviations are defined in the text

bacterium violaceum which has been reported to have a requirement for copper but not iron [27].

The copper ion, Cu^{2+} in the resting state, has several surprising features as compared to its iron-containing PAH counterparts. First of these is that the artificial cofactor, 6,7-dimethyltetrahydropterin [H_4DMP], is not able to reduce the copper atom in situ. Reduction prior to the catalytic event is evidently necessary, however, since dithiothreitol is effective in reducing the cupric ion to the cuprous state thereby accelerating the onset of enzyme turnover. Secondly a direct interaction between the cupric ion and [H_4DMP] was detected by EPR in the form of hyperfine coupling from a ^{15}N-labeled H_4DMP (at site N5) to Cu^{2+} [28]. EXAFS [29] and ESEEM [30] techniques also identified nitrogen ligands on the copper ion. Despite these data which support a direct copper-pterin interaction, the most recent work in this area calls into question whether the copper ion is needed for catalytic activity in atypical PAH from *C. violaceum* [31].

1.2.2
Oxo-Transferases

The enzymes in this class have been studied for over four decades. In contrast to the aromatic amino acid hydroxylases whose pterin dependence has been recognized since 1959, it is less than two decades since the first report appeared describing a possible pterin component in the oxo-molybdenum enzymes [32]. This class of molybdoenzymes is distinct from the celebrated molybdoenzyme nitrogenase and is referred to using the term oxo-transferases to emphasize the functional motif of this group – the transfer of oxygen atoms between substrates and water (Eq. 5).

$$\{E\text{-}Mo(6+)\} + S + H_2O \xrightarrow[\text{reductases}]{\text{oxidases}} \{E\text{-}Mo(4+)\} + S\text{-}O + 2H^+ + 2e^- \quad (5)$$

{E-Mo } = enzyme-bound Mo
S = substrate

There are now three dozen enzymes in this class, in addition to over a dozen related tungsten enzymes, but only three examples are detailed below (Eqs. (6)–(8) to illustrate typical substrate reactions. The reader is directed to several recent reviews for a more comprehensive discussion [33–35].

nitrate reductase

$$(6)$$

xanthine oxidase

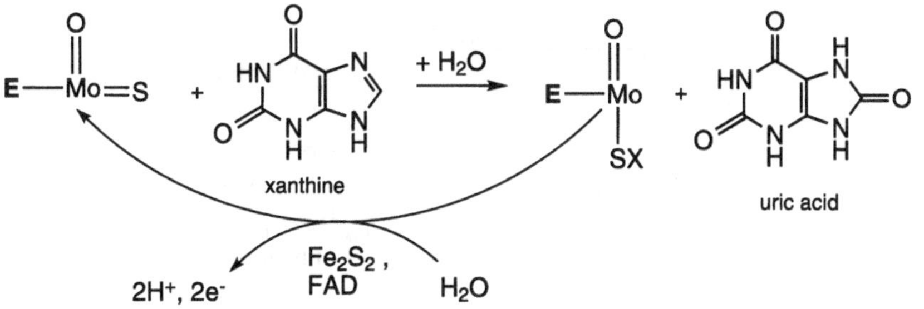

(7)

DMSO reductase

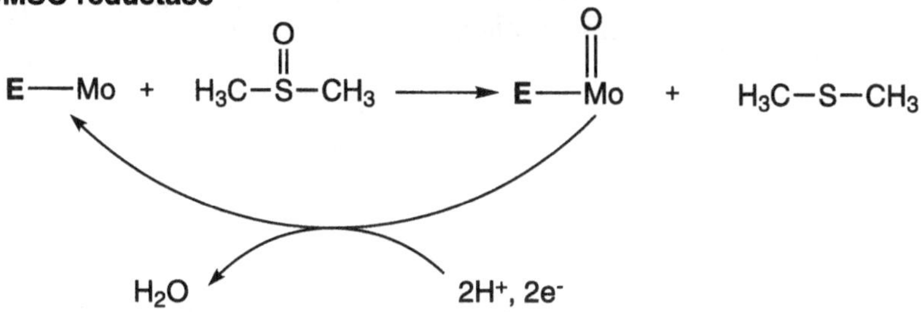

(8)

The dissociable catalytic center of these molybdenum enzymes is the molybdenum cofactor (Moco) and the discovery and characterization of the pterin component in Moco, which is now known by the term molybdopterin, has single-handedly been accomplished by Rajagopalan and co-workers in a long progression of remarkable and insightful papers [36]. Scheme 4 outlines the key steps in this progression. The characteristic fluorescence of the oxidized pterin ring structure appeared in oxidized samples of sulfite oxidase [36, 37]. Tracking the formation of this fluorescence and probing the oxidative conditions that generated it led to the crucial connection where urothione was identified as the natural metabolite of Moco [38]. The identity of the pterin side-chain was revealed by chemically trapping the dithiolene moiety as well as a combined strategy of NMR and FAB-MS methods [39]. The most recent efforts have been devoted to understanding the structure of the pterin precursor, Form Z, of Moco and its biochemical synthesis [40–42].

The recent X-ray crystal structures [43–47] of several oxo-molybdoenzymes have verified most of the aspects of Rajagopalan's proposed structure of Moco and he and his co-workers deserve considerable credit for this achievement. The crystal structures did, however, bring to light a few surprises. First is the 2:1 pterin:Mo ratio found in dimethylsulfoxide reductase [DMSO Red'ase] [46]. Second is the single oxo-ligand in the Mo(6+) oxidized form of the same enzyme [46]. The last is the tricyclic form of the pterin observed in every molybdoenzyme structure.

A description of the accepted changes at the Mo site is important to understanding the modeling strategy followed by bioinorganic chemists such

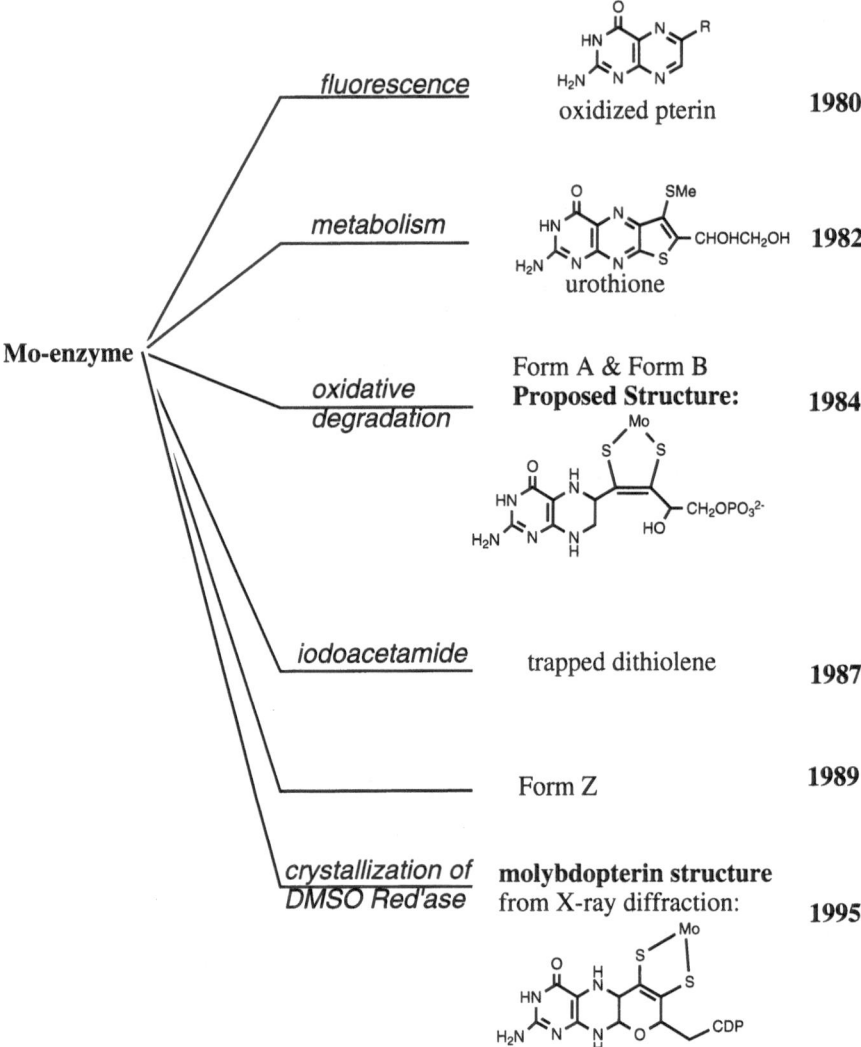

Scheme 4. Summary of key developments in the identity of pterin component, molybdopterin, of the molybdenum enzyme

as ourselves. From early on it was recognized from spectroscopic (primarily EPR and EXAFS) work that the molybdenum shuttles between the 4+ and 6+ oxidation states via 5+ with a parallel gain in oxo or sulfido ligands [33–35], as illustrated in Scheme 5. In certain cases the source of the second oxo ligand was proved to be water and this oxygen atom was eventually the one transferred to the substrate [48–49].

The role of the pterin was not clear. By analogy to the details known for the aromatic amino acid hydroxylases (Sect. 1.2.1) it seemed reasonable that the

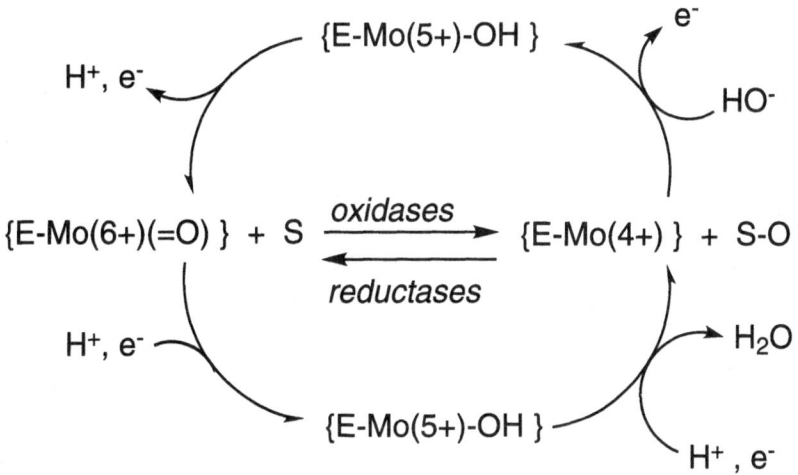

Scheme 5. A general schematic for all molybdenum and tungsten oxo-transferases showing the change in oxo ligation. {E-Mo(4+)} refers to holoenzyme with the molybdenum atom of the cofactor in the 4+ oxidation state

pterin was involved in a redox role. Confounding this hypothesis was the fact that redox titrations of several enzymes showed that the Mo, the flavin, and FeS clusters accounted for the eight electrons consumed, thereby prohibiting any net redox involvement by the pterin [50]. Clues indicating the pterin role have been provided by the recent X-ray crystal structures of both molybdenum and tungsten enzymes [43–47, 81]. In several cases the pterin is observed to link the molybdenum (or tungsten [81]) center to remote iron-sulfur clusters by hydrogen bonding through a pterinyl NH to a cysteinyl sulfur coordinated to Fe [43–45, 81] (structures A, B, and D in Fig. 1). The pterin effectively "hard-wires" the metal center to other electron transfer prosthetic groups, presumably facilitating electron flow out of the enzyme to the external acceptor. There are, however, still other structures [47] revealing no such system where the additional electron transfer group, such as a heme, is far removed from both the pterin and the Mo atom, or, other enzymes [46] without any direct linkage to additional prosthetic groups. Finally it should be mentioned that no experimental observation of pterin redox during any part of the substrate reaction has been made. The mystery of why nature selected the pterin as a partner for molybdenum in these enzymes has not been solved.

1.2.3
Flavins

The isoalloxazine ring structure is characteristic of the two important biomolecules FAD and FMN. Flavoproteins are common coenzymes in electron transport where they typically function to convert a two-electron process into two sequential one-electron steps, thereby serving as a critical link between the two-electron organic redox chemistry and the one-electron

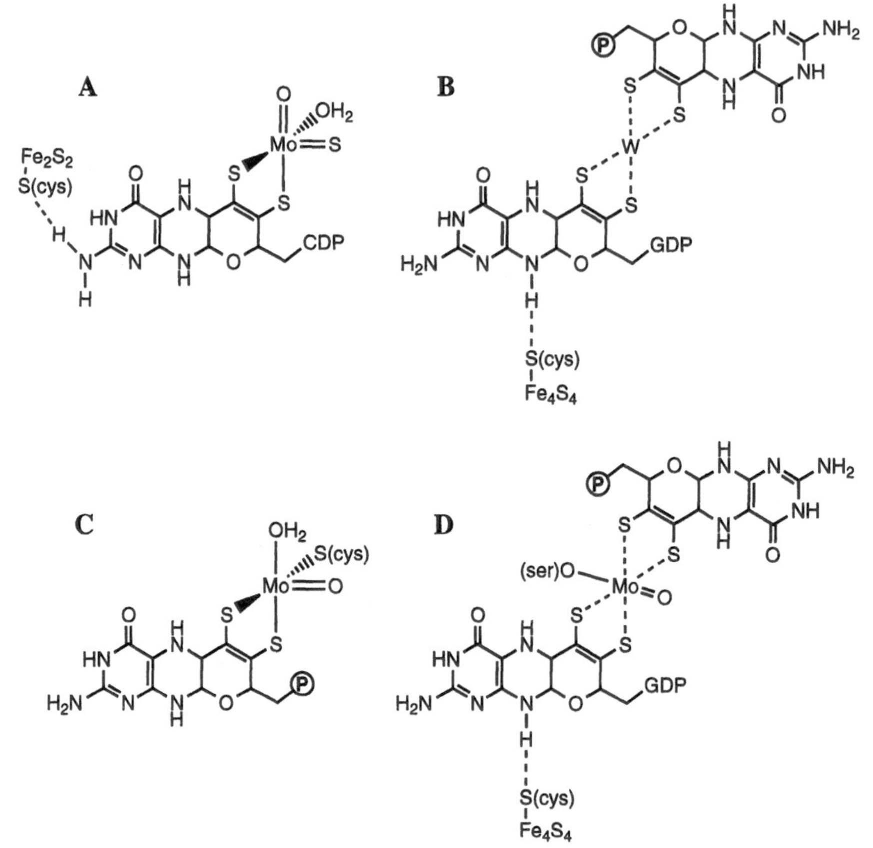

Fig. 1. The molybdenum site of several molybdoenzymes as determined by X-ray crystallography: Structure **A** is the cofactor as seen in *Desulfovibrio gigas* aldehyde oxidoreductase [44, 45]; structure **B** is the tungsten site from *Pyroccocus furious* aldehyde oxidoreductase [83]; structure **C** is the molybdenum site from chicken liver sulfite oxidase [47]; structure **D** is the oxidized cofactor in *Rhodobacter sphaeroides* dimethylsulfoxide reductase [46]

processes of cytochromes, for example [51]. The three ring structure of flavin is the key to its ability to absorb one- and two-electron reducing equivalents.

Unlike pterin redox chemistry which spans four electron transfers flavin has only three well-defined oxidation states: quinone, semi-quinone, and fully reduced. These three can exist in protonated and deprotonated forms. The entire assembly is shown in Scheme 6.

The diverse roles of flavins include simple electron transfer and substrate hydroxylation. The mechanism of the second function involves the reaction of reduced flavin with molecular oxygen and is well established. Dioxygen attacks the relatively electron-rich carbon C4a forming a C4a peroxyflavin [52], a species that is the prototype of the analogous C4a-peroxybiopterin postulated for PAH (see Eq. 4).

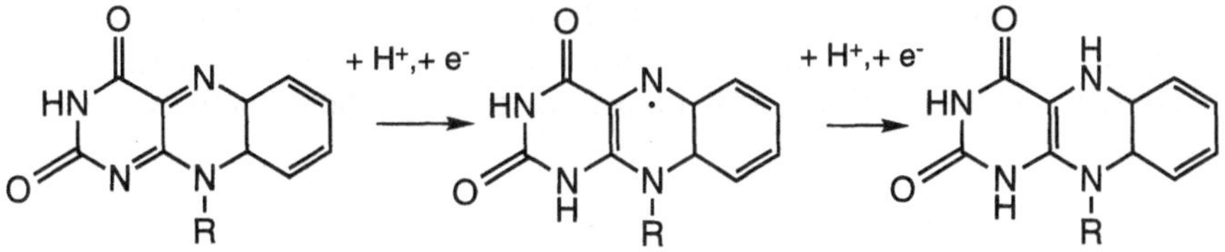

flavin quinone   flavin semi-quinone   dihydroflavin

$R = -CH_2(CHOH)_3CH_2OPO_3{}^{2-}$ in FMN
$R = -CH_2(CHOH)_3CH_2\text{-ADP}$ in FAD

Scheme 6. Redox reactions of dihydro-, semireduced and oxidized flavins

No metal requirement has been determined for normal FAD and FMN function.

1.2.4
Folates

Folic acid is a pterin substituted at the 6-position by a *p*-aminobenzoic acid esterified to mono- or polyglutamic acid groups [1, 2]. Folates refer more generally to the family of related structures including reduced forms (as tetrahydro and dihydro) and methylated structures. Some of these are illustrated below for a folate monoglutamate.

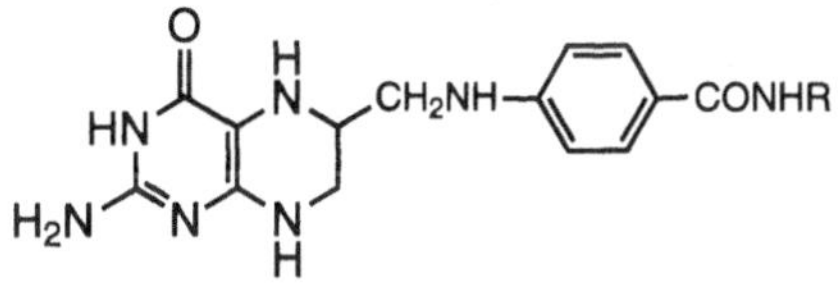

tetrahydrofolate

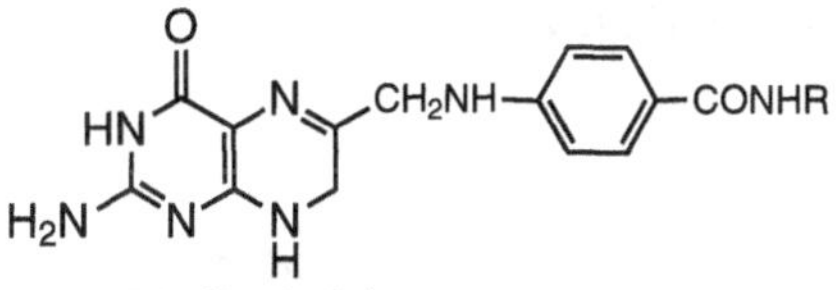

7,8-dihydrofolate

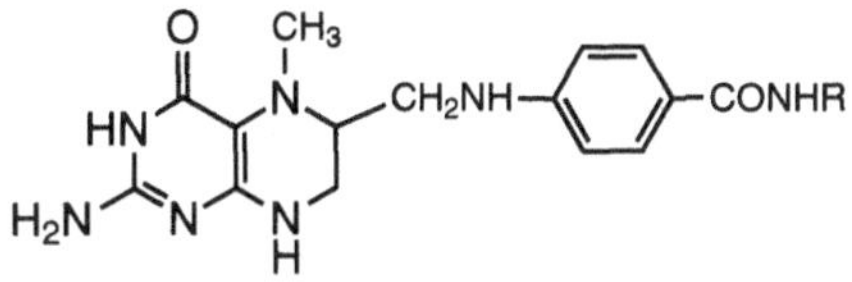

N5- methyltetrahydrofolate

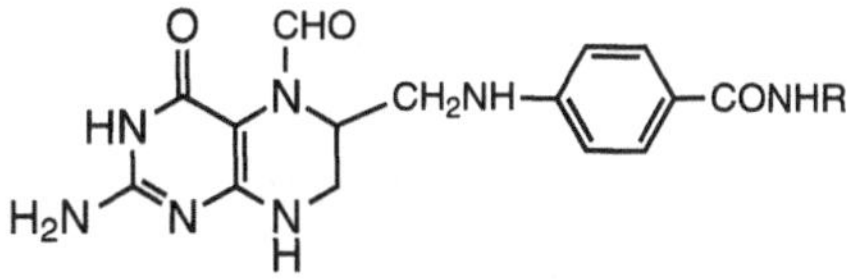

N5- formyltetrahydrofolate

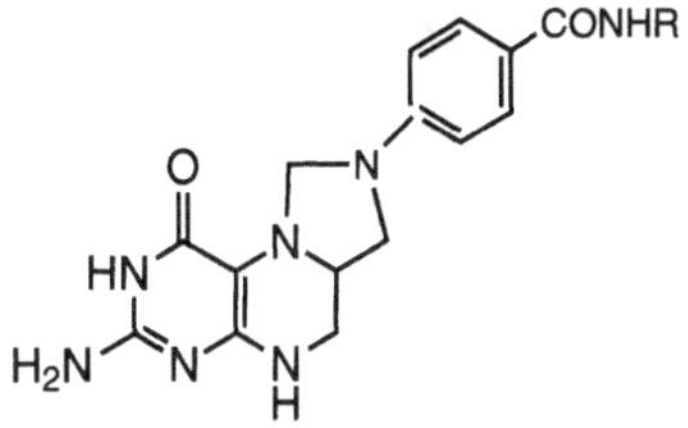

N5,10-methylene-tetrahydrofolate

$R = C(CO_2H)HCH_2CH_2CO_2H$

These are important molecules involved in methyl group transfer, often in conjunction with cobalamin, for amino acid biosynthesis [2, 53]. Like the flavins, no metal requirement is known for proper folate reactivity.

2
Synthetic Molybdenum-Pterin Systems

This section will cover two areas of research involving molybdenum and pterins. The first part recounts the development of molybdenum-tetrahydropterin systems and the difficulty in assigning formal oxidation states to the metal and reduced pterin in these systems. The second part presents model systems that illustrate a successful synthetic strategy to pterin-substituted dithiolenes coordinated to molybdenum.

2.1
Dioxo-Molybdenum(6+) Reactions with Tetrahydropterins

2.1.1
Hypotheses

The presence of a pterin requirement in Moco is intriguing – why would nature select such a complicated heterocyclic substituent for the dithiolene tether to molybdenum? Since the proposed molybdopterin structure presents several metal binding sites in addition to the dithiolene, the reason for the pterin might be to offer the molybdenum alternative coordination environments (Fig. 2).

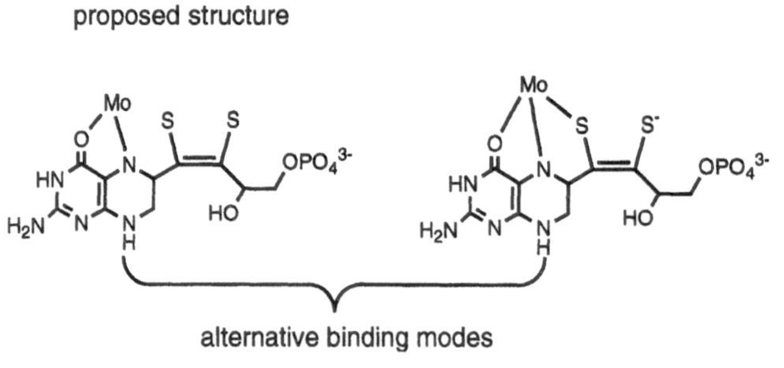

proposed structure

alternative binding modes

Fig. 2. The proposed structure of molybdenum-bound pterin portion of Moco after Rajagopalan et al. [37] and two alternative molybdenum binding modes

Another feature of Moco was provocative. Rajagopalan's proposed structure for Moco pairs an oxidized Mo(6+) center with a reduced tetrahydropterin. This combination seemed incongruous but suggestive of a possible redox reaction between the metal and the organic cofactor. Both hypotheses – alternative binding sites and likely redox reactions – were the motivation for initial studies of tetrahydropterin reactions with molybdenum(6+) complexes. Prior to these studies there was only one report of a reaction between reduced pterins and transition metals [54].

2.1.2
Preparation of Molybdenum Complexes of Reduced Pterins

The first redox investigation employed the known oxygen atom transfer agent $MoO_2(detc)_2$ and the easily prepared 6,7-dimethyltetrahydropterin [H_4DMP] (Eq. 9). These reactants were observed to undergo a rapid reaction signaled by the appearance of an intensely purple solution with λ_{max} at 505 nm. The observed color change from yellow ($MoO_2(detc)_2$) to purple typically indicates molybdenum reduction in molybdenum-oxo chemistry. Monitoring the reaction by ^{1}H NMR revealed that a new pterin species was formed. The characteristic proton resonances of this species resembled those of quinonoid-dihydropterin [18]. On the basis of these two pieces of spectroscopic data the outcome of Eq. (9) was initially interpreted as a two-electron redox reaction yielding a purple Mo(4+) complex bound to a quinonoid dihydropterin at atoms O4 and N5 [55]. Metal chelation at this site has been demonstrated for molybdenum [56] as well as other metals [57–62, 88, 94–100] when coordinated to oxidized pterins and pteridines. The importance of this interpretation was that it implied an unusual stability of coordinated quinonoid dihydropterin. This stability prevented the expected rearrangement of the unstable quinonoid dihydropterin to the 7,8-dihydro isomer (see Scheme 1). Unfortunately the product of Eq. (9) resisted isolation and its exact structure would remain unknown for seven years until a different preparative method was devised (see below).

$$MoO_2detc_2 \qquad H_4DMP \cdot 2HCl \qquad\qquad\qquad (9)$$

$$Mo^{4+}O(quin\text{-}H_2DMP)(detc)_x$$

In order to determine if stabilization of coordinated quinonoid dihydropterin was a general feature of metal ions, the performance of other dioxo-Mo(6+) reagents in reactions with tetrahydropterin was investigated [63].

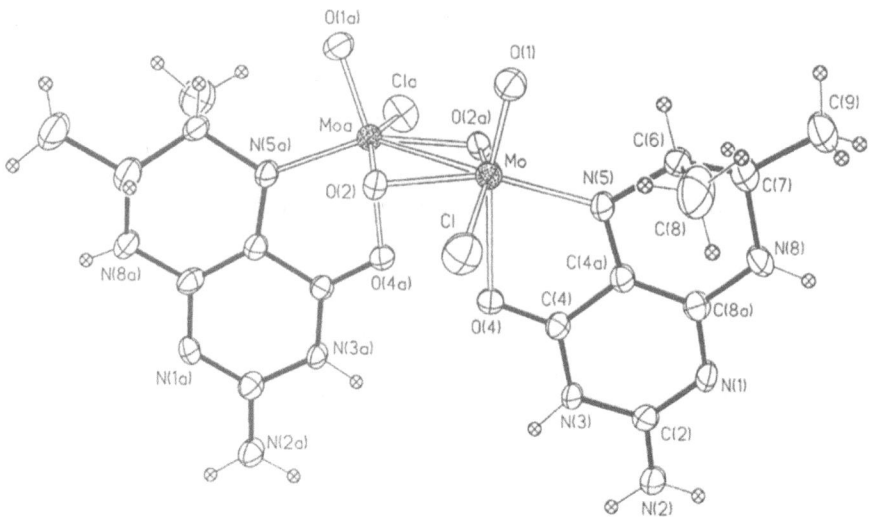

Fig. 3. ORTEP drawing of the crystal structure of $Mo_2O_4Cl_2(H_4DMP)_2$, **1** [63]

$MoO_2(acac)_2$ reacted with H_4DMP to produce results reminiscent of Eq. (9): formation of a purple solution exhibiting the characteristic downfield ^1H NMR resonances of quinonoid dihydropterin. The product of this reaction was in fact dimeric as proved through X-ray diffraction on crystals precipitated from the reaction solution. The X-ray structure confirmed pterin coordination and clearly showed saturation of the region C6,C7 as expected for a quinonoid isomer of dihydropterin (Fig. 3).

Thoughtful consideration of the metrical parameters around the molybdenum atoms led to the realization that this structure was *not* compatible with formal oxidation state assignments of Mo(4+) and *quin*-dihydropterin. Foremost in this reasoning is the fact that the long Mo-Mo distance of 3.01 Å precludes the Mo-Mo bond expected within a Mo(4+) dimer. Indeed, the dimensions of the Mo_2O_4-core observed in this structure were entirely consistent with a related Mo(6+) dimer [64]. It quickly followed that if Mo(6+) in the reagent had not in fact been reduced, tetrahydropterin had not been oxidized. Charge balance within the neutral dimer required that the coordinated tetrahydropterin exist in a deprotonated form (H_4pterinate); the dimer was hence formulated as $Mo(6+)_2O_4Cl_2(H_4DMP^-)_2$.

Skepticism increased towards the hypothesis that $Mo(4+)$-H_2pterin complexes were produced from tetrahydropterin reactions of dioxo-Mo(6+) reagents. A series of experiments were designed to reveal the correct formal oxidation state using the following reasoning. If the dimer were really a Mo(4+)-(quinonoid-H_2pterin) complex, dissociation of the unstable quinonoid-H_2pterin isomer would be expected to initiate rapid pterin isomerization to the stable 7,8-dihydropterin isomer (see Scheme 1). On the other hand, if the dimer was best considered a Mo(6+)-tetrahydropterinate complex then pterin dissociation would simply result in formation of free tetrahydropterin.

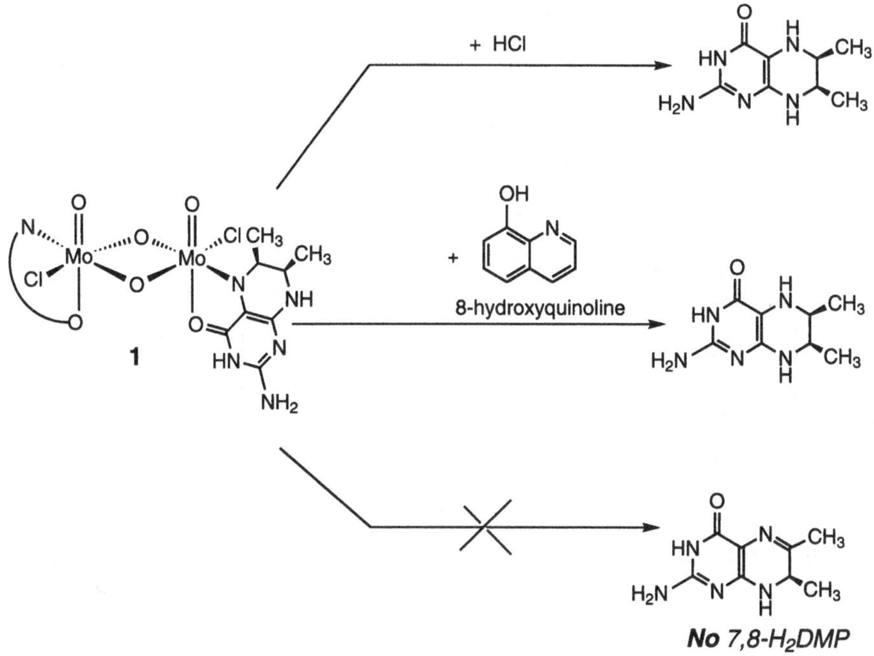

Scheme 7. Results of pterin dissociation reactions of $Mo_2O_4Cl_2(H_4DMP)_2$ **1** [63]

Distinguishing between these two scenarios was easily done using ^1H NMR which displays different signals for tetrahydropterin and 7,8-dihydropterin. Pterin dissociation was accomplished by the addition of either 8-hydroxy-quinoline or hydrochloric acid. The results (illustrated for $Mo_2O_4Cl_2(H_4DMP)_2$ **1** in Scheme 7) clearly showed *only* the presence of free tetrahydropterin; *no* 7,8-dihydropterin was observed. The same result was observed for other Mo-H$_4$pterin products (see below). It was therefore concluded that the best formal oxidation state assignment was Mo(6+)-(H$_4$pterinate$^-$) in these complexes [63].

Simultaneous to our attempts to explore molybdenum-tetrahydropterin redox, another research team was pursuing studies also based on the idea that pterin might directly ligate Mo in Moco. They reported a reaction between a reduced pterin, tetrahydrobiopterin (H$_4$B), and the molybdenum(6+) reagent MoO_2Cl_2 [65]. The product of this reaction was a red-purple material (λ_{max} at 487 nm) having $\nu_{Mo=O}$ at 985 cm^{-1}. X-ray diffraction was the basis for the structure drawn in Eq. (10). From spectroscopic and structural data these authors concluded that the product was a Mo(4+) complex of quinonoid-H$_2$biopterin which they formulated as MoOCl$_3$(quin-H$_2$biopterinH$^+$) with the pterin in a protonated quinonoid form on a Mo(4+) center. Facts consistent with this assignment are the high stretching frequency observed for the Mo=O bond, the intense visible absorption near 500 nm, and the bond distances within the pterin suggestive of localized -C=C- and -C=N- bonds.

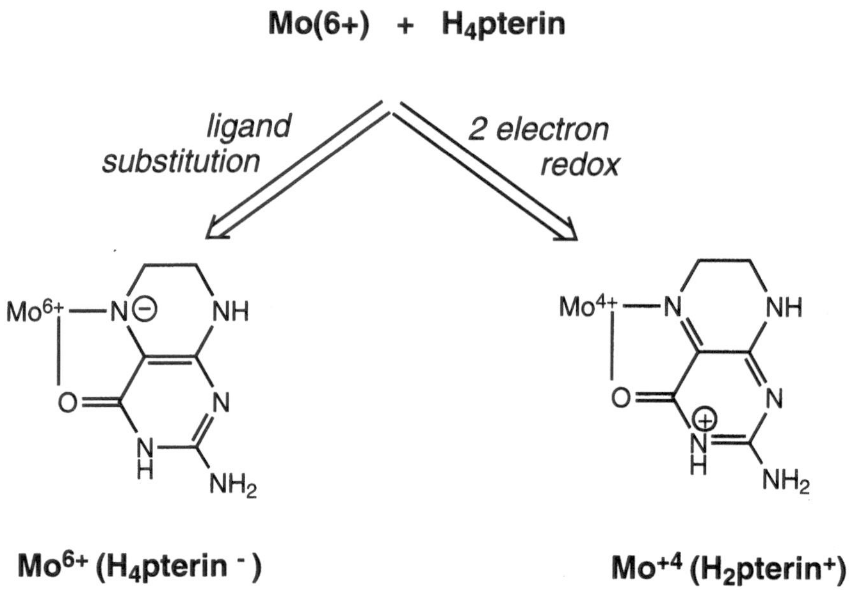

Mo(6+) + H₄pterin

Scheme 8. Two views of oxidation state assignments for molybdenum complexes of reduced pterins

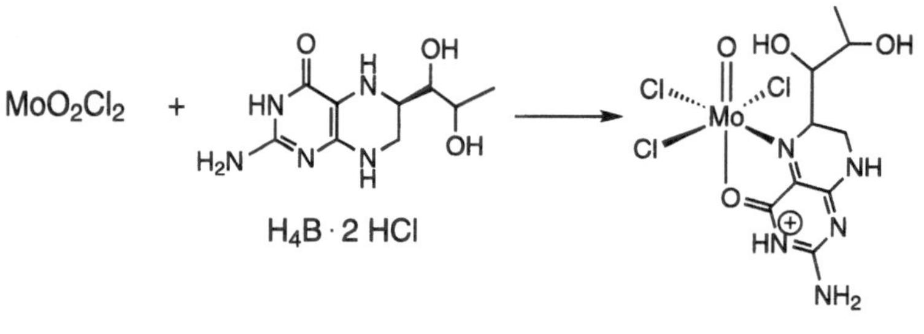

(10)

There were now three examples of Mo(6+) reactions with tetrahydropterins (summarized in Scheme 8): two described as redox and one as ligand substitution. These contradictory interpretations begged for resolution and prompted us to further study.

Scheme 9 summarizes the repertoire of molybdenum reagents known to react with a variety of tetrahydropterins. The product observed in the initial studies (Eq. 9) using MoO₂detc₂ was eventually proved to have the stoichiometry [MoOCl(H₃DMP)(detc)]Cl **3a** and has been fully characterized. The best preparation of product **3a** occurs by a different route where a Mo(6+)-mono-oxo reagent MoOCl₂(detc)₂ reacts with H₄pterin [66].

Each of the products **1-3** shown in Scheme 9 were subjected to the same pterin dissociation experiments described earlier (Scheme 7). Addition of HCl or hydroxyquinoline to each of the complexes **1-3** caused pterin dissociation and ¹H NMR detected the appearance of tetrahydropterin, but not dihydropterin. Therefore, each molybdenum complex behaves as a

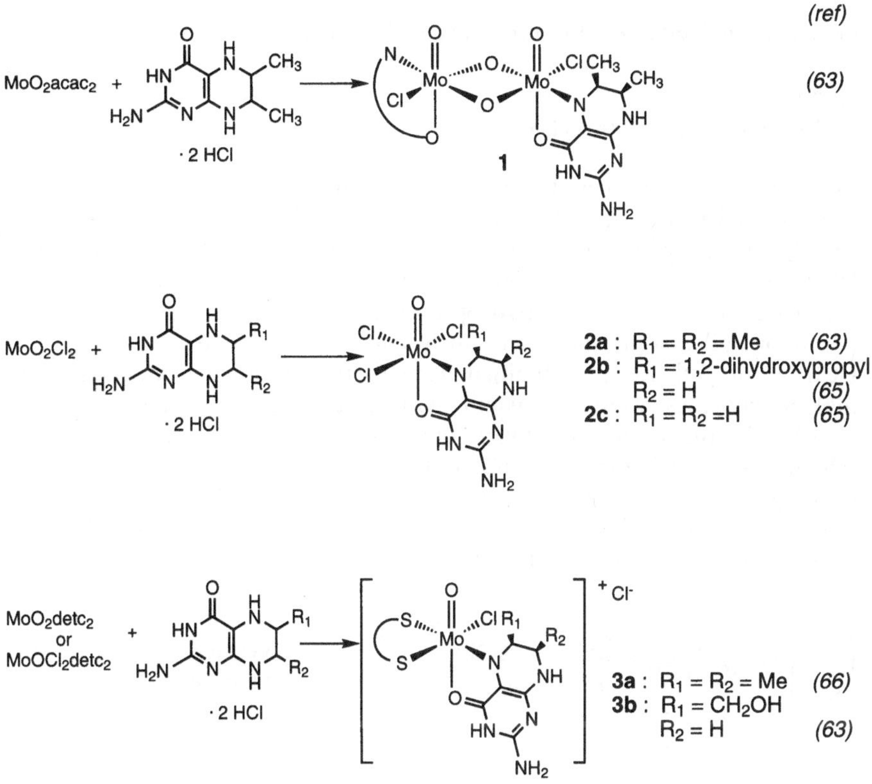

(ref)

(63)

2a : $R_1 = R_2 = Me$ (63)
2b : $R_1 = $ 1,2-dihydroxypropyl
     $R_2 = H$ (65)
2c : $R_1 = R_2 = H$ (65)

3a : $R_1 = R_2 = Me$ (66)
3b : $R_1 = CH_2OH$
     $R_2 = H$ (63)

Scheme 9. A summary of the variety of tetrahydropterin reactions with dioxomolybdenum(6+) reagents

Mo(6+)(H$_4$pterinate$^-$) complex. The data are consistent with the idea that *no* complete transfer of 2 electrons from pterin to molybdenum has occurred.

Adopting the formal oxidation state assignment Mo(6+)-(H$_4$pterinate$^-$) offered a different interpretation of a peculiar observation [65]. It was reported that MoOCl$_3$(H$_2$biopterinH$^+$) **2b** was unstable in methanol where decomposition was marked by the appearance of free tetrahydrobiopterin. This observation was explained by a proposed reversal of the synthetic redox reaction (Eqs. 11 and 12):

synthesis: Mo(6+)O$_2$Cl$_2$ + H$_4$biopterin · 2HCl
   → Mo(4+)OCl$_3$(H$_2$biopterinH$^+$) **2b**   (11)

decomposition: Mo(4+)OCl$_3$(H$_2$biopterinH$^+$) + MeOH
   → Mo(6+)OCl$_3$(OMe) + H$_4$biopterin   (12)

In contrast, an alternative interpretation of this instability was simple pterin dissociation promoted by the protic solvent methanol (Eqs. 13 and 14):

synthesis: MoO$_2$Cl$_2$ + H$_4$biopterin · 2HCl → Mo(6+)OCl$_3$(H$_4$biopterinate$^-$)   (13)

decomposition: Mo(6+)OCl$_3$(H$_4$biopterinate$^-$) + MeOH
   → Mo(6+)OCl$_3$(OMe) + H$_4$biopterin   (14)

2.1.3
Reactivity of Molybdenum Complexes of Reduced Pterins

2.1.3.1
Dichlorophenolindophenol Oxidation

The reduction state of molybdopterin (i.e., the pterin portion of Moco) had been probed using the redox dye dichlorophenolindophenol (DCIP) [36, 67, 68]. The intensely blue solution of oxidized DCIP is bleached to colorless when DCIP is reduced by two electrons, providing a convenient visual and spectroscopic monitor. Tetrahydropterins reduce DCIP instantaneously while quinonoid dihydropterins react slowly and 7,8-dihydropterin do not reduce DCIP at all [69]. Results obtained from DCIP additions to Moco in several molybdoenzymes were consistent with a dihydropterin reduction state, possibly in the quinonoid form, but eliminated the possibility of a 7,8-dihydropterin structure [36, 67, 68]. It seemed appropriate to use this technique on our new molybdenum pterin complexes. Stoichiometric additions of DCIP to molybdenum complexes of reduced pterins gave the results summarized in Eqs (15)–(18) [63]:

$$\text{1} + \text{DCIP} \xrightarrow{\text{DMF}} \text{no bleaching, no reaction} \qquad (15)$$

$$\text{3a, 3b} + \text{DCIP} \xrightarrow{\text{DMF}} \text{no bleaching, no reaction} \qquad (16)$$

$$\text{2a} + \text{DCIP} \nearrow \xrightarrow{\text{DMF}} \text{red solution, proton transfer to DCIP} \qquad (17)$$

$$\searrow \xrightarrow{\text{MeOH}} \text{bleaching, DCIP reduction}$$

$$\xrightarrow{\text{MeOH}} \text{free H4DMP} \xrightarrow{+ \text{ DCIP}} \text{bleaching, DCIP reduction} \qquad (18)$$

None of the complexes in dimethylformamide solution reduced DCIP, though in one case (Eq. 17), proton transfer occurred. Only in the case of MoOCl$_3$(H$_4$DMP) in methanol solution was DCIP reduction observed. This contradictory result can be integrated with the remaining data when one remembers that its analog, MoOCl$_3$(H$_4$biopterin), is unstable towards dissociation of free tetrahydropterin in methanol solution (Eqs. (11–14)) [65]. Therefore the true sequence of reactions is shown in Eq. (18): tetrahydropterin first dissociates from molybdenum in methanol prior to the reaction of free tetrahydropterin with DCIP.

The uniform lack of reactivity of any intact molybdenum complex with DCIP could be explained by several arguments: (a) coordination of tetrahydropterinate stabilizes the pterin towards oxidation by DCIP; (b) the coordinated pterin is, in fact, a partially oxidized dihydropterin, or (c) the coordinated pterin is bound as a *partially* oxidized pterin where ownership of the two electrons is *shared* by both the molybdenum and the pterin. Later experiments will show that option (c) is the most accurate explanation.

2.1.3.2
DMSO Reduction

A successful model for the molybdenum cofactor should be capable of mimicking biological activity. The ability of synthetic model complexes to demonstrate oxygen atom transfer by reduction of dimethylsulfoxide (DMSO) to dimethyl sulfide (DMS) is a frequently used criterion [74, 75]. A specific illustration of this reaction for the enzyme DMSO reductase was presented earlier in this review (Eq. 8).

The ambiguity surrounding the molybdenum oxidation state in the pterin complexes 1–3 has been addressed by testing each complex for its ability to reduce DMSO. The results are summarized in Eqs. (19)–(22). Dimer 1 is unreactive towards DMSO and shows no change after two days in DMSO solution [63]. Complexes 2a and 2b demonstrate ligand substitution with DMSO but not further oxygen atom transfer [63, 65] (Eq. 20). In contrast, complex 2c reduces two equivalents DMSO to yield an oxidized Mo(6+)(=O)$_2$ species and fully oxidized pterin [70] (Eq. 21). Complexes 3a and 3b react slowly with DMSO [66, 71] (Eq. 22). 3a consumes 3 equivalents DMSO to produce oxidized pterin, a Mo(6+)(=O)$_2$ core and oxidized dithiocarbamate. The precise outcome of DMSO oxidation of complex 3b is still unknown but a Mo(6+)(=O)$_2$ core is formed in addition to oxidized pterin. Oxidation of the pterin side chain is indicated as well.

$$1 + \text{DMSO} \longrightarrow \text{no reaction} \tag{19}$$

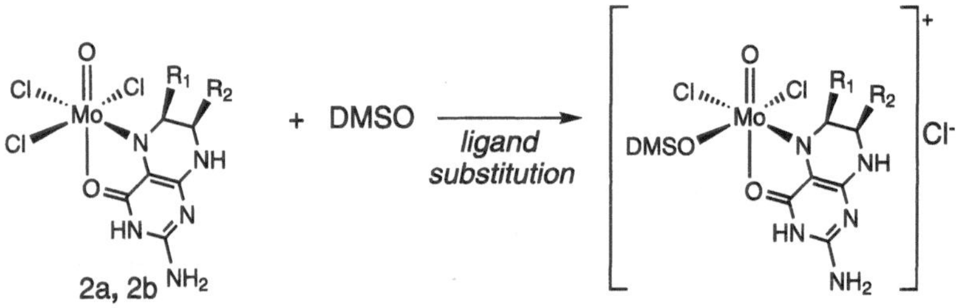

(20)

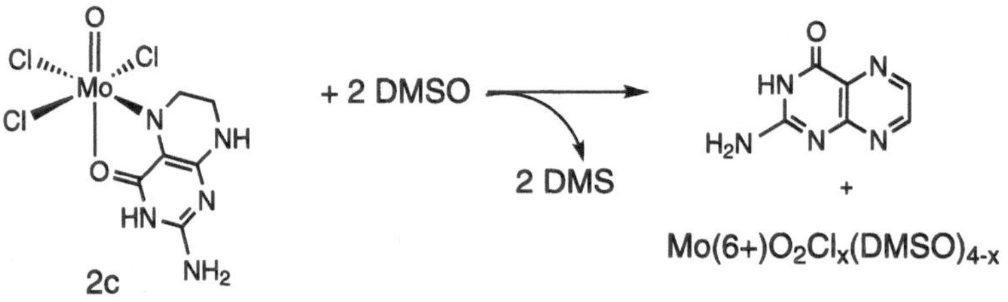

2a, 2b + DMSO →(ligand substitution) [product]$^+$ Cl$^-$

(21)

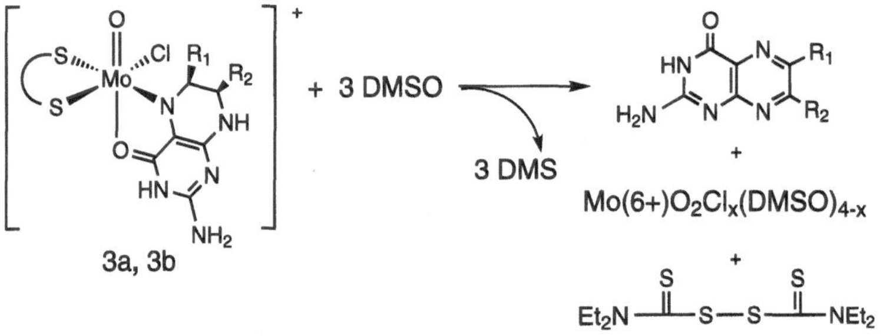

2c + 2 DMSO → (2 DMS) pterin + Mo(6+)O$_2$Cl$_x$(DMSO)$_{4-x}$

(22)

3a, 3b + 3 DMSO → (3 DMS) pterin + Mo(6+)O$_2$Cl$_x$(DMSO)$_{4-x}$ + Et$_2$N–C(=S)–S–S–C(=S)–NEt$_2$

The ability of both **3a** and **3b** to reduce DMSO but not complexes **2a**, **2b**, and **2c** is readily understandable in view of previous studies demonstrating an easier oxidation of Mo(4+) when complexed to sulfur ligands [72, 73]. The singular reduction effected by **2c**, but not **2a** or **2b**, is considerably more challenging to explain. There is no apparent difference in the bond distances between **2c** and **2a** or **2b** that would offer a simple solution to the problem. The oxidized pterin product released from **2c** is undoubtedly much more insoluble than oxidized biopterin or dimethylpterin but this difference would seem insufficient to explain the observed chemistry.

One possible mechanism for DMSO reduction is offered in Scheme 10 as detailed for complex **2c**.

The first step (a) is undoubtedly DMSO substitution for chloride, a fact proved by spectroscopic and conductivity data observed for **3a** and **2a** in DMSO solution [63]. The most likely site for introduction of DMSO into the molybdenum coordination sphere is at the trans position to the pterin nitrogen N5 based on the observed weakening of a bond *trans* to a double bond involving oxygen or nitrogen [74]. However, the X-ray structure of MoOCl$_3$(H$_3$dmp) shows that the longest, and presumably weakest, Mo-Cl bond

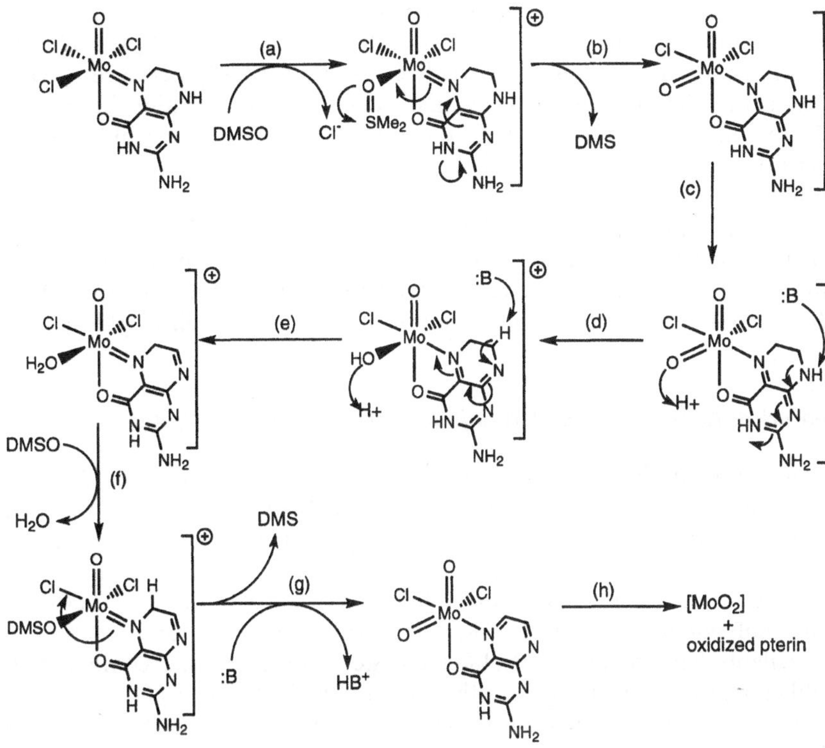

Scheme 10. A possible mechanism for the reduction of two molecules of DMSO by one molecule $MoOCl_3(H_4pterin)$

is the substitution site cis to N5 illustrated in Scheme 10. The transfer of an oxygen atom (b) from DMSO to Mo is the point of formal oxidation of the complex which may be viewed as the transfer of two electrons from the Mo=N5 bond to the incipient Mo=O bond resulting from oxygen atom transfer. Prior to further oxidation, the pterin must lose at least two protons to a base designated :B in Scheme 10 (c, d). Solution species capable of proton abstraction are either another pterin molecule or a Mo=O group. The Mo=O group is used in Scheme 10 where the sequential protonation (d, e) of a Mo=O bond has the illusion of resulting from intramolecular proton transfer between pterin and Mo, but note that this event could also be intermolecular. As a result, the Mo=O group is converted from an oxo ligand via a hydroxo to an aquo group on Mo(6+) simultaneous with the further oxidation of pterin via electron transfer to molybdenum. Note that this now formally regenerates a Mo(4+) oxidation state; an alternative view is that this step makes an electron pair (localized between Mo and pterin) available for a second DMSO reduction. Substitution of water by a second DMSO molecule (f) precedes the reduction (g) of a second DMSO. Overall the reaction is a four-electron oxidation of the molybdenum-pterin complex **2a** coupled to the four electron reduction of two molecules DMSO (h).

The six electron oxidation observed for complex **3a** can be imagined to proceed similarly to the reactions (a–h) in Scheme 10. However, at some point two dithiocarbamate ligands are oxidized to a disulfide. There are as yet no experimental data to suggest mechanistic details of this additional redox step.

Before proceeding to describe theoretical efforts to determine the best description of the oxidation and reduction states in the molybdenum-pterin complexes **1–3**, it might be useful to summarize the results of the three experimental approaches. Complexes **1–3** all react with acid or hydroxyquinoline to dissociate free tetrahydropterin. There is no reaction of complexes **1–3** with the oxidant DCIP. There is variable ability among the complexes **1–3** to reduce DMSO to produce DMS, oxidized Mo(6+) and oxidized pterin.

2.1.4
Theoretical Approaches to Determining Formal Oxidation States

In concert with the experimental approaches described in Sect. 2.1.3, two computational methods were employed to resolve the issue of molybdenum and pterin oxidation states by revealing charge distribution between the metal and the redox-active ligand [63].

Extended Huckel molecular orbital (EHMO) calculations were performed on all molybdenum pterin complexes for which crystallographic parameters are available. The calculated Mulliken charge on the metal was compared with Mulliken charges calculated for other oxo-molybdenum complexes for which unambiguous formal oxidation states can be made. The results are listed in Table 1.

The bond valence sum (BVS) method was also applied to the set of complexes listed in Table 1. This empirical method has been most recently used to estimate oxidation states on metal ions in enzymes [75, 76]. The BVS method is based on the observation that, as the unit charge on the metal increases, the bond length to ligated atoms decreases [77]. The best illustration of this general trend is in the series of neutral fragments M-OH$_2$, M-OH, M=O where the metal bears no charge, +1 and +2 units of charge, respectively, in parallel with decreasing Mo-O distances to the neutral, anionic and dianionic ligands. Table 1 also lists molybdenum charges calculated using the BVS method for molybdenum pterin complexes as well as for the molybdenum complexes that constitute the yardstick for comparison. Figure 4 presents a graphical display of the BVS results. It is observed that compounds conventionally assigned Mo formal charges of 4+, 5+, and 6+ are clustered in well-separated groups. When the calculated Mo charges in molybdenum pterin complexes are superimposed on the graph in Fig. 4 it is seen that they span the range of charges calculated for authentic Mo(5+) compounds.

Both the EHMO and BVS methods, although different in approach, converge on the same conclusion: the molybdenum pterin complexes, as a group, are most consistent with an oxidation state assignment of 5+ for molybdenum. To understand the pterin redox state most appropriate to pair with a Mo(5+) center, it is helpful to recall the pterin partners for Mo(4+) and Mo(6+) cations. For example, a formal molybdenum oxidation state of 6+

Table 1. Formal oxidation states and Mulliken atomic charges

Molecule	Formal Mo oxidation state	Calculated Mulliken Charge	BVS Calculated Mo Charge
$MoO_2(5\text{-}t\text{-Busap})(MeOH)$	6	4.42	5.99
$MoO(naph\text{-}cat)(sap)$	6	4.41	5.87
$MoO_2(HB(Me_2prz)_3)(NCS)$	6	3.98	6.12
$MoO_2(HB(Me_2prz)_3)(S_2P(OEt)_2)$	6	3.96	5.79
$MoO_2(N_2S_2)$	6	3.82	5.73
$MoO(detc)_2Cl_2$	6	3.70	5.32
$MoO(HB(Me_2prz)_3)(N_3)_2$	5	3.95	5.15
$MoO(HB(Me_2prz)_3)(NCS)_2$	5	3.98	5.03
$Mo_2O_3(3\text{-}OEt\text{-Busap})_2(dmf)_2$	5	3.60	–
$MoOCl(HB(Me_2prz)_3)(NCS)$	5	3.58	–
$Mo_2O_3(3\text{-}t\text{-Busap})_2(dmf)_2$	5	3.50	5.43
$MoO(HB(Me_2prz)_3)(SPh)_2$	5	–	5.09
$MoO(HB(Me_2prz)_3)(S_2P(OEt)_2)$	4	4.05	4.57
$MoO(HB(Me_2prz)_3)(detc)$	4	3.88	4.55
$MoS(HB(Me_2prz)_3)(detc)$	4	3.87	4.22
$MoO(3\text{-}t\text{-Busap})(bipy)$	4	3.36	4.78
$MoOCl(HB(Me_2prz)_3)(py)$	4	2.90	4.57
$Mo_2O_4Cl_2(H_4DMP)_2$ **1**		4.25	5.42,[a] 5.52,[b] 5.05[c]
$MoOCl_3(H_4biopterin)$ **2b**		3.9	5.18,[a] 5.24,[b] 4.84[c]
$[MoOCl(H_4DMP)(detc)]^+$ **3a**		3.9	5.02,[a] 4.92,[b] 4.58[c]

[a] Calculated using BVS parameters for Mo(6+).
[b] Calculated using BVS parameters for Mo(5+).
[c] Calculated using BVS parameters for Mo(4+).
Complexes used for comparison are:
(a) $MoO_2(ssp)$, $Mo_2O_3(ssp)_2$, $MoO(ssp)(bipy)$: Inorg Chem (1989) 28:2082
(b) $MoO(ssp)(catecholate)$: Inorg Chem (1988) 27:3950;
(c) $MoO_2(NCS)(L\text{-}N_3)$, $MoOCl(NCS)(L\text{-}N_3)$, $MoOCl(py)(L\text{-}N_3)$: Inorg Chem(1990) 29: 3650
(d) $MoO(S_2P(OR)_2)(L\text{-}N_3)$, $MoO_2(S_2P(OR)_2)(L\text{-}N_3)$: Inorg Chem (1988) 27:3044;
(e) $MoO(N_3)_2(L\text{-}N_3)$, $MoO(NCS)_2(L\text{-}N_3)$: Acta Cryst (1987)C43:51;
(f) $MoO(detc)(L\text{-}N_3)$, $MoS(detc)(L\text{-}N_3)$: J Am Chem Soc (1987) 109:2938;
(g) $MoO_2(L\text{-}S_2N_2)$: J Am Chem Soc (1987) 10:5655
(h) $MoO(SPh)_2(L\text{-}N_3)$: Inorg Chem (1987) 26:1017.

requires the H_4pterin to have an anionic charge to account for the overall neutral charge on the complex $MoOCl_3$(reduced pterin), that is, any of the complexes **2a**, **2b**, or **2c**. Likewise, the choice of Mo(4+) demands a protonated H_2pterin to attain neutrality. The intermediate oxidation state of 5+ then corresponds to a neutral, *trihydropterin*. The set of three Mo/reduced pterin partners is diagrammatically illustrated in Fig. 5.

If one must adhere to formal oxidation state labels then Mo(5+), a compromise between the extreme choices of Mo(6+) vs Mo(4+), most accurately conveys the extent of charge delocalization between the Mo and the pterin. A more flexible view of the electronic picture can be invoked that seems a better choice. This view discards the confines of formal oxidation states and regards the second bond between the atoms Mo and N5 as an electron pair that remains available for use by *both* the Mo and the pterin. This view accommodates all the experimental data. For example, tetrahydropterin ligand dissociation (Scheme 7) suggests a heterolytic cleavage of the Mo-N5 bond

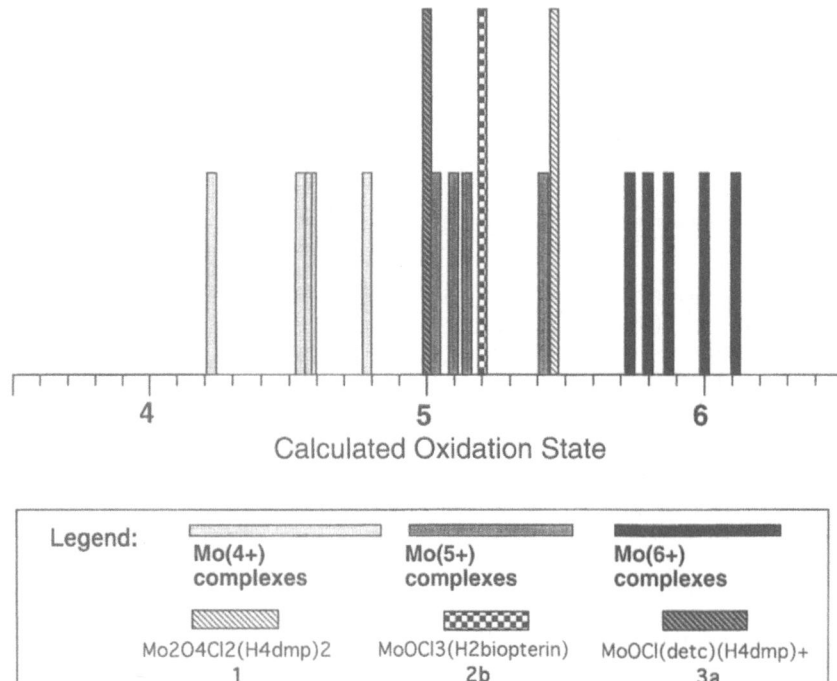

Fig. 4. A graphical display of the results of Bond Valence Sum calculations. It is observed that compounds conventionally assigned Mo formal charges of 4+, 5+ and 6+ are clustered in well-separated groups and this is emphasized with different bar shading. The legend identifies the bar shading corresponding to these Mo(6+), Mo(5+), and Mo(4+) complexes used for comparison and the Mo complexes of reduced pterins, **1, 2b** and **3a**

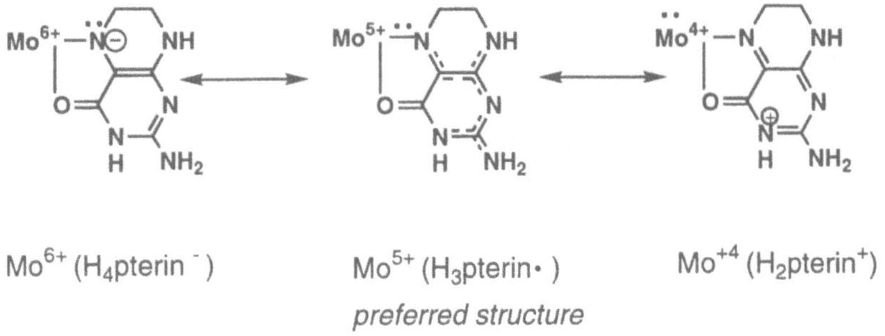

Mo^{6+} (H_4pterin$^-$) Mo^{5+} (H_3pterin·) Mo^{+4} (H_2pterin$^+$)

preferred structure

Fig. 5. Resonance structures for molybdenum coordinated to reduced pterin. Note how the location of the pair of electrons depicted near Mo and N5 shifts in the three structures

with the pterin regaining its original lone pair electrons. The lack of reactivity of molybdenum pterin complexes 1–3 with the mild oxidant DCIP is consistent with a deactivated tetrahydropterin through partial electron delocalization onto the molybdenum. The variable ability of molybdenum pterin complexes to reduce DMSO can be interpreted as the result of a subtle shift of the electron pair between Mo and N5 in response to changing ligand fields of ancillary ligands on Mo.

2.1.5
Requirements for the Synthesis of Molybdenum Complexes of Reduced Pterins

This account of tetrahydropterin reactions of dioxo-molybdenum reagents will conclude with comments on several aspects of the syntheses that shed light on the conditions necessary to form these complexes.

The stability of these complexes is closely tied to the formation of a Mo=N5 bond. Conversely, if reaction conditions prevent formation of this bond, no pterin coordination is observed. Several pieces of experimental data corroborate this idea [63]. First is the observation that the syntheses in Scheme 9 fail if performed in basic media. For example, if two equivalents triethylamine are added to the reaction mixture to neutralize the two equivalents HCl associated with each type of pterin, no molybdenum-pterin complexes can be isolated and no pterin coordination is observed spectroscopically. Secondly, when the synthesis is attempted in neutral media a highly reactive $MoO_2(H_4pterin)$ complex forms that rapidly decomposes in any coordinating solvent. Neutral, aprotic reaction conditions are attained by using the solubilized pterin shown in Eq. (23).

$$MoO_2Cl_2 \ + \quad \xrightarrow{\text{CHCl}_3} \qquad (23)$$

In both of these examples the failure to form a stable oxo-Mo-(H_4pterin) complex can be attributed to competition between the two extant Mo=O groups on the molybdenum reagent with the incipient Mo=N5 bond to the pterin. The competition is relieved in protic environments where one oxo ligand can be protonated and removed as water. One may consider that the Mo(=O)(=N5) unit of the molybdenum pterin complexes substitutes for the common Mo(=O)$_2$ core frequently observed in Mo(6+) complexes [74–76].

2.2
Complexes of Molybdenum Coordinated by Pterin-Substituted Dithiolene Ligands

Molybdenum and pterin have been combined in another type of system. The goal of this work is the synthesis of molybdopterin as it appears in the proposed structure of Rajagopalan. Though there are only a few points within this area having relevance to the topic of this review, it is briefly discussed to give a complete picture of molybdenum-pterin systems.

The attachment of a pterin to a dithiolene chelate was accomplished using the method shown in Eq. (24). This strategy was based on known reactivity trends of molybdenum tetrasulfides and initially proved for quinoxalyl alkynes [78].

$$\text{(24)}$$

A pterin-substituted dithiolene was formed using the strategy in Scheme 11 [79, 80]. The reaction proceeds through a so-called trithiolene chelate that can be converted to the dithiolene by phosphine reduction.

Oxidation of the dithiolene complex product to a Mo(5+) species (Scheme 12) allowed spectroscopic analysis by EPR. Small molybdenum hyperfine values indicate that substantial spin density is delocalized onto the pterinyl-dithiolene ligands in these molecules. Further evidence of the delocalized nature of the metal dithiolene system is a chemical transformation of the heterocyclic ring. In the case of the quinoxalyl-dithiolene complex, a one-electron oxidation induces electrophilic attack of the side chain keto-carbon at one of the ring nitrogen atoms, possibly accompanying hydrogen atom abstraction from the solvent (Scheme 12).

The authors remark that the same dithiolene transformation is induced by addition of acid and reversed by subsequent addition of base, illustrating acid-base reactivity coupled to a redox process. This observation is a recurring theme in enzymes as well as in the previous section of this review.

2.3
Pterin Function in Molybdenum and Tungsten Enzymes

The above studies were completed prior to several recent X-ray crystallographic structure determinations of molybdenum [43–47] and tungsten [81] enzymes. Certain recurring features of these X-ray structures will be described, followed by a discussion that focuses on the relationship of the enzyme structures to the results from previous studies on molybdenum-pterin systems.

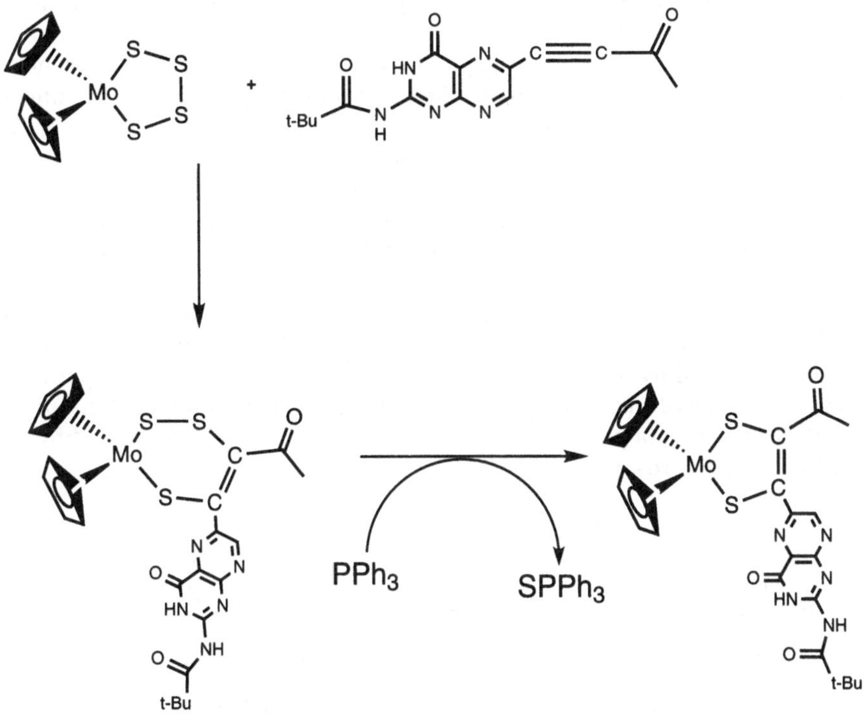

Scheme 11. One synthetic scheme forming a molybdenum-bound pterin-substituted dithiolene [79, 80]

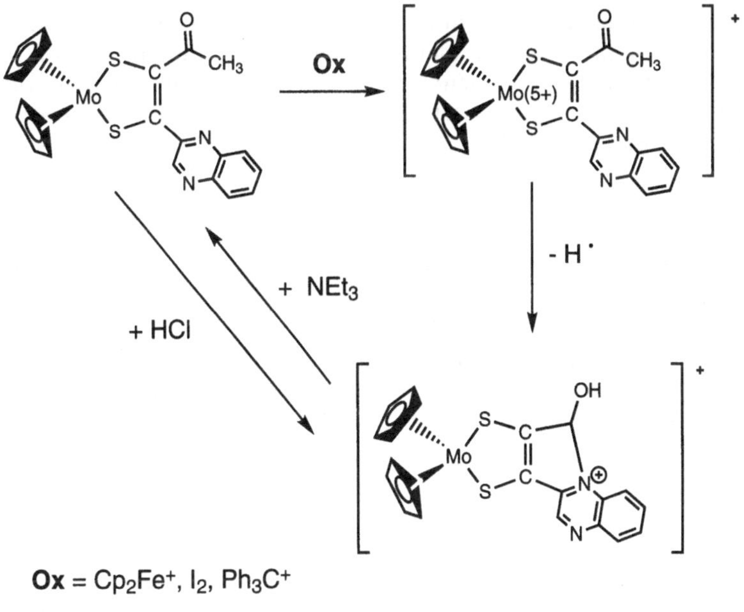

Ox = Cp$_2$Fe$^+$, I$_2$, Ph$_3$C$^+$

Scheme 12. Molybdenum oxidation in a pterin-substituted dithiolene [79]

From the perspective of this review, perhaps the most surprising aspect of the molybdopterin structure is the cyclization of the γ-hydroxyl on the pterin side chain to form a pyrano ring. There was no experimental evidence to predict such a structure, though computer modeling suggested an uncomfortably close interaction of the γ-hydroxyl with the proton at C7 [82]. The interpretation of this structure as a protected form of a 5,6-dihydropterin is intriguing (Eq. 25), especially in light of Rajagopalan's results obtained during attempts to define the reduction state of the pterin in Moco. From these studies he concluded that Moco contained a dihydro-, not tetrahydro-, pterin based on reactivity patterns with the redox dye DCIP [36, 67, 68]. He further speculated that the isomeric form of the dihydropterin was likely quinonoid or another unstable isomer but definitely not the 7,8-isomer. With hindsight it seems that the unknown dihydropterin isomer detected by Rajagopalan's experiments was the 5,6-dihydromolybdopterin. Recent experiments conducted in our labs confirm that pyranopterins such as that drawn in Eq. (25) react as dihydro-, but not tetrahydro-, pterins [83].

5,6-dihydropterin   cyclized molybdopterin
                                     (pyrano-molybdopterin)

(25)

With the known Moco structure in hand, the value of the modeling efforts above can be evaluated. The dithiolene binding site for the metal observed in the X-ray structures at first glance seems to discount the results from the Mo(6+) – tetrahydropterin studies. However, given the proclivity of molybdenum for strongly binding tetrahydropterins at the O4,N5 chelate demonstrated in those investigations, one may conclude that one job of the protein is to constrain the pterin via a complex hydrogen bonding network to prohibit metal access to this site. The extent of the H-bonding is illustrated in Fig. 6 for formate dehydrogenase [43]. Further speculation suggests that in dysfunctional enzymes or inactive Moco samples the site of Mo (or W) binding may have changed.

A redox role for the pterin is clear from the intimate interactions, again through H-bonds, between the pterin and other electron carriers that will conduct the electron flow to or from the ultimate acceptor or donor. Specific examples were illustrated in Fig. 1 in Sect. 1.2.2 which showed how the flexibility of the conjugation and multiple donor/acceptor sites in the pterin system are exploited by different protein structures. Certainly in this regard the pterin is functioning as part of the 'circuit' to deliver electrons to and from the metal. This pterin-mediated electron flow was expected from the results of both the Mo(6+) – H$_4$pterin studies and the pterinyl-dithiolene complexes of molybdenum. It is curious that communication between molybdenum and

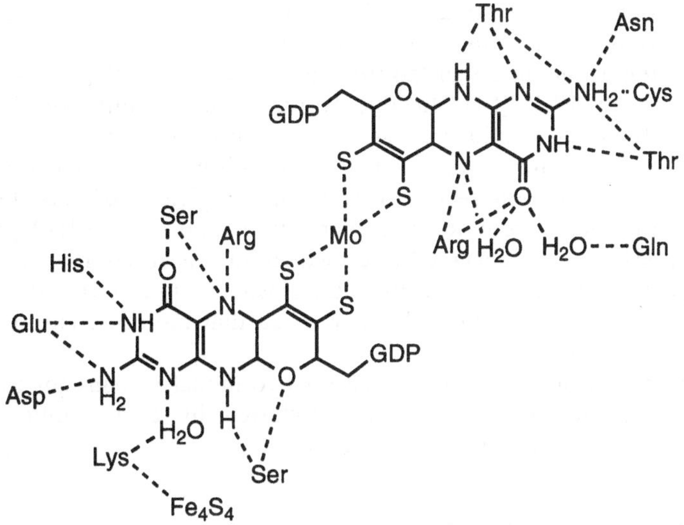

Fig. 6. A diagram of the extensive hydrogen bonding around Moco in formate dehydrogenase [43]

pterin via the conjugation of the dithiolene unit seems blocked by the saturated region C6,C7 in the pyranopterin as observed in the enzyme X-ray structures and by the molecules in Eq. (25). An alternative view is that the cyclized structure represents a protected resting state and substrate turnover involves a different electronic description at the pterin. Opening the pyrano ring in a different way from that shown in Eq. (25) produces a 5,8-dihydropterin (Eq. 26), an intriguing structure which maintains pi-conjugation throughout the pterin.

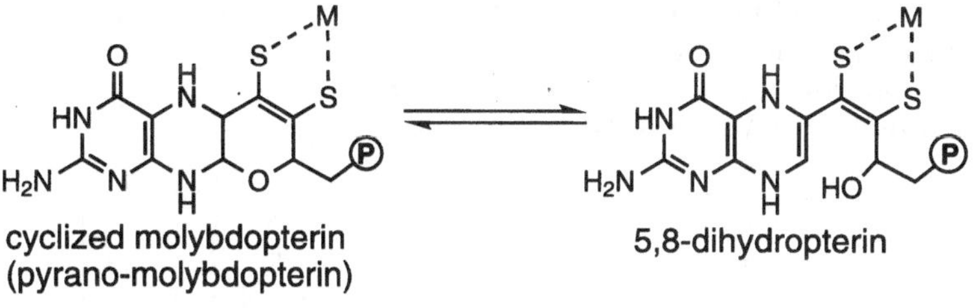

(26)

cyclized molybdopterin
(pyrano-molybdopterin)

5,8-dihydropterin

3
Model Systems for the Metal Site in Phenylalanine Hydroxylase

There is scant evidence to indicate the nature of the metal-pterin interaction in the aromatic amino acid hydroxylases PAH, TyrH, and TrpH. While direct copper-pterin binding is inferred from spectroscopic results on bacterial PAH (see Sect. 1.2.1), no information exists either supporting or discounting a similar interaction in the iron-dependent enzymes. Hence, the field was ripe for designing model systems to determine the characteristics which favor metal coordination and the requirements for electron transfer. Prior to 1989 the only report of tetrahydropterin redox reactions with transition metals dated back to 1967 [54]. In this work Vonderschmitt and Scrimgeour described their observations on aqueous reactions of ferric and cupric ions with tetrahydropterins. They observed that both metals reacted with tetrahydropterin with different results. Ferric ion completely oxidized tetrahydropterin to produce ferrous ion and pterin in a ratio of 4 Fe(II):1 pterin. 7,8-Dihydropterin was an observed intermediate in this reaction. In contrast, no 7,8-dihydropterin was detected in solution when copper (II) and tetrahydropterin reacted, but a purple color developed that indicated cupric complexation by H_4pterin.

The first part of this section discusses two model systems for the copper-pterin active site in bacterial PAH followed by a description of brief investigations of several iron-pterin systems.

3.1
Copper Reactions with Tetrahydropterins

3.1.1
Functional Model for the Copper Site in Chromobacterium Violaceum

The observation made by Vonderschmitt and Scrimgeour of a purple complex from copper(II) and tetrahydropterin spurred several groups to reinvestigate this reaction. Yamauchi and coworkers explored the behavior of cupric complexes with 6,7-dimethyltetrahydropterin [H_4DMP] under a variety of conditions [84]. When equimolar amounts of simple cupric salts, $Cu(NO_3)_2$ or TEA_2CuCl_4, were combined with H_4DMP in a methanol/acetonitrile solvent mixture, a colorless solution resulted (Eq. 27). The protonated trihydropterin radical H_4DMP$^+$ was observed by EPR and the pterin radical identity is supported by successful EPR spectral simulation of experiments incorporating both protonated and deuterated pterins. The rapidly formed trihydropterin radical disproportionates to a 1:1 mixture of H_4DMP and H_2DMP as demonstrated by both UV and NMR spectroscopy.

$$Cu(l) + 0.5 \text{ eq } 7,8\text{-}H_2DMP + 0.5 \text{ eq } H_4DMP$$

When the simple cupric salts were replaced by a cupric dipeptide complex, Cu(Gly-Gly), the reaction with aqueous H_4DMP at pH 7 proceeded much more slowly and without formation of a radical intermediate. Instead a new cupric species was detected by EPR and was postulated to be the ternary complex Cu(II)(Gly-Gly)- H_4DMP (Eq. 28).

$$\text{Cu(Gly-Gly)} \quad + \qquad\qquad\qquad\qquad\qquad \longrightarrow \qquad\qquad\qquad\qquad\qquad (28)$$

H₄DMP

In a third system they observed formation of hydroxyl radical when a 1:1 mixture of H_4DMP and $[Cu(II)bipy]^{2+}$ in aqueous solution at pH 7 was bubbled with O_2. This system was capable of hydroxylating unactivated phenyl groups as demonstrated by the conversion of phenylalanine ethyl ester to *o*-, *m*- and *p*-tyrosine esters (Eq. 29), albeit in low (4%) yields [85]. This observation is significant when compared to the complete lack of tyrosine formation when $[Cu(II)bipy]^{2+}$ was omitted from the system.

$$[Cu(bipy)]^{2+} \; + \; H_4DMP \; + \; O_2 \qquad\qquad\qquad \text{4\%} \qquad (29)$$

3.1.2
Formation of a Cupric Complex of Tetrahydropterin

Results obtained on a different cupric system complement and corroborate the conclusions of Yamauchi [57, 86, 87]. In addition, these investigations illustrate how the stability of intermediates can be altered by use of different solvent media and different (i.e., bidentate vs tridentate) ancillary ligands. Scheme 13 diagrammatically depicts the various reactions studied in the course of this work.

Reduction of Cu(II) by H_4DMP yielding Cu(I) and 7,8-H_2DMP in each of the reactions is seen in Scheme 13. The Cu(I) can be isolated as the perchlorate salt and the H_2DMP is identified by UV spectroscopy. Some of these reactions proceed through a visible intermediate characterized by an intense blue color. The stability of this blue intermediate is greatly enhanced when the Cu(II) is coordinated to the tridentate tris(pyrazolyl)borate ligand Tp* whereas no intermediate is detected in tetrahydropterin reactions of $Cu(bipy)_2^{2+}$ or $Cu(phen)_2^{2+}$. The λ_{max} of this blue intermediate as well as its lifetime vary as a function of both the pterin used in the reaction (i.e., H_4DMP vs H_4HMP) and the substitution of the tridentate Tp* ligand (Tp* = tppb, tbpb, or tippb). Data summarizing the effects of these variations are listed in Table 2. The extinction coefficient of the 'blue' absorption band is estimated to be 2500 $M^{-1}cm^{-1}$.

EPR spectra obtained from reaction solutions frozen immediately after mixing show signals from a new Cu(II) complex and this signal degrades in parallel with the fading of the blue color. No signal characteristic of a pterin radical is observed by EPR. The absence of radical nature in the blue intermediate is further consistent with its lack of reactivity to molecular oxygen or to a radical trap (2,4,6-tri(t-butyl)phenol). For example, adding pure dioxygen gas via syringe to the blue intermediate has no effect on the intensity of the characteristic band near 600 nm. The blue intermediate is extremely sensitive to solvent conditions. Addition of acetonitrile to the blue intermediate causes immediate bleaching of the 600 nm absorption. Generating the blue intermediate in a mixed dichloromethane/methanol solvent environment increases its stability. The conclusion is made from these data that the blue

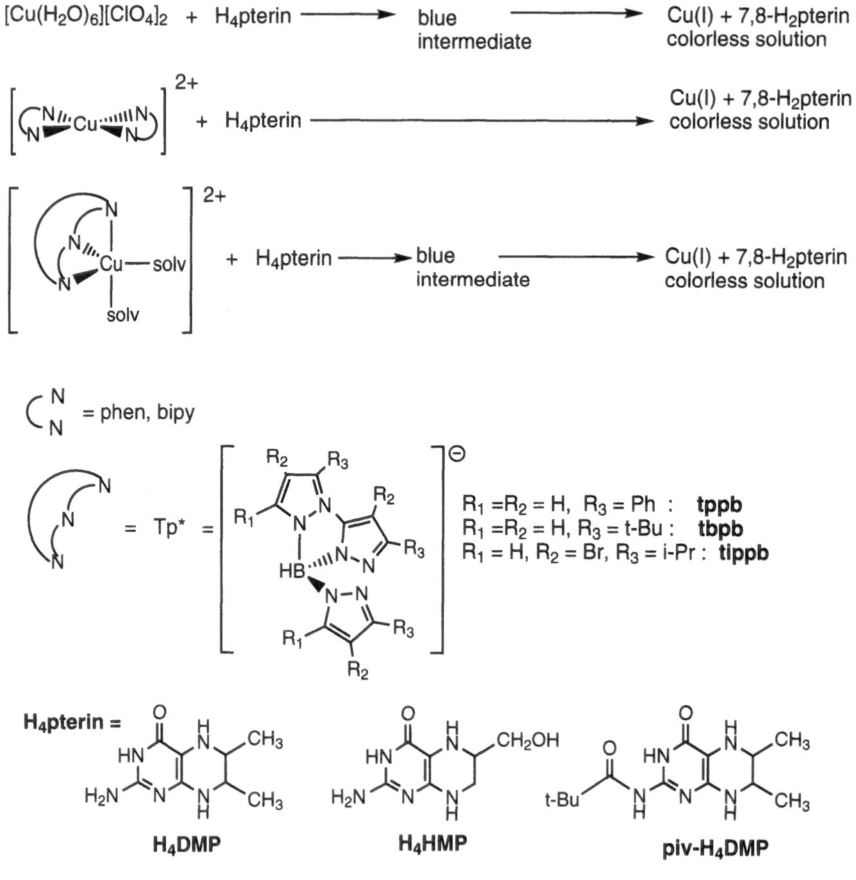

Scheme 13. Cupric complex reactions with tetrahydropterins [57, 86]

intermediate has the formulation [Cu(II)(H$_4$DMP)Tp*], that is, a five-coordinate ternary cupric complex containing both tetrahydropterin and the tridentate Tp* ligand [86, 87]. Based on Yamauchi's results, the speculation is made that the decay of the blue intermediate corresponds to the one-electron transfer from H$_4$pterin to Cu(II) forming the trihydropterin radical which subsequently disproportionates to 7,8-H$_2$pterin and H$_4$pterin.

The above studies serve to model several different aspects of possible Cu-pterin interactions in PAH. The nature of the ancillary ligands apparently determines the relative rate of electron transfer. Specifically it is observed that Cu(II) coordinated by polydentate Gly-Gly or tridentate tris(pyrazolyl)hydroborate ligands formed a transient but observable complex with tetrahydropterin (Eq. 28 and Scheme 13). In contrast, Cu(II) with monodentate ligands undergoes rapid electron transfer yielding a transient trihydropterin radical (Eq. 27). Cu(bipy)$_2$$^{2+}$ and Cu(phen)$_2$$^{2+}$ apparently undergo immediate electron transfer to H$_4$pterin with no observable Cu(II)-H$_4$pterin intermediate (Scheme 13). When Cu(II) is coordinated to bidentate bipyridine, it reacts with tetrahydropterin and O$_2$ to produce small amounts of hydroxyl radical and hydroxylates a model substrate (Eq. 29). The ancillary ligands may modulate Cu-pterin interactions by occupying certain coordination sites, thereby restricting both the geometry of the copper and the metal orbitals available for pterin bonding. In addition, the lifetime of the Cu(II)-H$_4$pterin intermediate is observed to decrease with increasing steric pressure of Tp* (see Table 2), suggesting that, in addition to coordination mode, second coordination sphere effects determine the electron transfer rate. Ring stacking interactions may also play a role in stabilizing or directing the substrate in the quaternary copper-pterin-substrate-O$_2$ complex. Such interactions are present in the crystal structures of copper-pterin complexes [57] as well as copper-tyrosinate compounds [85].

In the context of this section, it will briefly be noted that a cupric-pterin complex, bound only at pterin N5, undergoes one-electron reduction at the metal without apparent change of the pterin moiety (Eq. 30) [88].

Table 2. Electronic spectral data and lifetimes of Cu(H$_4$pterin)L, the "Blue Intermediate"

Cupric reagent	H$_4$pterin	λ_{max}, nm	Lifetime[a]
Cu(tppb)(DMF)$^+$	H$_4$DMP	610	1 min
Cu(tppb)(DMF)$^+$	H$_4$HMP	630	20 min
Cu(tppb)(DMF)$^+$	2-piv-H$_4$DMP	626	3 min
Cu(tppb)(DMSO)$^{2+}$	H$_4$HMP	626	30 min
Cu(tbpb)(MeOH)$^+$	H$_4$DMP	578	10 s
Cu(tbpb)(MeOH)$^+$	H$_4$HMP	582	30 s
Cu(tippb)(DMF)$^+$	H$_4$HMP	602	2 min
Cu(ClO$_4$)$_2$	H$_4$DMP	578	10 s
Cu(ClO$_4$)$_2$	H$_4$HMP	578	10 s

[a] The relative stability, "lifetime", is estimated by the time required to bleach reaction solutions having a cupric ion concentration of 2 mM in methanol.

$$\text{Eq. 30}$$

3.2
Iron Reactions with Tetrahydropterins

Except for the initial work of Vonderschmitt and Scrimgeour, only one other effort has been made to study ferric interactions of reduced pterins [89, 90]. This seems to be due to the extreme reactivity of the ferric-tetrahydropterin pair. Preliminary investigations in the author's labs indicated rapid reaction yielding dihydropterins and (presumably) ferrous products but no spectroscopically observable intermediates were detected.

Fischer et al. have probed this problem successfully by utilizing the relatively new technique of electrospray-ionization mass spectrometry (ESI-MS) [89, 90]. As they explain, this technique lends itself to systems involving polar and non-volatile molecules and is especially useful for detecting transient species of high reactivity.

The system they studied involves acetylacetonate iron, or $Fe(III)(acac)_3$, and two tetrahydropterins. In the first study depicted in Scheme 14, an acidic solution of $Fe(III)(acac)_3$ and 5-methyl-tetrahydropterin in a 1:1 ratio generated a signal in the ESI-MS having a mass consistent with the formulation $Fe(MeH_4P)(acac)_2{}^{2+}$. This species was postulated to be an equilibrium mixture of Fe(III) bound to the tetrahydropterin MeH_4P and the one-electron redox product Fe(II) coordinated by the radical cation of protonated trihydropterin of MeH_3P^+. The presence of a trihydropterin radical was inferred from EPR which showed the characteristic seven-line pattern previously observed for trihydropterin radicals. No EPR signal for the putative $Fe(II)(MeH_3P^{+\cdot})(acac)_2$ complex could be obtained, ostensibly for reasons of anisotropic coupling between the paramagnetic Fe(II) and the pterin radical. The hypothesis that the tetrahydro- and trihydro-pterins are coordinated to the iron is consistent with the observation that excess acetylacetonate inhibits formation of the species as indicated by a lower intensity signal. The evidence supporting an equilibrium mixture of these two complexes, rather that a single product, is that free tetrahydropterin MeH_4P *and* free trihydropterin radical MeH_3P^+ are detected, presumably according to the pterin dissociation reaction specified in Scheme 14.

In a second study, $Fe(acac)_3$ was allowed to react with tetrahydropterin (H_4P) [91]. Again pterin oxidation and iron reduction is observed (Scheme 15), although there are two significant differences from the former experiment. First, the reaction solution degrades within minutes in contrast to

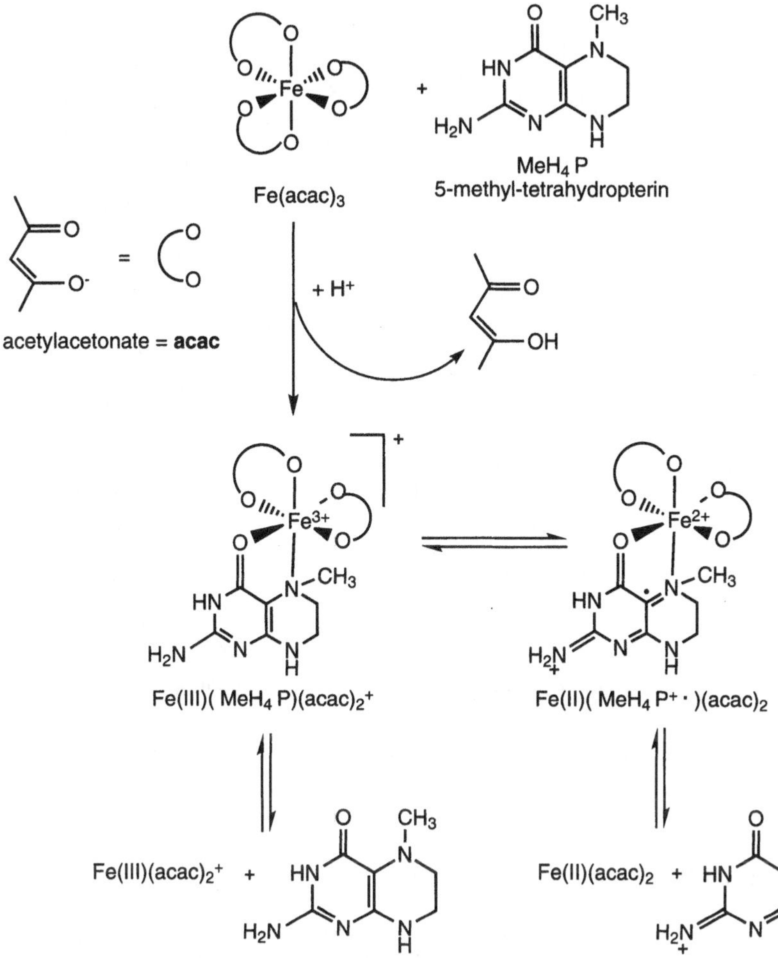

Scheme 14. Products of acetylacetonate iron(3+) with N5-methyl-tetrahydropterin as identified by ESI-MS and postulated in [89, 90]

the $Fe(MTHP)(acac)_{2+}$ species which is stable over several weeks. Second, the redox reaction is reported to yield a ferrous complex of dihydropterin as the quinonoid isomer.

Viewed through the lens of other pterin complexes described in this review, the pterin coordination to ferric and ferrous ions shown in Schemes 14 and 15 is not surprising. The extreme reactivity of these species, especially in the case of unsubstituted pterin H_4P, is consistent with the biological use of a ferrous-pterin couple as an efficient catalyst and the inability to obtain spectroscopic proof of metal-pterin interactions in the iron-dependent enzymes [23].

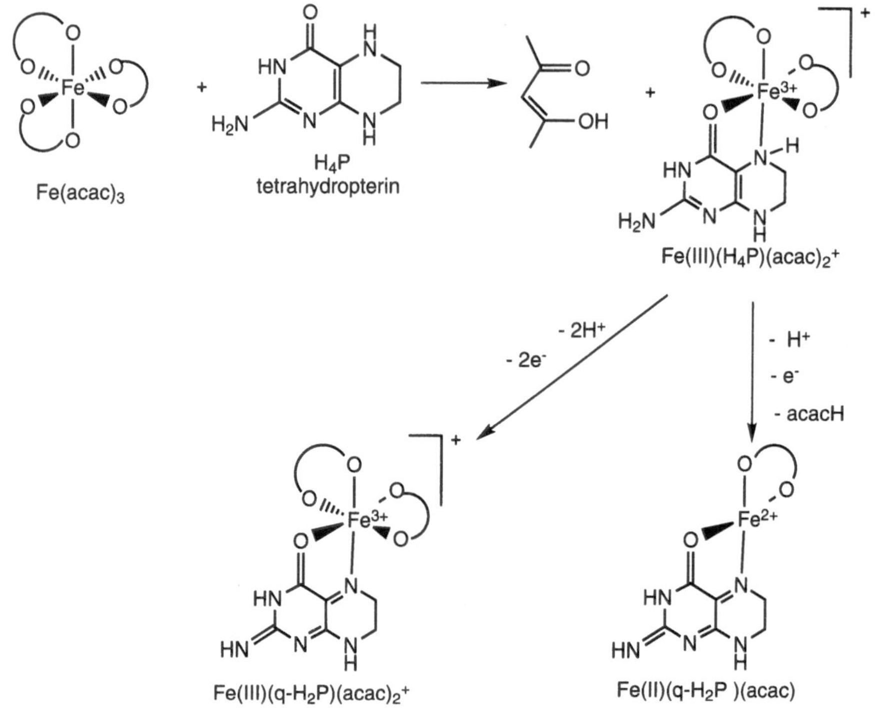

Scheme 15. Products of acetylacetonate iron(3+) with tetrahydropterin as identified by ESI-MS [91]

4
Transition Metal Complexes of Flavins

Studies of flavin interactions with transition metals were primarily motivated by the many examples of metal-containing flavoproteins. The majority of early studies were limited to oxidized flavins and isoalloxazines. Adrian Albert was first to attempt to quantify transition metal complexation by flavin as well as oxidized pterins [92, 93]. The work of Hemmerich et al., though now several decades old, is notable for its comprehensive investigation of metal reactions with all three flavin oxidation states [94]. This section will begin with a synopsis of the findings of the classic Hemmerich studies. Next, the first structurally characterized example of a ruthenium-flavin semiquinone will be discussed as background for presenting the more recently accomplished work in this area. Finally, an account of molybdenum(4+) chelation by flavin, previously discredited but now proved correct, will close this section.

4.1
Early Flavin Studies

The decidedly weak affinity of flavin for divalent redox-active metals of biological relevance was recognized early by Hemmerich and others [94, 95].

This behavior is in contrast to the favorable formation constants for pteridine coordination. These conflicting observations were confusing since pteridines and the isoalloxazine core of flavin both possess the oxinate chelate structure.

pterin   isoalloxazine   oxine
                                                        (8-hydroxyquinoline)

Hemmerich et al. identified the many factors that contributed towards these observations [94, 95]. First, since many of the studies were conducted in aqueous solution, the competing events of metal hydrolysis and flavin deprotonation were typically resolved in favor of modest complexation as quantified by formation constants, K_f, between 10 and 100. Second, it was concluded that flavin had little basicity at N5. The exception to flavin ambivalence towards metals was demonstrated in reactions of silver [96] and copper monocations [97]. The interpretation of this observation was that soft, low valent Ag^+ and Cu^+ ions were capable of significant π-backbonding to the empty π-antibonding orbitals of the flavin. In contrast to oxidized flavin, the one-electron reduced flavin radical (or flavin semiquinone), has significantly higher metal affinity. This was traced to its higher basicity at N5 [98]. Reduced, or dihydro-, flavins had little affinity to bind metals, a fact attributed to the low basicity of N5 and the awkward geometry presented to the metal by a bent dihydropterin [95].

One of Hemmerich's studies of flavin-metal redox reaction will be described. Ferrous ion is oxidized to the ferric ion (isolated as the insoluble $Fe(OH)_3$) as flavin is reduced to flavin semiquinone. The flavin semiquinone then disproportionates to oxidized flavin and dihydroflavin. Under different conditions the Fe(II) chelate of flavin can be spectroscopically characterized where the most distinctive feature is a bathochromic shift of the low energy flavin absorption. Hemmerich considered this bathochromic shift indicative of significant charge transfer from the ferrous ion to the pterin. From this experiment Hemmerich et al. realized that a Fe(II)-flavin "charge-transfer" complex and a Fe(III)-flavin semiquinone complex constituted a redox-related pair and speculated that they might exist in equilibrium (Scheme 16).

4.2
Ruthenium-Flavin Chemistry

A thorough study of the electrochemistry and spectroscopy of ruthenium pteridines has been recently completed in the lab of Clarke. The cornerstone of this work was laid much earlier when the structure and characterization of a ruthenium flavin product indicated substantial electron delocalization onto the heterocyclic ring [99, 100] (Scheme 17). The reaction of diaquo-tetraam-mine-ruthenium(II) with 10-methylisoalloxazine proceeded similar to other metal-pteridine reactions to yield a ruthenium complex coordinated to the flavin at the usual O4,N5 chelation site. The structure determined by X-ray diffraction was the first evidence to suggest that π-backbonding from the Ru(II) was more extensive than previously observed in silver or copper (I) complexes [96, 97]. Close scrutiny of bond distances and angles led these investigators to conclude that the complex could be formulated as a Ru(III)-flavinsemiquinone. This alternative formulation was supported by UV/vis, ^{1}H NMR, and spectroelectrochemistry. In particular, NMR indicated a significant electron density increase at N10 and spectroelectrochemistry produced a one-electron reduced species, formulated as Ru(II)-flavin-radical, having an intense low energy absorption remarkably similar to that observed for an Fe(II)- flavin-radical [94, 95].

Recent work from this lab has been devoted to electrochemical and spectroscopic study of ruthenium complexes of lumazine and methylated pterins [101]. Pterin and pteridine coordination to ruthenium(II) uniformly causes the site of protonation in the complexes to shift from N1 or N3 in the free ligands to N8 (compare Eqs. 31 and 32).

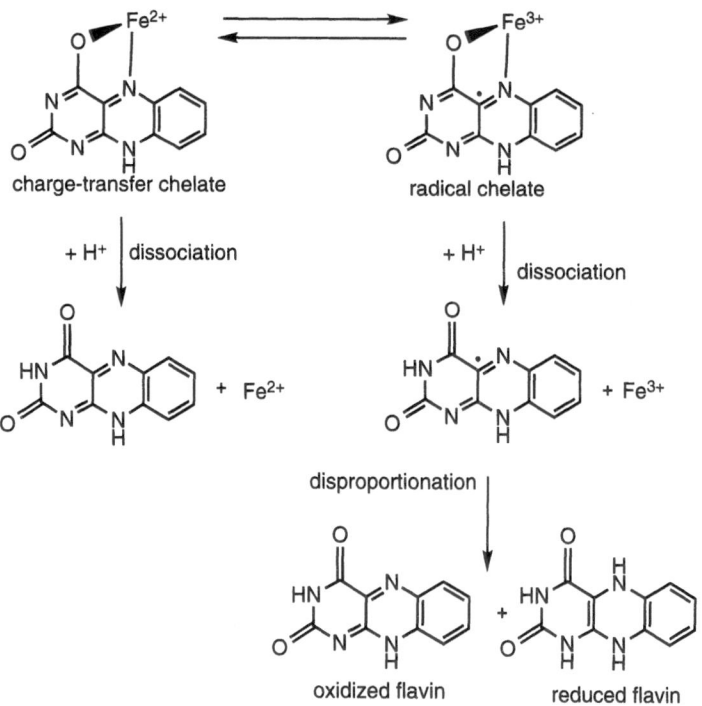

Scheme 16. Equilibrium between a charge-transfer complex and a radical chelate complex according to Hemmerich et al. [94]

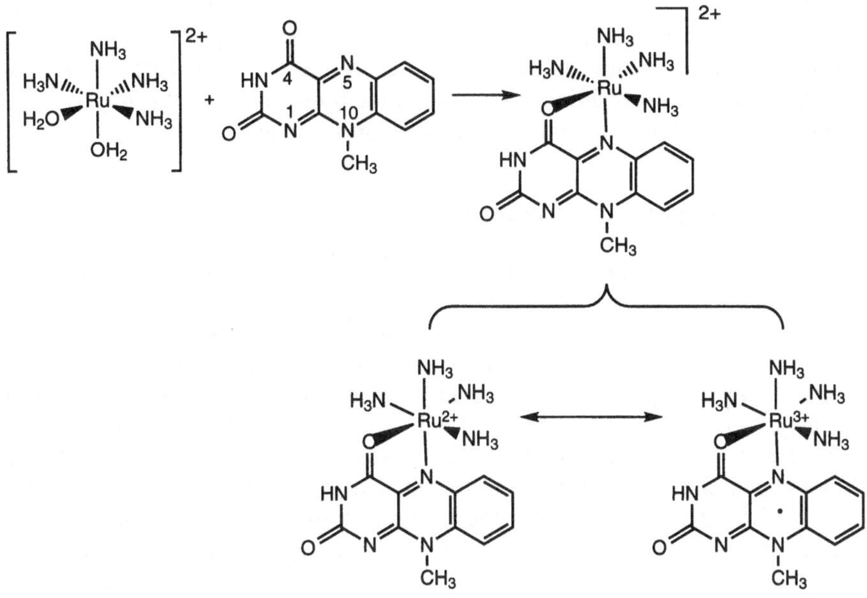

Scheme 17. Formation of a ruthenium coordinated by flavin and two resonance structures suitable for the product [99, 100]

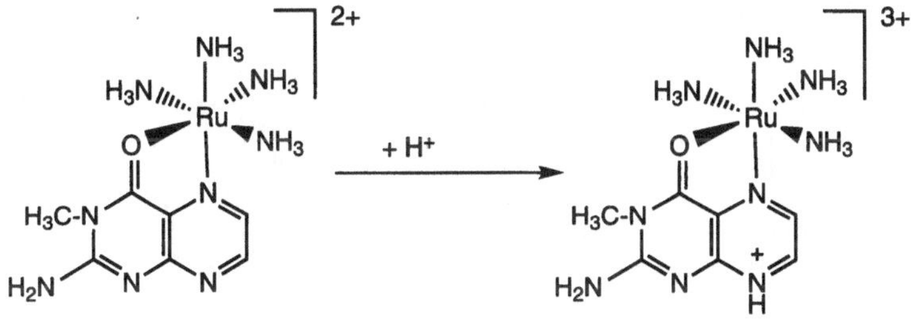

$$+ H^+ \qquad\qquad (31)$$

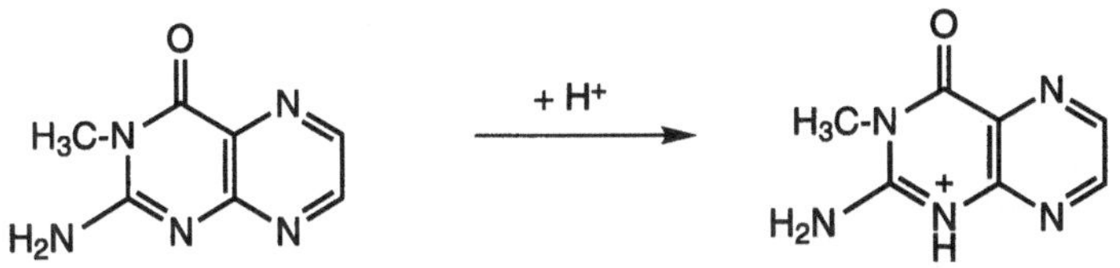

$$+ H^+ \qquad\qquad (32)$$

This is essentially the same effect noted earlier for the Ru-flavin complex, i.e., that Ru π-backbonding into the electron deficient heterocycle produces excess electron density localized at N8 in the pyrazine ring of the pteridine structure. The additional basicity at N8 is also verified by pK_a titrations [101]. In the case of lumazine, for example, the acidic pK_a for [lumazineH$^+$] is –3 but this is decreased to –0.9 in the complex [Ru(lumazineH$^+$)(NH$_3$)$_4$]$^{3+}$. A significant finding with respect to the role of electron transfer within metalloflavoenzymes is that the $2e^-/2H^+$ reduction normally observed for free pterin was separated into two one-electron transfer steps in the ruthenium pterin compounds. Pourbaix plots exhibiting slopes of 59 mV/pH are obtained for the ruthenium-pteridine complexes and these provide the experimental evidence in support of single proton/electron transfer.

4.3
Molybdenum Complexes of Flavin and Pterin

In 1974 Selbin et al. reported that the reaction of molybdenum(4+) chloride with various flavin derivatives led to the first isolable Mo(4+)-flavin products [102]. This is the only extant report concerning molybdenum and flavins since Hemmerich's studies, a surprising result given the numerous examples of molybdenum flavoproteins. Within the year the Selbin work was put into doubt by Sawyer who believed the Mo(IV)-flavin interpretation was incorrect. Instead, Sawyer and Doub claimed that the observed products were likely to be Mo(V) dimers, an assignment felt to be more consistent with the observed data [103]. This rebuttal may have contributed to the lack of further investigation in the area of molybdenum-flavin compounds. The case put forth by Selbin has been recently re-opened and it was found that his interpretation was precisely correct.

The reactions reported by Selbin et al. are shown in Eqs. (33) and (34). The products were characterized by IR, UV/vis, and NMR spectroscopy, microanalysis, and thermal gravimetric analysis. Selected data has been extracted from their paper [102] and combined with the recent data [66, 104] in Table 3.

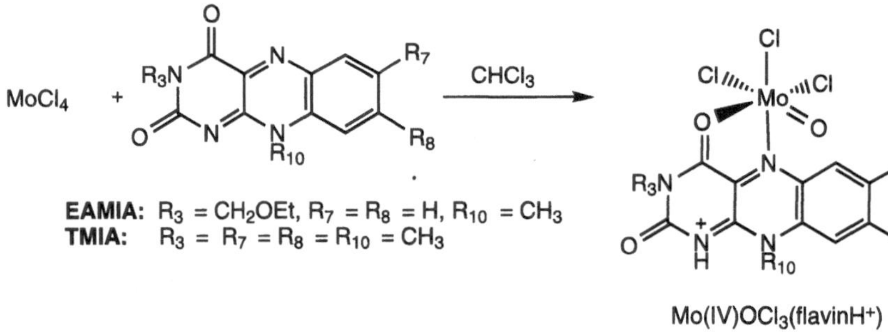

EAMIA: $R_3 = CH_2OEt$, $R_7 = R_8 = H$, $R_{10} = CH_3$
TMIA: $R_3 = R_7 = R_8 = R_{10} = CH_3$

$$Mo(IV)OCl_3(flavinH^+)$$

(33)

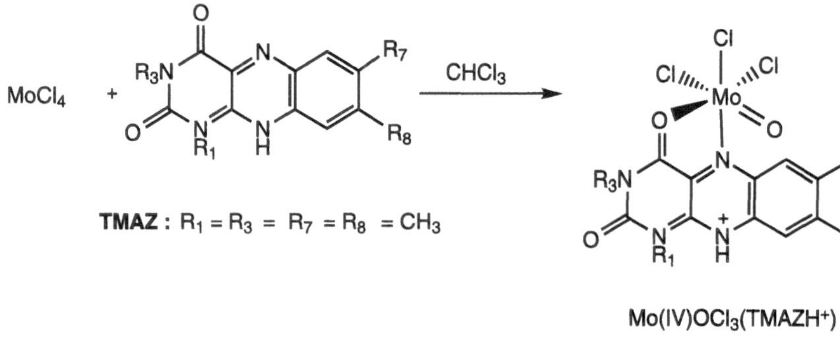

TMAZ : $R_1 = R_3 = R_7 = R_8 = CH_3$

$$Mo(IV)OCl_3(TMAZH^+)$$

(34)

Table 3. Spectral data for molybdenum(IV) flavin and pterin complexes

MoOCl$_3$(pteridine)	λ_{max}, nm	$\nu_{Mo=O}$, cm^{-1}	$\nu_{C=O}, \nu_{C=N}$, cm^{-1}
MoOCl$_3$(HTMAZ)·1.5 HCl[a]	375,650,755	982	1728,1625
MoOCl$_3$(HTMAZ)[a]	388,530	982	1730,1609
MoOCl$_3$(HTMAZ)[b]	384,626,754	982	1730,1609
MoOCl$_3$(HTMAZ)[c]	386,626,758	982	1730,1609
MoOCl$_3$(HTMIA)	412,525	980	1752,1595
MoOCl$_3$(HEAMIA)	393,509	984	1730,1605
MoOCl$_3$(Hpiv-DMP)[a]	394,564	992	1638,1609
MoOCl$_3$(Hpiv-DMP)[b]	406,630,768	992	1638,1609
MoOCl$_3$(Hpiv-DMP)[d]	398,544	992	1638,1609
MoOCl$_3$(DMP)[a]	394,570	982	1643,1599

[a] in acetonitrile.
[b] in DMF.
[c] in DMSO.
[d] in acetone.

In this two decade-old report it was the protonated form of the products (Eqs. 33 and 34) that aroused interest. These protonated pteridine structures are reminiscent of the protonated pterin of the resonance structures (structure A below) presented for the molybdenum-reduced pterin system described earlier in Sect. 2.1. The oxidation state ambiguity described for the molybdenum complexes of reduced pterins in Sect. 2.1 concerned the alternative formulations of tetrahydropterin vs quinonoid-dihydropterin coordinated to oxidized and reduced molybdenum, respectively.

Mo(4+)OCl$_3$(H$_2$pterinH$^+$)   Mo(6+)OCl$_3$(H$_4$pterinate$^-$)

A   **B**

Recognizing this parallel quickly leads one to draw an analogous pair of redox-related structures (C and D) for the Mo-flavin complexes starting from the suggested formulation of Selbin.

Mo(4+)OCl₃(flavinH⁺)   Mo(6+)OCl₃(H₂flavin⁻)

C   **D**

Taking these redox resonance structures one step further suggests a similar redox pair consisting of Mo(4+)-oxidized pterin and Mo(6+)-5,8-dihydropterin (E and F).

Mo(4+)OCl₃(pterinH⁺)   Mo(6+)OCl₃(5,8-H₂pterinate⁻)

E   **F**

This line of thought initiated investigations into the reactions of Mo(4+) with oxidized pterins and with the flavin TMAZ [66]. First the synthesis of TMAZ and the formation of MoOCl₃(TMAZH) were duplicated using the procedures reported by Selbin. Spectroscopic data (Table 3) verified that the same Mo species was isolated and an X-ray structure determination of MoOCl₃(TMAZH) (Fig. 7) provides the ultimate proof of the veracity of Selbin's formulation.

Happily, identical conditions were successful for reactions of MoCl₄ with two oxidized pterins (piv-DMP and DMP) (Eq. 35). The Mo(4+)-pterin complexes are spectroscopically similar to the Mo(4+)-TMAZ complex and are characterized by strong MLCT absorptions in the visible region, by downfield shifts of pterin ring substituents and by the familiar pattern of νC=O, νC=N, absorptions for O4, N5-chelated pterin in the infrared spectrum. One of these, MoOCl₃(piv-DMPH), was also structurally characterized by X-ray diffraction and its ORTEP is shown in Fig. 8.

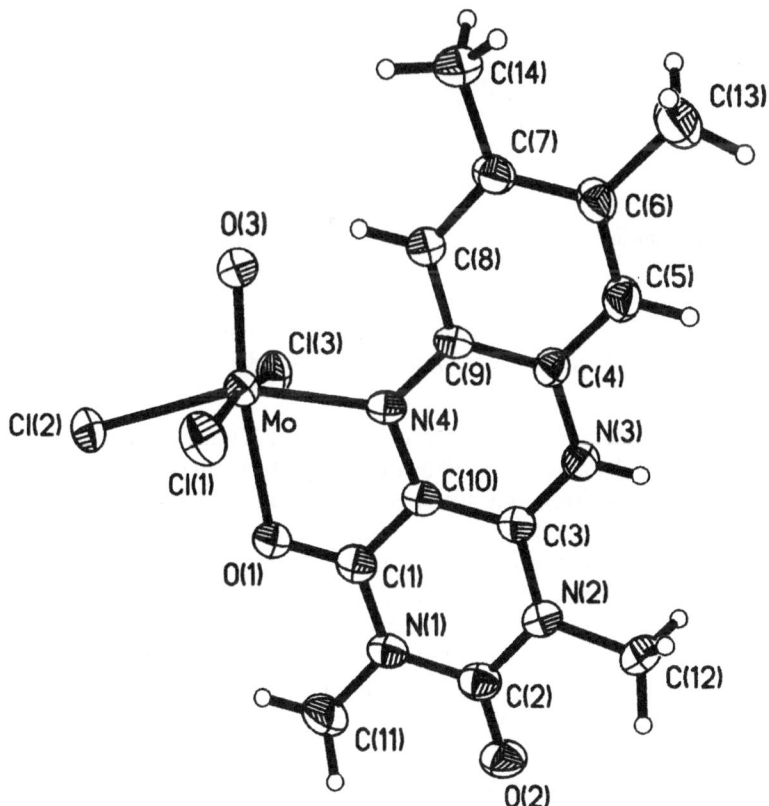

$$(35)$$

R = H : **DMP**

R = -C(=O)(t-Bu): **piv-DMP**

Mo(IV)OCl₃(pterinH⁺)

Beyond providing confirmation of Mo(4+) coordination by flavin (and pterin), the X-ray crystal structures highlight several key features of these complexes. First, a distinct bending of the flavin along the N5-N10 axis is observed in MoOCl₃(TMAZH) (Fig. 9). An uncoordinated dihydroflavin is expected to be bent along the N5-N10 axis [95]. Flavin bending had been previously observed in the ruthenium-flavin complex [101, 102] and used as one of the criteria to

Fig. 7. ORTEP drawing of the crystal structure of MoOCl₃(TMAZH) [66]

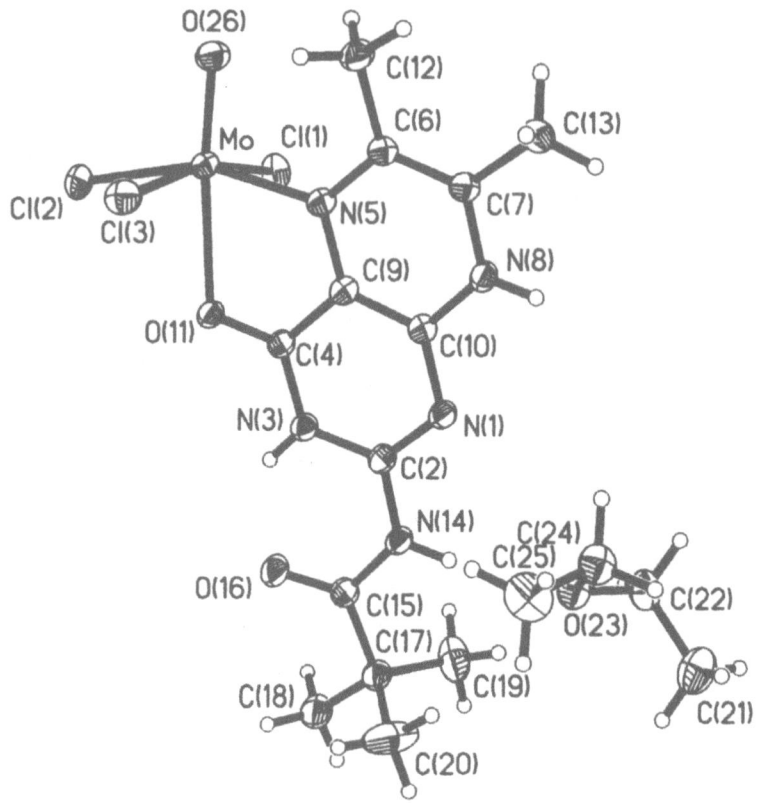

Fig. 8. ORTEP drawing of the crystal structure of MoOCl₃(piv-DMP)·EtOEt [66]

support the notion that coordinated flavin was partially reduced [99, 100]. This suggests that Mo(4+), like Ru(2+), is partially oxidized when coordinated by a flavin.

Second, the structures of both MoOCl₃(TMAZH) and MoOCl₃(piv-DMPH) display Mo-N5 distances of 2.080 and 2.082 Å, respectively. This distance indicates partial multiple bonding between Mo and N5 but to a lesser degree than observed for the reduced pterin complexes of Mo(6+) discussed in Sect. 2 which have Mo-N5 distances between 1.997 and 2.027 Å [63, 65, 66, 70]. Third, pteridine coordination is coupled to protonation. In the flavin complex protonation occurs at N3 while coordinated pterin accepts a proton at N8. This protonation is additional evidence that electron density has been increased in the pteridine system, that is, the pteridine is effectively reduced from the partial oxidation of the metal. The idea that the electron delocalization and the protonation are coupled is supported by the solution behavior of the molybdenum flavin and pterin complexes. While the molybdenum complexes produce relatively stable blue solutions in acetonitrile or acetone, dissolution in a more basic solvent like DMF causes an immediate color change to green followed by rapid bleaching. This series of reactions is interpreted as pterin deprotonation at site N8 by DMF to produce the green species followed by the dissociation of the neutral pteridine as signaled by complete bleaching of the solution [66]. This hypothesis then predicts that flavin or pterin coordination will be favored under acidic conditions.

5
Epilogue

In terms of electron transfer, metal-pteridine systems behave in a variety of ways. There are examples where metal reduction by pterin is fast, such as observed in the reactions of iron and certain copper compounds, and there are examples where the electrons are shared and available for use by metal and pterin, such as shown by the oxo-molybdenum complexes. In cases where reduced metals coordinate oxidized pteridines, structural evidence and facile pteridine protonation point to significant electron delocalization from the metal to the ligated pteridine. So the electron flow can be bi-directional, that is, both metal-to-pteridine and pteridine-to-metal, depending on the system.

There exists a distinct preference of metals in certain oxidation states for pterins in specific oxidation states. Highly oxidized metals such as Mo(6+) readily react with fully reduced tetrahydropterins, but are reluctant to coordinate oxidized pterins. In contrast to the many examples of Mo(6+)-tetrahydropterin reactions covered in Sect. 2, there is only one report of molybdenum(6+) bound to an oxidized pteridine [56]. Section 4 described instead how oxidized pteridines react easily with Mo(4+). The same preference is observed in a different context using copper. Cu(2+) reacts and binds to tetrahydropterins (albeit with eventual reduction) while monovalent copper binds oxidized pteridines with conspicuously high stability [94, 95]. Noticing the trend that an oxidized pterin prefers a more reduced metal partner, the

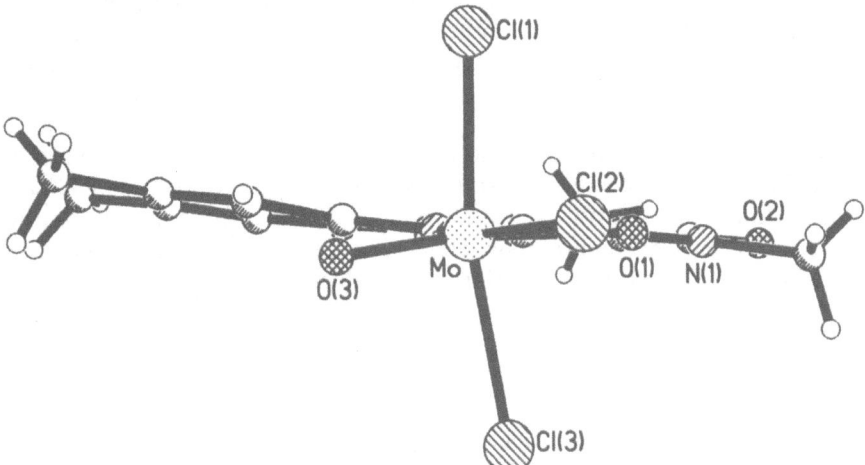

Fig. 9. An illustration of flavin bending when coordinated to Mo(4+) in MoOCl$_3$(TMAZH)

reactions of oxidized pteridines with zero-valent molybdenum have been investigated (Eq. 36) [66].

$$\text{Mo(CO)}_6 + \quad \underset{\substack{\textbf{PUP}\\\textbf{3 eq}}}{\text{(t-Bu...pteridine)}} \quad \xrightarrow[\text{relux}]{\text{acetonitrile}} \quad \text{(Mo complex)} \quad + \; 6\,\text{CO}_{(g)} \qquad (36)$$

Mo(0)(PUP)$_3$

Indeed, these tris(pteridine)Mo(0) complexes possess a much enhanced solution stability as compared to the pteridine complexes of tetravalent molybdenum, thereby emphasizing the importance of electron delocalization over the metal-pteridine structure to the stability of the complex.

The conclusion that reduced pteridines prefer to react with highly oxidized metal ions and oxidized pteridines pair with relatively reduced metals is neither novel nor surprising [63]. The same trend has been established for reactions of transition metals with the redox-active orthoquinones, semiquinones, and catecholates [106].

In 1965 Hemmerich et al. summarized their studies on metal-flavin complexes by stating "... for any 'charge transfer' chelate, two independent dissociation reactions have to be considered, yielding oxidized metal plus reduced ligand and reduced metal plus oxidized ligand, respectively. At the same time the metal and the ligand valences in the complex are not distinguishable" [94]. This statement serves as an apt summary of the whole of this review where many examples have illustrated the highly delocalized nature of the metal-pterin unit. To advocate consideration of the metal-pteridine unit as a separate chemical species rather than merely the sum of its parts with inaccurately assigned oxidation states echoes a useful view of hemes. It has been suggested that, since the delocalized metal-porphyrin moiety functions differently from the isolated metal or free porphyrin, the unit is almost like a new element [105]. The well-known difficulties associated with assigning oxidation state to the reactive ferryl species such as in cytochrome P450 also foreshadow this oxidation state dilemma within transition metal-pteridine systems. It seems that nature enjoys enhancing the covalent bonding capabilities of redox-active metals by combining them with compatible ligands.

6
References

1. Brown DJ (1988) Fused pyrimidines, pt 3. pteridines. Wiley, New York
2. Hurst DT (1980) Chemistry and biochemistry of pyrimidines, purines and pteridines. Wiley, New York, p 64
3. Pfleiderer W (1992) J Heterocycl Chem 29: 583

4. The name pyrimido[4,5-*b*]pyrazine was adopted by Chemical Abstracts for a period of time and sometimes appears in older literature
5. Albert A (1976) Adv Heterocycl Chem 20: 117
6. Bieri JH, Viscontini M (1977) Helv Chim Acta 60: 447
7. Bieri JH, Viscontini M (1977) Helv Chim Acta 60: 1926
8. Armarego WLF, Waring P (1980) J Chem Res 318
9. Armarego WLF, Schou H (1977) J Chem Soc Perkin Trans 1: 2529
10. Ganguly AN, Sengupta PK, Bieri JH, Viscontini M (1980) Helv Chim Acta 63: 395
11. Weber R, Viscontini M (1975) Helv Chim Acta 58: 1772
12. Benkovic SJ (1980) Annu Rev Biochem 49: 227
13. Williams TC, Storm CB (1985) Biochem 24: 458
14. Dryhurst G (1982) Electrochemistry of reduced pterin cofactors. In: Kadish KM (ed) Electrochemical and spectrochemical studies of biological redox components. American Chemical Society, Washington p 457
15. Dryhurst G (1977) In: Electrochemistry of biological molecules. Academic Press, New York, p 320
16. Pfleiderer W, Gottlieb R (1985) Electrolysis of pteridines. In: Wachter H, Curtius H, Pfleiderer W (eds) Biochemical and clinical aspects of pteridines. DeGruyter, Berlin, p 3
17. Kaufman S (1961) J Biol Chem 236:804
18. Lazarus RA, DeBrosse CW, Benkovic SJ (1982) J Am Chem Soc 104: 6871
19. Benkovic SJ, Sammons D, Armarego WLF, Waring P, Inners R (1985) J Am Chem Soc 107:3706
20. Archer MC, Scrimgeour KG (1970) Can J Biochem 48: 278
21. Bailey SW, Ayling JE (1983) Biochemistry 1790
22. Blair JA, Pearson AJ (1975) J Chem Soc Perkin Trans 2
23. Kappock TJ, Caradonna JP (1996) Chem Rev 96: 2659
24. For example, common artificial cofactors are 6,7-dimethyl tetrahydropterin and 6-methyltetrahydropterin
25. Massey V, Palmer G, Ballou D (1971) On the reaction of flavins and flavoproteins with molecular oxygen. In: Kamin H (ed) Flavins and flavoproteins. University Park Press, Baltimore, p 349
26. Mager HIX, Addink R, Berends W (1967) Recueil 86: 833
27. Pember, SO, Villafranca JJ, Benkovic SJ (1986) Biochemistry 25:6611
28. Pember, SO, Benkovic SJ, Villafranca JJ, Pasenkiewicz-Gierula M, Antholine WE (1986) Biochemistry 26: 4477
29. Blackburn N, Strange RW, Carr RT, Benkovic SJ (1992) Biochemistry 313: 5298
30. McCracken J, Pember SO, Benkovic SJ, Villafranca JJ, Miller RJ, Peisach J, (1988) J Am Chem Soc 110: 1068
31. Carr RT, Benkovic SJ (1993) Biochemistry 32: 14,132
32. Johnson JL, Hainline BE, Rajagopalan KV (1980) J Biol Chem 255: 1783
33. Pilato RS, Stiefel EI (1993) Catalysis by molybdenum-cofactor enzymes. In: Reedjik J (ed) Bioinorganic catalysis. Dekker, New York, p 131
34. Enemark JH, Young CG (1993) Bioinorganic chemistry of pterin-containing molybdenum and tungsten enzymes. In: Advances in inorganic chemistry, vol 40, p 1
35. Hille R (1996) Chem Rev 96: 2757
36. Rajagopalan KV (1991) Novel aspects of the biochemistry of the molybdenum cofactor. In: Meister A (ed) Advances in enzymology and related areas of molecular biology, vol 64. Wiley, New York, p 215
37. Johnson JL, Hainline BE, Rajagopalan KV, Arison BH (1984) J Biol Chem 259: 5414
38. Johnson JL, Rajagopalan KV (1982) Proc Natl Acad Sci USA 79:6856
39. Kramer SP, Johnson JL, Ribeiro AA, Millington DS, Rajagopalan KV (1987) J Biol Chem 262: 16,357
40. Wuebbens MW, Rajagopalan KV (1993) J Biol Chem 268: 13,493
41. Johnson JL, Wuebbens MW, Rajagopalan KV (1989) J Biol Chem 264: 13,440
42. Johnson ME, Rajagopalan KV (1987) J Bacteriol 169:110

43. Boyington JC, Gladyshev VN, Khangulov SV, Stadtman TC, Sun PD (1997) *E. coli* formate dehydrogenase. Science 275: 1305
44. Romao MJ, Archer M, Moura I, Moura JJG, LeGall J, Engh R, Schneider M, Hof P, Huber R (1995) *Desulfovibrio gigas* aldehyde oxidoreductase. Science 270: 1170
45. Huber R, Hof P, Duarte RO, Moura JJG, Moura I, Liu M-Y, LeGall J, Hille R, Archer M, Romao MJ (1996) Proc Natl Acad Sci USA 93: 8846
46. Schindelin H, Kisker C, Hilton J, Rajagopalan KV, Rees DC (1996) *Rhodobacter sphaeroides* DMSO reductase. Science 272: 1615
47. Schindelin H, Kisker C, Rees DC Chicken liver sulfite oxidase (personal communication)
48. Schultz B, Hille R, Holm RH (1995) J Am Chem Soc 117: 827
49. Hille R, Sprecher H (1987) J Biol Chem 262: 10,914
50. Hille R, Massey V (1982) J Biol Chem 257: 8898
51. Dolphin D (1980) Model studies and the biochemical function of coenzymes. In: Biomimetic chemistry. American Chemical Society, Washington, DC, p 65
52. Bruice TC (1980) Acc Chem Res 13: 256
53. Blakely RL, Benkovic SJ (1984) Folates and pterins. Wiley, New York
54. Vonderschmitt DJ, Scrimgeour KG (1967) Biochem Biophys Res Comm 28: 302
55. Burgmayer SJN, Baruch A, Kerr K, Yoon K (1989) J Am Chem Soc 111: 4982
56. Burgmayer SJN, Stiefel EI (1986) J Am Chem Soc 108: 8310
57. Perkinson J, Brodie S, Yoon K,Mosny K, Carroll PJ, Burgmayer SJN (1991) Inorg Chem 30: 719
58. Bessenmacher C, Vogler C, Kaim, W (1989) Inorg Chem 28: 4645
59. Kohzuma T, Masuda H, Yamauchi O (1989) J Am Chem Soc 111: 3431
60. Kohzuma T, Odani A, Morita Y, Takani M, Yamauchi O (1988) Inorg Chem 27: 3854
61. Mitsumi M, Toyoda J, Nakasuji K (1995) Inorg Chem 34:3367
62. Hueso-Urena F, Jimenez-Pulido SB, Moreno Carretero MN, Quiros-Olozabal M, Salas-Peregrin JM (1997) Polyhedron 16: 607
63. Burgmayer SJN, Arkin MR, Bostick L, Dempster S, Everett KM, Layton HL, Paul KE, Rogge C, Rheingold AL (1995) J Am Chem Soc 117: 5812
64. Wieghardt K, Hahn M, Swiridoff W, Weiss J (1984) Inorg Chem 23: 94
65. Fischer B, Strahle J, Viscontini M (1991) Helv Chim Acta 74: 1544
66. Kaufmann HL (1977) PhD dissertation. Bryn Mawr College, Bryn Mawr, PA
67. Gardlik S, Rajagopalan KV (1990) J Biol Chem 265: 13,047
68. Gardlik S, Barber MJ, Rajagopalan KV (1987) Arch Biochem Biophys 259: 363
69. Kaufman S (1961) J Biol Chem 236: 804
70. Fischer B, Schmalle H, Dubler E, Schafer A, Viscontini M (1995) Inorg Chem 34: 5726
71. Kaufmann HL, Burgmayer SJN (submitted to Inorg Chem)
72. Holm RH (1990) Coord Chem Rev 100: 183
73. Holm RH (1987) Chem Rev 87: 1401
74. Stiefel EI (1977) The coordination and bioinorganic chemistry of molybdenum. In: Lippard SJ (ed) Progess in inorganic chemistry, vol 22. Wiley, New York, p 3
75. Liu W, Thorp HH (1993) Inorg Chem 32: 4102
76. Thorp HH (1992) Inorg Chem 31: 1585
77. Brown ID, Altermatt D (1985) Acta Crystallogr B41: 244
78. Soricelli CL, Szalai VA, Burgmayer SJN (1991) J Am Chem Soc 113: 9877
79. Pilato RS, Eriksen K, Greaney MA, Gea Y, Taylor EC, Goswami S, Kilpatrick TG, Spiro TG, Rheingold AL, Stiefel EI (1993) Pterins, quinoxalines, and metallo-ene-dithiolates: synthetic approach to the molybdenum cofactor. In: Stiefel EI, Coucouvanis D, Newton WE (eds) Molybdenum enzymes, cofactors, and model systems. American Chemical Society, Washington, DC, p 83
80. Pilato RS, Eriksen K, Greaney MA, Gea Y, Taylor EC, Goswami S, Kilpatrick TG, Spiro TG, Rheingold AL, Stiefel EI (1991) J Am Chem Soc 113: 9372
81. Chan MK, Mukund S, Kletzin A, Adams MWW, Rees DC (1995) Science 1463
82. Burgmayer SJN (unpublished results)

83. Burgmayer SJN, Pearsall D, manuscript to be submitted to J Am Chem Soc.
84. Funhashi Y, Kohzuma T, Odani A, Yamauchi O (1994) Chem Lett 385
85. Yamauchi O (1995) Pure Appl Chem 67: 297
86. Burgmayer SJN, Everett KM, Arkin MA, Mosny K (1991) J Bioinorg Chem 43: 581
87. Burgmayer SJN, Bharwani L, Mosny K (unpublished results)
88. Nasir M, Karlin K, Chen Q, Zubieta J (1992) J Am Chem Soc 114: 2264
89. Schafer A, Fischer B, Paul H, Bosshard R, Hesse M, Viscontini M (1992) Helv Chim Acta 75: 1955
90. Fischer B, Schafer A, Paul H, Bosshard R, Hesse M, Viscontini M (1993) Pteridines 4: 206
91. Schafer A, Fischer B, Bosshard R, Hesse M, Viscontini M (1993) In: Ayling JE, Gopal Nair, M, Baugh CM, Viscontini M (eds) Chemistry and biology of pteridines. Proceedings of the 10th International Symposium on Pteridines and Folates. Plenum Press, New York, p 29
92. Albert A (1950) Biochem J 47:xxvii
93. Albert A (1953) Biochem J 54: 646
94. Hemmerich P, Muller F, Ehrenberg A (1965) The chemistry of flavin-metal interactions. In: King TE, Mason HS, Morrison M (eds) Oxidases and related redox systems, vol 1. Wiley, New York, p 157
95. Hemmerich P, Lauterwein J The structure and reactivity of flavin-metal complexes. In: Eichhorn GL (ed) Bioinorganic chemistry, chap 32, p 1198
96. Wade TD, Fritchie CJ (1973) J Biol Chem 248: 2337
97. Yu MY, Fritchie CJ (1975) J Biol Chem 250: 946
98. Muller F, Hemmerich P, Ehrenberg A, Palmer G, Massey V (1970) Eur J Biochem 14: 185
99. Clarke MJ, Dowling MG, Garafalo AR, Brennan TF (1980) J Biol Chem 255: 4372
100. Clarke MJ, Dowling MG (1981) Inorg Chem 20: 3506
101. Abelleira A, Galang R, Clarke MJ (1990) Inorg Chem 29: 633
102. Selbin J, Sherrill J, Bigger CH (1974) Inorg Chem 13: 2544
103. Sawyer DT, Doub WH (1975) Inorg Chem 14: 1736
104. Kaufmann HL, Burgmayer SJN (submitted to Inorg Chem)
105. Frausto da Silva JJR, Williams RJP (1991) The bioinorganic chemistry of the elements. Clarendon Press, Oxford
106. Pierpont C (1994) Prog Inorg Chem 41: 331

Binding and Transport of Nonferrous Metals by Serum Transferrin

Wesley R. Harris

Department of Chemistry, University of Missouri-St. Louis, St. Louis, MO 63121
E-mail: wharris@umsl.edu

Transferrin is the serum iron transport protein in a wide variety of species. Its normal function is to transport iron through the blood between sites of uptake, utilization and storage. The protein consists of two distinct lobes, each of which contains a high affinity metal binding site. This review describes the thermodynamics of metal binding at these two sites, including the unique requirement of a synergistic carbonate anion for strong metal complexation. Normally the transferrin in serum is only partially saturated with iron, so that the protein provides a significant concentration of vacant binding sites which can sequester other metal ions that enter the blood. This review emphasizes metals for which transferrin appears to play a significant role in serum transport. There is strong evidence that transferrin is the primary serum transport agent for Mn^{3+}, Hf^{4+}, the group 13 metal ions Al^{3+}, Ga^{3+}, and In^{3+}, and the tetravalent actinides Pu^{4+}, Th^{4+}, Np^{4+}, and Pa^{4+}. Transferrin appears to play a lesser role in the binding and transport of the trivalent actinides Cm^{3+} and Am^{3+}, the oxocations VO^{2+}, VO_2^+, and UO_2^{2+}, and several of the lanthanides. There is also evidence that some of the metal-transferrin complexes are recognized and transported across cell membranes by the transferrin receptor.

Keywords: Transferrin; metal binding gallium; aluminum; indium manganese; actinides; lanthanides

List of Symbols and Abbreviations

apoTf	apotransferrin
CD	circular dichroism
cta	citric acid
DFO	desferrioxamine B
ESR	electron spin resonance
hepes	N-(2-hydroxyethyl)piperazine-N'-(2-ethanesulfonic acid)
LFER	linear free energy relationship
LMW	low molecular weight
MRI	magnetic resonance imaging
NHE	normal hydrogen electrode
nta	nitrilotriacetic acid
PAC	perturbed angular correlation spectroscopy
Tf	transferrin

1
Introduction

1.1
Biological Functions of the Transferrins

Serum transferrin (Tf) is one member of a small family of iron-binding proteins which also includes lactoferrin and ovotransferrin [1–6]. These three proteins have very similar structures [6], but appear to serve different functions. Transferrin is responsible for the transport of iron as ferric ion through the blood among sites of uptake, utilization, and storage. Such a carrier is necessary in part because the strong tendency of the ferric ion to form insoluble $Fe(OH)_3$ at physiological pH makes it impossible to transport significant amounts of unchelated Fe^{3+} in blood. Transport of iron as the more soluble Fe^{2+} ion is precluded by the oxidation of free divalent iron by dissolved oxygen. This produces insoluble Fe^{3+} as well as generating toxic species such as the superoxide anion and the hydroxyl radical [7, 8].

Lactoferrin is found in tears, milk, and other fluids. It binds iron more tightly than transferrin [9] and is thought to act as a bacteriostatic agent by ensuring that there is no free iron available to support the growth of microorganisms. The lactoferrin in milk may also play a role in the absorption of iron by the nursing infant. Ovotransferrin is found in avian egg whites, and it also appears to function primarily as a bacteriostatic agent.

To facilitate the delivery of iron to specific tissues, the diferric-transferrin complex is recognized by transferrin receptors [10]. The transferrin receptor is a protein bound to the outer surface of the cell membrane which selectively binds the diferric form of serum transferrin. The receptor-diferric transferrin complex is internalized by a process called endocytosis, in which the cell membrane invaginates, and the resulting cavity eventually pinches off to form an intracellular vesicle. This vesicle contains the receptor-diferric transferrin complex on its internal membrane surface. An ATP-dependent proton pump increases the acidity within the vesicle to about pH 5.5. This promotes the release of iron from the transferrin, but at this low pH the resulting apotransferrin remains bound to the receptor. After the iron has been removed, the vesicle fuses back into the cell membrane, returning the receptor-apotransferrin complex to the outer membrane surface. At the higher extracellular pH of 7.4, the apoTf dissociates from the receptor and returns to the serum to participate in additional cycles of iron transport and uptake.

1.2
Transferrin Structure

Transferrin is a glycoprotein of approximately 80,000 MW. Although it consists of a single polypeptide chain, the protein folds into two distinct, homologous lobes connected by a short polypeptide tether [11, 12]. Essentially the same overall structure has been reported for the closely related iron-binding proteins lactoferrin [6, 13–15] and ovotransferrin [16, 17]. Each lobe (usually identified as either N-terminal or C-terminal) is further divided into two domains by a cleft. A single, high-affinity metal binding site is contained within each cleft. The crystal structures for apo- and diferric lactoferrin show that the two domains pivot on a hinge region at the base of the cleft, so that the cleft is open wider in the apoprotein than it is in the ferric-protein complex [6, 18]. This conformational change has been detected in solution by low angle X-ray scattering studies for lactoferrin, ovotransferrin, and serum transferrin [19].

It has been known for years that the strong binding of a metal ion to apoTf requires the concomitant binding of a synergistic anion. Under biological conditions carbonate functions as this synergistic anion, although a number of small carboxylic acids can form weaker Fe-anion-Tf ternary complexes under carbonate-free conditions [20]. The crystal structures now show that the first coordination sphere around each ferric ion includes the phenolic oxygens of two tyrosine residues, the imidazole group of a histidine residue, and a monodentate carboxylate side chain of an aspartic acid residue (Fig. 1) [6, 11–14]. The fifth and sixth positions in the distorted six-coordinate structure are occupied by two oxygens of the bidentate carbonate synergistic anion. The carbonate anion also interacts with the positive dipole associated with the terminus of an alpha helix and hydrogen bonds to a threonine side chain.

The transferrin crystal structures do not show any significant difference between the ferric ion coordination geometries of the C-terminal and N-terminal binding sites. However, small chemical and spectroscopic differences between the C- and N-terminal binding sites are observed fairly routinely. The two sites often have different rate constants and equilibrium constants for

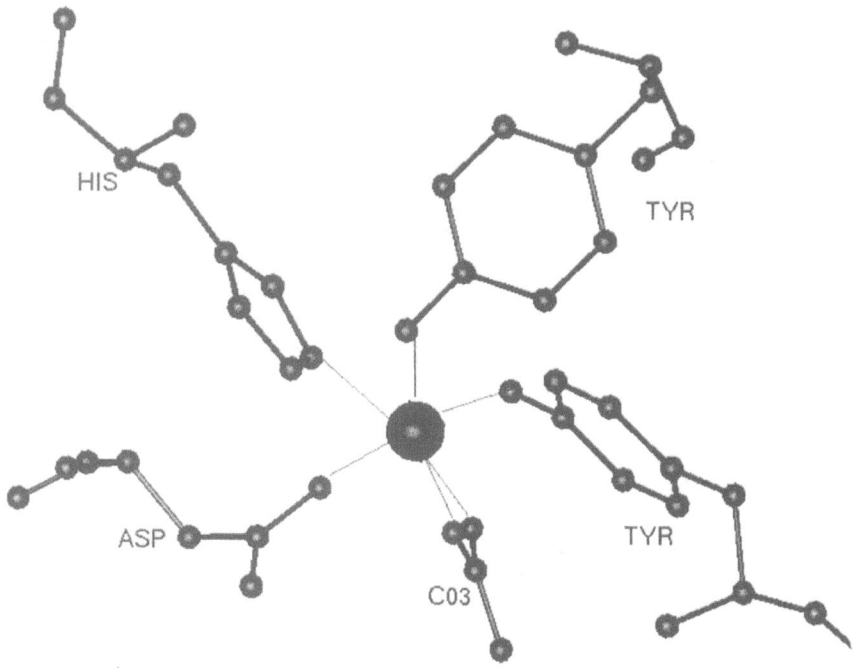

Fig. 1. *N*-terminal iron binding site of human serum transferrin

metal binding, and metals bound at the two sites often give slightly different ESR or NMR spectra [1–3, 21–23].

2
General Description of Metal Binding

The tyrosine side chains in the metal binding site are strong, hard Lewis bases, so Tf preferentially binds hard metal ions such as Fe^{3+}, Ga^{3+}, and Pu^{4+}. Similar selectivity is observed for low molecular weight ligands which contain phenolate coordinating groups [24–27]. Since there are few trivalent metal ions other than iron found in normal serum, transferrin often appears to be highly selective for iron. In fact, the protein will bind a wide variety of metal ions. Table 1 lists approximately 40 metal ions that have been reported to bind (usually in vitro) to either transferrin or ovotransferrin. Whether these metal ions bind to transferrin in vivo is a more difficult question. This review will focus on metals for which there is evidence that Tf is actively involved in serum transport.

One general feature to keep in mind is that the transferrin in normal serum is only about 30% saturated with ferric ion [4, 28]. Given that the serum transferrin concentration ranges from 25 to 50 µmol/l [4], normal serum will contain about 50 µmol/l vacant transferrin binding sites. Thus while it is true

Table 1. Metal ions reported to bind to the transferrins

Transition Metal Ions	V^{3+}, VO_2^+, VO^{2+}, Cr^{3+}, $Mn^{2+/3+}$, $Fe^{2+/3+}$, $Co^{2+/3+}$, Ni^{2+}, Cu^{2+}, Zn^{2+}, Cd^{2+}, Hf^{4+}, Sc^{3+}
Lanthanides	$Ce^{3+/4+}$, Pr^{3+}, Nd^{3+}, Sm^{3+}, Eu^{3+}, Gd^{3+}, Tb^{3+}, Ho^{3+}, Er^{3+}, Yb^{3+}, Lu^{3+}
Actinides	Th^{4+}, U^{4+}, UO_2^{2+}, Np^{4+}, Pa^{4+}, Pu^{4+}, Am^{3+}, Cm^{3+}
P-Block Metal Ions	Al^{3+}, Ga^{3+}, In^{3+}, Tl^{3+}, Bi^{3+}

that ferric ion is bound more tightly than other metal ions, it is usually the case that other metal ions, even at concentrations of several μmol/l, need not compete directly with ferric ion for binding sites on the protein. Instead, the binding of other metal ions to transferrin depends primarily on competition for the metal ion from other ligands in serum, including proteins, such as albumin, organic ligands such as cysteine, histidine, and citrate, and inorganic ligands such as phosphate, carbonate, and hydroxide.

For metal ions that bind to Tf very tightly, e.g., Mn^{3+} or Pu^{4+}, protein binding can be easily demonstrated by the fractionation of serum containing an appropriate radiolabel by methods such as electrophoresis or column chromatography. However, for many metal ions of interest, including Al^{3+}, Ga^{3+}, and Zn^{2+}, the binding may not be strong enough to survive some of the fractionation methods. For these metal ions, it is useful to estimate the extent of transferrin binding under biological conditions by the use of equilibrium speciation calculations. Both approaches to speciation are discussed in the following sections on individual metal ions.

2.1
Metal Complexation Equilibria

Equilibrium constants have been reported for the transferrin complexes of 15 metal ions ranging in oxidation state from II to V: Zn^{2+} [29, 30], Cu^{2+} [31], Cd^{2+} [32], Mn^{2+} [33], Fe^{2+} [34, 35], Ni^{2+} [35], Fe^{3+} [36, 37], Al^{3+} [36, 38–40], Ga^{3+} [41–44], In^{3+} [42, 45, 46], Gd^{3+} [47, 48], Yb^{3+} [49], Nd^{3+} [50], Sm^{3+} [50], Bi^{3+} [51], Pu^{4+} [49], and VO_2^+ [52]. Most studies report effective binding constants for a specific pH and carbonate concentration, usually defined as

$$K_1^* = \frac{[\text{M-CO}_3\text{-TF}]}{[\text{Tf}][\text{M}]} \tag{1}$$

$$K_2^* = \frac{[\text{M-CO}_3\text{-TF-CO}_3\text{-M}]}{[\text{M-CO}_3\text{-Tf}][\text{M}]} \tag{2}$$

where the free bicarbonate anion is not explicitly shown as a reactant in the complexation equilibria. As a general rule, the binding constants for the two Tf binding sites differ by about one order of magnitude [21, 22]. In most cases in which site selectivity has been examined, the larger binding constant is associated primarily with the *C*-terminal binding site [31, 33, 38, 47, 50, 53, 54].

In contrast, ovotransferrin typically shows either stronger binding at the N-terminal site [23, 55, 56] or no site selectivity [31, 57].

Metal and anion binding to transferrin has traditionally been regarded as an all-or-nothing event, with neither the metal ion nor the anion binding in the absence of the other. It is now clear from difference UV [58–60] and ^{13}C NMR [23, 61, 62] studies that several anions, including bicarbonate, will bind to apotransferrin in the absence of metal ions. Anion-apoTf binding constants range from $10^{2.5}$ for bicarbonate to about 10^6 for pyrophosphate [58–60]. Furthermore, kinetic studies on the formation of ferric transferrin by the binding and air-oxidation of Fe^{2+} have shown that the binary carbonate-Tf species forms first, followed by complexation and finally oxidation of the ferrous ion [34]. Since the bicarbonate-apoTf equilibria have been characterized, one can easily treat the overall complexation reaction as two sequential equilibria

$$HCO_3^- + apoTf \xrightleftharpoons{K_C} HCO_3\text{-}Tf \tag{3}$$

$$HCO_3\text{-}Tf + M \xrightleftharpoons{K_M} M\text{-}CO_3\text{-}Tf \tag{4}$$

The advantage of this approach is that the value of K_M for a given metal ion is independent of the solution bicarbonate concentration, and this provides a common frame of reference for comparing results from experiments conducted under different conditions. One should note, however, that both K_C and K_M are still conditional constants with respect to the $[H^+]$ and are valid only at pH 7.4.

For trivalent metal ions such as Fe^{3+}, a total of three protons per iron are released by Eqs. (3) and (4). The free anion is depicted as HCO_3 in Eq. (3) because that is the predominant species in solution at pH 7.4, but both structural and spectroscopic data indicate that in M^{3+}-Tf complexes, the bound synergistic anion has been deprotonated to carbonate [12, 13, 23, 55, 63, 64]. Whether the deprotonation of bicarbonate occurs in Eq. (3) or (4) is not known. The other two protons would come from the deprotonation of the phenolic oxygens of the two coordinated tyrosine residues.

Complexation of divalent cations such as Zn^{2+} releases only two protons [65], both of which can be accounted for by the tyrosine residues. This would imply that the anion remains protonated in complexes of divalent cations, and the crystal structure of the Cu^{2+}-lactoferrin complex shows monodentate anion coordination at one of the two sites [66]. However, ^{13}C-NMR studies on Zn^{2+}-transferrin are more consistent with carbonate than with bicarbonate [63]. Thus there is still some uncertainty as to the state of protonation of the synergistic anion in complexes with divalent anions.

Except under carbonate-free conditions, the metal-free sites of apotransferrin consist of a mixture of truly vacant sites and those occupied by the carbonate anion as a product of Eq. (3). The magnitude of the effective metal-binding constant (K^*) is directly proportional to the degree of saturation of the metal-free sites with bicarbonate [29]. Thus the constants K_M and K_1^* are related by the equation

$$\log K_1^* = \log K_M + \log \alpha \tag{5}$$

where α is defined as

$$\alpha = \frac{K_C[HCO_3]}{1 + K_C[HCO_3]} \tag{6}$$

A similar set of equations can be used for K_2^*. Given that the value of $\log K_C$ is 2.5 for both Tf binding sites [58], one can easily convert experimental values of K^* which have been measured for different solution bicarbonate concentrations into carbonate-independent K_M's.

There is no fundamental reason that metals should not bind to apotransferrin in the absence of carbonate. However, in the absence of carbonate the binding of many metal ions, particularly Fe^{3+}, is weakened to the point that protein-binding can no longer compete with the strong tendency of the metal ion to hydrolyze. Thus, as a practical matter, the presence of carbonate usually acts as an effective trigger to enable metal binding. Two exceptions to this rule involve the complexes between transferrin and VO_2^+ [59] and between VO^{2+} and ovotransferrin [67]. In each case, the metal ion presumably brings its own synergistic anion into the binding site in the form of an oxo ligand.

2.2
Difference Ultraviolet Titrations

The ferric transferrin complex has a reasonably strong charge transfer band at 465 nm [1–3]. Thus one can measure ferric transferrin binding constants by allowing the protein to equilibrate with a competitive chelating agent and then measuring the distribution of the iron between the protein and the competitive ligand based on the visible absorbance [36, 37]. However, many metal ions of interest lack this type of charge-transfer band, and the most widely used method for measuring metal-transferrin binding constants is difference UV spectroscopy [41, 50]. In this method, a sample of the protein is split between the reference and sample cuvettes. After a baseline of protein vs protein is recorded, an aliquot of a metal titrant is added to the sample cuvette, while an equal aliquot of water is added to the reference cuvette to maintain the same protein concentration. The coordination of the metal to the phenolic oxygen of the two tyrosine residues perturbs the UV spectrum of the aromatic rings. This results in a characteristic difference spectrum, with a major peak near 245 nm and a broader, smaller peak around 290 nm. The sequential addition of titrant produces a family of spectra such as that shown in Fig. 2 for the titration with Al^{3+} [38].

To normalize the data from different experiments, the absorbance at 245 nm is converted to an absorptivity by dividing by the analytical concentration of transferrin. Titration curves are prepared by plotting absorptivity vs the equivalents of metal added. By repeating the titration using a series of titrants which contain the metal ion plus varying

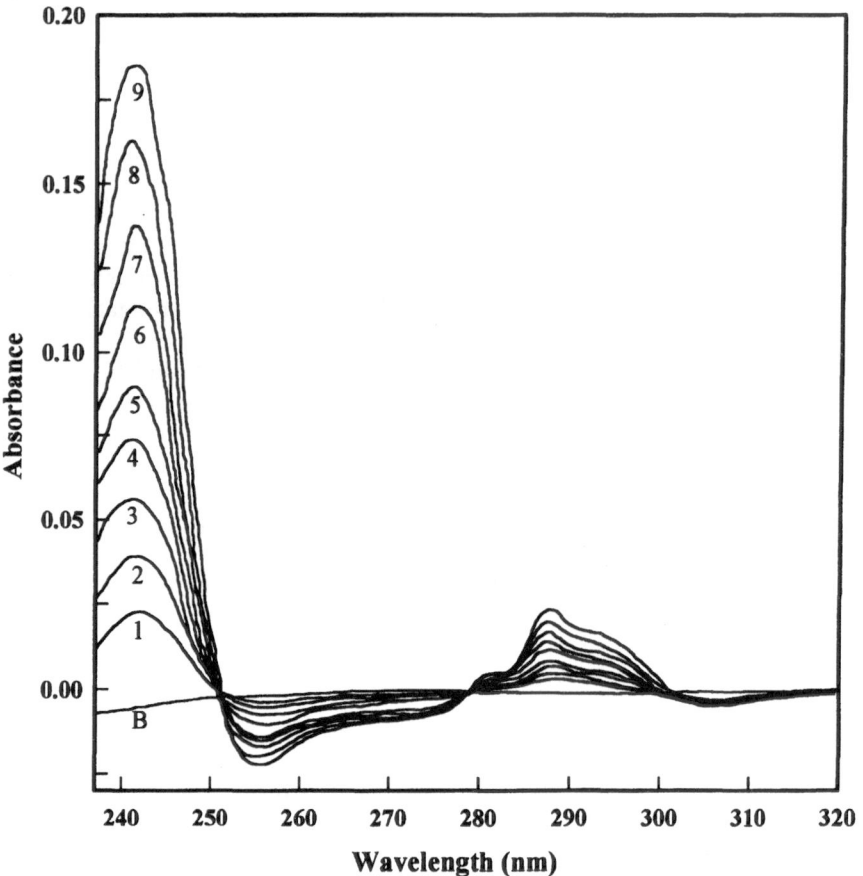

Fig. 2. Difference UV spectra produced by the titration of apotransferrin with Al^{3+}. *Curve B* is a baseline of protein vs protein. *Curve 1* – 0.24 equiv. of Al^{3+}; *curve 2* – 0.36; *curve 3* – 0.48; *curve 4* – 0.60; *curve 5* – 0.78; *curve 6* – 1.02; *curve 7* – 1.26; *curve 8* – 1.50; *curve 9* – 1.86. Data have been taken from [38]

concentrations of a competitive ligand, one can generate a family of titration curves as shown in Fig. 3. These data show the titration of apoTf with Al^{3+} titrants containing nitriliotriacetic acid (NTA) as the competitive ligand. The molar absorptivity of the Al-Tf complex can be calculated as the initial slope of the titration curve in the absence of NTA. Metal-transferrin equilibrium constants can be calculated from non-linear least squares fits of the titration curves using K_1^* and K_2^* as adjustable parameters [21].

Transferrin metal-binding constants (K_M's) typically range from 10^4 to 10^{12} for trivalent lanthanides and divalent transition metals and from 10^{13} to 10^{22} for other trivalent metal ions [21, 22, 31, 51]. Although these binding constants are reasonably large, there is often significant competition for the metal ion from hydroxide and/or carbonate anions in the buffer. This is illustrated by the Al^{3+} data shown in Fig. 3. Early in the titration, when the occupancy of the

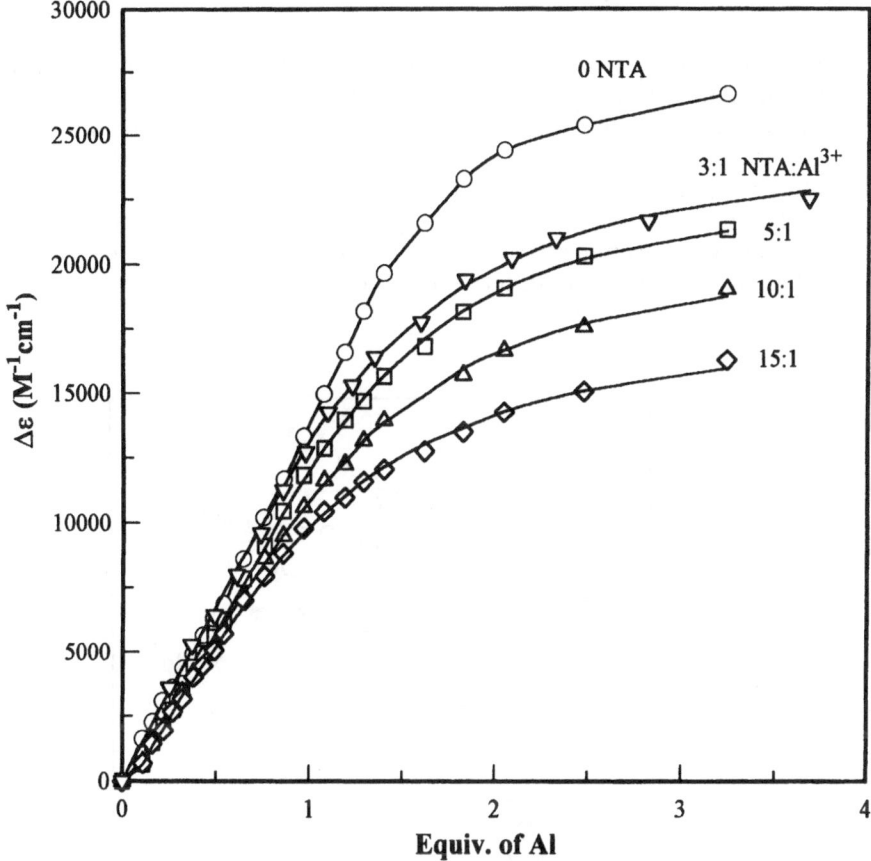

Fig. 3. Titrations of apoTf with Al^{3+} titrants which contain varying concentrations of nitrilotriacetic acid (NTA). The curves are identified by the NTA: Al^{3+} ratio. The data have been taken from [38]

stronger C-terminal binding site is relatively low, there is a linear increase in the absorptivity. From the slope of this initial linear segment, one can calculate a molar absorptivity of 14,800 l/mol cm^{-1} for the Al-CO_3-Tf complex. However, as the titration proceeds and the stronger C-terminal site is saturated, the binding of the Al^{3+} by hydroxide begins to compete with binding to the remaining, weaker N-terminal site. Later in the titration one observes a curvature in the plot, which eventually levels off at about 27,000 l/ mol cm^{-1}. This corresponds to only 1.8 Al^{3+} ions bound per transferrin molecule. Results from ^{13}C and ^{27}Al NMR studies indicate that both serum Tf and ovotransferrin bind about 1.5 Al^{3+} ions per molecule [55].

The apparent saturation in the binding at less than 2.0 Al^{3+} ions per molecule arises because the K_{SP} for the metal hydroxide has been exceeded. Even though no precipitation is observed at low metal concentrations, the formation of polymeric metal hydroxo complexes essentially caps the concentration of free metal ion in solution at a level below that needed to achieve saturation of both Tf binding sites.

Such competition from hydroxide also affects the binding of Ga^{3+} [41], while competition from carbonate is a very strong factor in titrations of apoTf with lanthanides [47, 50] and zinc [30]. This phenomenon can be illustrated more clearly by using the observed absorptivities from difference UV titrations to calculate n, the average number of metal ions bound per protein molecule. Plots of n vs equivalent of added metal for several metal ions are shown in Fig. 4. It needs to be emphasized that the very tight binding of two equivalents of metal ion, as is observed for ferric ion, is the exception rather than the rule

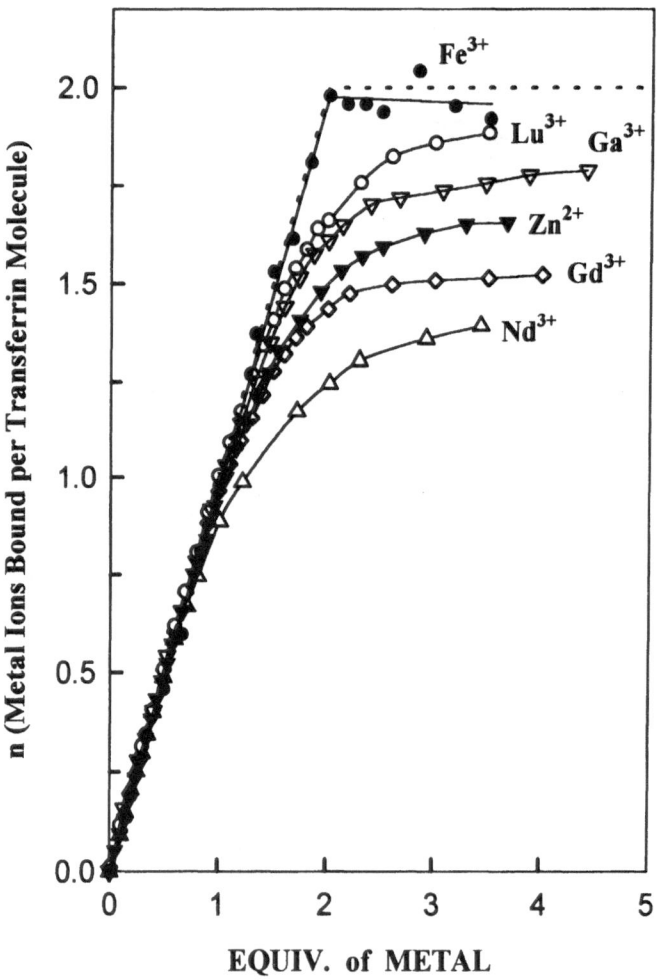

Fig. 4. Plots of n, the average number of metal ions bound per transferrin molecule, vs the equiv. of metal added to the solution

in transferrin chemistry, and that the addition of excess metal ion to apoTf in no way guarantees the formation of the saturated M_2-Tf complex.

3
Binding and Transport of Individual Metal Ions

3.1
Gallium

Both ^{67}Ga and ^{68}Ga are used as radiopharmaceuticals [68, 69]. The largest current application is the use of ^{67}Ga for soft tumors and abscesses [68]. However, ^{68}Ga is of growing interest because it is suitable for three dimensional imaging by positron emission tomography.

Gallium isotopes are usually injected in a solution of citric acid, which is added to prevent the formation of insoluble gallium hydroxide complexes. However, early studies demonstrated that once in the circulation, the metal ion dissociates from the citrate and binds to serum transferrin [70–73]. This binding is easily understood in terms of the very strong similarities between Ga^{3+} and Fe^{3+} [74]. These two trivalent ions are almost the same size. In addition, the electron configurations of the high spin d^5 Fe^{3+} ion and the d^{10} Ga^{3+} ion are such that their complexes lack any crystal field stabilization energy. Thus neither metal ion shows a strong preference for strict octahedral geometry, and each can easily adapt to the distorted six-coordinate geometry of the Tf binding site.

It is not surprising that Ga^{3+} binds not only to transferrin and lactoferrin [75–78], but also to the iron storage protein ferritin [77, 79–82], since ferritin also binds iron in its +3 oxidation state. The major chemical difference between Fe^{3+} and Ga^{3+} is the absence of a stable divalent oxidation state for Ga. Thus one does not observe incorporation of Ga into hemoglobin and other heme proteins that utilize divalent iron.

A very wide range of Ga-Tf binding constants have been reported [41–44]. The initial attempt at measuring the Ga-Tf binding constants used an excess of Ga in a competitive assay vs sephadex G-100 [44]. Analysis of the data by a Scatchard's plot indicated that there were 14 binding sites with a binding constant of log K=0.25. Larson et al. [43] measured the Ga binding constants of Tf using equilibrium dialysis with a large excess of Tf and reported a binding constant of log K=5.4.

Neither of these studies took into consideration the hydrolysis of the free gallium ion. The small, highly charged Ga^{3+} ion hydrolyzes extensively, even in acidic solutions [83]. The hydrolysis constants for Ga are defined as

$$\beta_n = \frac{[Ga(OH)_n^{3-n}][H^+]^n}{[Ga^{3+}]} \tag{7}$$

with log β_n values of −2.9, −6.6, −11.0, and −16.5 for n=1–4 [83]. Thus the Ga^{3+} ion acts as a polyprotic acid with four stepwise pK_a's ranging from 2.9 to 5.5. At physiological pH, monomeric Ga^{3+} exists in solution as 98% $Ga(OH)_4$ and 2%

$Ga(OH)_3$. At higher concentrations of the metal ion, polynuclear hydroxo complexes begin to form, and an accurate description of the hydrolysis equilibria becomes virtually impossible. Because of this extensive hydrolysis, protein binding studies such as that of Clausen et al. [44], which involve a large excess of the metal over the protein, should be avoided.

Harris and Pecoraro used difference UV spectroscopy to measure Ga-Tf binding constants [41]. The titration of apoTf with Ga^{3+} produces a family of spectra that are very similar to the aluminum-Tf spectra shown in Fig. 2. Nitrilotriacetic acid and ethylenediaminediacetic acid were included in the Ga titrations as competitive chelating agents. This was the first study on gallium transferrin to account explicitly for hydrolysis of the free Ga^{3+} ion. The effective binding constants (defined in Eqs. (1) and (2)) were log $K_1^* = 19.5$ and log $K_2^* = 18.6$ for pH 7.4 and 5 mmol/l bicarbonate.

The actual chemical equilibrium between Tf and Ga at physiological pH corresponds essentially to

$$Ga(OH)_4^- + apoTf \xrightleftharpoons{K_{obs}} Ga\text{-}Tf \qquad (8)$$

Thus the binding constant of $10^{5.4}$ reported by Larson et al. [43] would correspond to Eq. (8). If this binding constant is corrected for hydrolysis of the Ga and adjusted to 5 mmol/l bicarbonate, the resulting value is log $K_1^* = 19.6$, in excellent agreement with the difference UV results.

One other Ga-Tf binding constant of log K = 23.7 has been reported [42], but this value is clearly too high. The accepted ferric transferrin binding constants are only log $K_1^* = 20.7$ and log $K_2^* = 19.4$ [37]. Because of the strong chemical similarity between Fe^{3+} and Ga^{3+}, there is a very strong linear correlation between the binding constants of iron and gallium for a common set of ligands. Such a linear free energy relationship (LFER) is shown in Fig. 5. This plot is typical of several such LFERs that have been reported for the complexation of Fe^{3+} and Ga^{3+} [41–44, 75, 84]. The log K^* values for Fe^{3+} [37] and Ga^{3+} [41] have been converted to K_M values to account for different experimental conditions, and the Tf values are plotted as the solid triangles in Fig. 5. These points are in excellent agreement with the LFER. Conversely, the higher Ga-Tf binding constants from Kulprathipanja et al. [42] would move these points to about three log units above the LFER and thus would be inconsistent with the normal relationship between the relative metal-ligand binding constants for Fe^{3+} and Ga^{3+}.

Serum fractionation studies indicate that at low gallium concentrations, between 85 and 100% of serum gallium binds to protein [70, 71, 85–90]. There is no evidence that proteins other than transferrin are involved in gallium binding, and there are specific data which indicate that serum albumin does *not* bind Ga [71, 78, 90]. It is difficult to get a more precise measure for the percentage of gallium bound to Tf in serum. The complexation equilibrium in serum between the gallate anion and apoTf is labile and easily shifted by the physical methods used to separate macromolecules from low-molecular-weight complexes. In addition, the effective Ga-Tf binding constant drops off rather sharply when the bicarbonate concentration drops below ~10 mmol/l

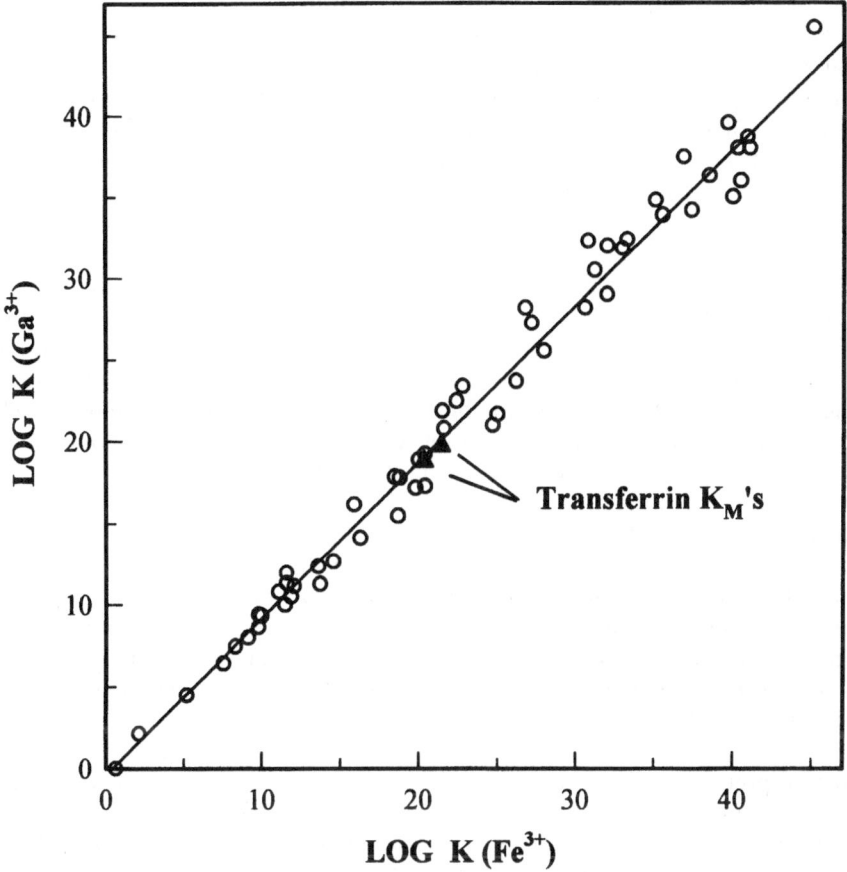

Fig. 5. Linear free energy relationship between Fe^{3+} and Ga^{3+}. Each data point represents a ligand. The x-coordinate is equal to the log K value of that ligand with Fe^{3+}, and the y-coordinate is the log K value of that ligand with Ga^{3+}. The log K_{M1} and log K_{M2} values for serum transferrin are shown as the *filled triangles*. Data on the LMW ligands used to construct the plot were taken from [109]

[22, 80, 87], so that the problem of dissociation of Ga-Tf during the physical separation of transferrin from the other serum proteins can become quite serious unless one takes care to maintain the bicarbonate concentration throughout the separation process [73, 85, 87, 89, 91]. It has even been suggested that the lower intracellular bicarbonate concentration (1–5 mmol/l) could be a factor in the intracellular release of gallium from transferrin [80].

Although gallium binds to transferrin rather tightly in serum, Ga clears from the blood more quickly than either Fe^{3+} or I-labeled albumin [85–87]. This has been attributed to a small but kinetically significant labile pool of low molecular weight (LMW) gallium. The chemical composition of this LMW pool is still not known. Citrate is an obvious possibility as a Ga ligand. However, spiking serum with 500 μmol/l [89] or even 1000 μmol/l citrate [92] removes very little Ga from transferrin. Thus it seems unlikely that the 100 μmol/l citrate in normal serum could bind a significant fraction of serum gallium. Computer models of gallium speciation in serum [22, 93] indicate that the LMW pool consists primarily of $Ga(OH)_4^-$ and $Ga(OH)_3$, with only trace amounts of Ga-citrate complexes. A number of phosphate compounds are more effective than citrate as mediators of gallium exchange between transferrin and either lactoferrin or ferritin [77, 79, 80]. Thus it is possible that phosphates may also contribute to the pool of kinetically active Ga in serum.

The clearance of gallium from blood is biexponential, consisting of two roughly equal components with clearance half-lives of around 10 and 150 min [85]. Organ uptake is highest in bone, followed by liver, kidney, spleen, and lymph nodes [94, 95]. The bone uptake is not transferrin-dependent, and the metal appears to be bound to the bone mineral surface, not to erythroid cells in the marrow [96]. When relatively low doses of Ga are injected, about 25% of the injected dose is excreted through the kidneys within a few days, but injection of doses that exceed the transferrin binding capacity leads to higher urinary excretion and kidney damage [94].

There has been some controversy regarding the role of Tf in the uptake of Ga into soft tissues and tumors. Larson and co-workers [72, 97] originally proposed the "transferrin receptor hypothesis", in which Ga follows the receptor-mediated pathway described in Sect. 1.1 for the cellular uptake of iron. Receptor-mediated uptake of Ga in a variety of tumor cell lines has been reported [72, 82, 97–99].

Data on tumor uptake of Ga in whole animals has been somewhat difficult to interpret. Tissue and tumor uptake tend to decrease if serum transferrin levels are low or if the transferrin is saturated with iron, which is consistent with receptor-mediated uptake [96, 98, 100]. However, it has been argued that it is actually LMW Ga which is taken up, and apoTf enhances tissue uptake simply by binding Ga, which retards Ga excretion and maintains a buffered free Ga concentration in the serum [73, 101, 102]. Proponents of this view point out that lowering the pH, which increases the free Ga concentration [92], tends to enhance the cellular uptake of Ga [73, 103]. It has also been shown that intracellular Ga can traverse back across the cell membrane into the extracellular fluid [82, 99].

Receptor-mediated uptake of Ga has now been demonstrated conclusively by studies which show that antibodies to the transferrin receptor block Ga uptake, even in the presence of normal levels of apoTf [82, 99]. But there are cell lines for which Ga uptake is neither enhanced by apoTf nor blocked by antibodies to the transferrin receptor [82, 101]. In addition, Sohn et al. [96] implanted two tumor types into hypotransferrinemic mice and found that the reduced transferrin levels markedly reduced Ga uptake in one type of tumor, but had very little effect on the other type. Thus it now seems clear that there are two pathways for Ga uptake, one which is transferrin-dependent and one which is transferrin-independent [82, 96, 98]. Weiner [98] has estimated that in most cell lines, the transferrin-dependent pathway accounts for 75–90% of Ga uptake. The relative importance of the two mechanisms for any particular cell type may depend on factors such as the density of transferrin receptors in that cell line. Weiner [98] has speculated that the importance of the receptor-mediated pathway might also depend on the extent to which the cell relies on reduction of Fe^{3+} to Fe^{2+} to release iron from Tf intracellularly, since Ga^{3+}-Tf cannot be reduced.

Receptor-mediated cellular uptake of Ga-Tf appears to be less important in the localization of Ga in purulent abscesses [98, 104]. Instead, this process appears to depend more on the transfer of Ga from serum transferrin to lactoferrin, which is released into the extracellular fluid in inflammatory lesions [76, 78, 104, 105]. Such a transfer is thermodynamically favorable. This was originally demonstrated by direct competition studies between the two proteins [78]. The higher binding affinity for lactoferrin has been confirmed by independent determinations of the binding constants of Ga with both transferrin and lactoferrin by difference UV spectroscopy [41, 75]. These studies reported log K_1^* values of 19.4 for transferrin and 21.4 for lactoferrin. The Tf binding of Ga^{3+} is also more sensitive to pH than that of lactoferrin, so that lowering the pH shifts the relative gallium binding affinities even further in favor of lactoferrin [78].

It has also been suggested that gallium localization in abscesses might involve the binding of Ga by siderophores, which are powerful LMW iron chelators excreted by the invading microorganism [104, 105]. It is clear that the siderophores are capable of binding Ga^{3+}. The stability constant of Ga with the trihydroxamate desferrioxamine B is $10^{28.5}$ [106], which is more than enough to remove Ga from transferrin. This Ga binding constant is still substantially smaller than the Fe^{3+}-desferrioxamine stability constant of $10^{30.99}$, and it is very unlikely that Ga^{3+} can displace Fe^{3+} from siderophores. However, Emery has shown that Ga^{3+} can effectively replace iron under reducing conditions, where Ga-binding to the siderophore facilitates reduction of the Fe^{3+} to the very weakly bound Fe^{2+} [107]. Emery and Hoffer have confirmed that the Ga complex of the siderophore ferrichrome is taken up by *Ustilago sphaerogena* at the same rate as the ferric complex [108].

3.2
Indium

Like Ga^{3+}, the In^{3+} ion has been investigated intensively because of the widespread interest in its use in radiopharmaceuticals. Two γ-emitting isotopes, ^{111}In ($t_{1/2} = 2.8$ d) and ^{113m}In ($t_{1/2} = 1.7$ h) are of interest [69]. There are a number of chemical similarities between Ga^{3+} and In^{3+} [74]. Both are filled shell d^{10} ions lacking any crystal field stabilization energies, and for each metal only the +3 oxidation state is found under biological conditions. Both are extensively hydrolyzed at physiological pH. However, it is important that In^{3+} is much less amphoteric. The fourth pK_a describing the hydrolysis of $In(OH)_3$ to $In(OH)_4^-$ is 9.7, compared to 5.5 for Ga^{3+} [109]. Thus at neutral pH, the In^{3+} ion exists in aqueous solution almost exclusively as the insoluble $In(OH)_3$ ($K_{SP} = 10^{-36.9}$). In this respect In^{3+} more closely resembles Fe^{3+} than Ga^{3+}. The binding of unchelated In^{3+} or Fe^{3+} to apoTf at neutral pH is very slow, presumably because of the very rapid formation of unreactive hydroxy polymers of the metal hydroxides [45]. In contrast, the addition of unchelated Ga^{3+} to apoTf proceeds smoothly and rather quickly because of the predominance of the soluble, monomeric $Ga(OH)_4^-$ anion at neutral pH [41].

Studies on the In-Tf binding constants have lagged behind those on Al and Ga. The initial values for the In-Tf binding constants of almost 10^{30} were reported by Kulprathipanja et al. [42]. However, as discussed above for the Ga-Tf binding constants, these In-Tf constants are much too high and should be disregarded. Lurie et al. [46] reported an In-Tf binding constant of 10^{24}, which was determined by using perturbed angular correlation spectroscopy to measure the distribution of In between Tf and a series of bidentate competitive chelating agents.

More recently, the In-Tf system has been studied by use of difference UV spectroscopy [45]. The binding of In^{3+} to apoTf was much slower than anticipated, and a new batchwise titration method was developed to allow time for equilibration. In-Tf binding constants determined by these difference UV studies titrations are log $K_1^* = 18.52$ and log $K_2^* = 16.64$ in 5 mmol/l HCO_3^-. The evaluation of these indium transferrin binding constants by use of a LFER between Fe^{3+} and In^{3+} is more complicated than was the case for Ga^{3+}. The low molecular weight data for aminocarboxylic acids such as EDTA fall on a separate line from the binding constants for ligands which coordinate via phenolic oxygens [25]. The In-Tf values are consistent with the LFER based on the phenolate ligands. Based on the currently accepted binding constant for Fe-Tf and the phenolate LFER, one obtains an estimate of 18.8 for the log K_1^* value for In-Tf, which is in good agreement with the results from the difference UV studies [45].

Transferrin appears to be the only significant In-binding protein in serum. If In is injected either as an acidic solution or as a weak chelate, e.g., In-citrate, then >95% binds to macromolecular ligands, and essentially all of this appears to be associated with transferrin [88–90, 110–112]. In contrast to the slow binding of macroscopic amounts of In^{3+} to apoTf which was observed in difference UV studies [45], the binding of trace amounts of In^{3+} in plasma appears to be quite rapid [86, 112]. Either the higher dilution retards polymerization, or serum ligands such as citrate serve to deliver monomeric In to Tf.

Compared with Ga, the clearance of In-Tf from the blood is relatively slow, with a half-life in the range of 4–10 h [86, 88, 113]. The slow release of In from Tf and slow blood clearance rates for In make it possible to use In-Tf as a marker for vascular permeability [86, 88, 110]. If the pH of the indium stock solution is raised to neutral or basic values prior to injection, colloidal indium-hydroxide polymers form which do not donate In to Tf and are cleared from the serum within about 5 min [69, 90].

There are several reports that In-Tf is more stable in serum than is Ga-Tf [86, 88–90, 114]. In addition, In-Tf is more stable than Ga-Tf to dissociation during urea/polyacryamide electrophoresis [115]. This at first seems incompatible with the reported stability constants of log $K_1^* = 19.4$ for Ga-Tf and only 18.5 for In-Tf [41, 45]. To understand the behavior in serum, one must also consider the impact of metal hydrolysis, which lowers the *effective* binding constants at pH 7.4 to log $K_1 = 7$ for Ga and around 10 for In [45]. It is the stronger competition from $Ga(OH)_4^-$ relative to $In(OH)_3$ which leads to a higher degree of dissociation of Ga from transferrin in vivo.

The clearance of In-Tf from blood is even slower than the clearance of iron, which suggests that the receptor mediated processes involved in iron clearance are less effective for In [86]. The binding affinities of In-Tf and Fe-Tf to the Tf receptors on reticulocytes are very similar [116]. However, whereas iron is successfully internalized and subsequently appears in hemoglobin in circulating erythrocytes, the In-Tf remains membrane bound and almost no In appears with the mature erythrocytes. Instead, there is a slow translocation of In from the reticulocyte membrane to the mineral surface of the nearby bone. The lack of effective receptor mediated uptake of In may reflect a difficulty in dissociating In^{3+} from the complex between In-Tf and the transferrin receptor. Other studies have shown that In-Tf can bind to the transferrin receptors at the placenta, but again it appears that there is no actual transport of In across the membrane [94]. Thus aside from the binding to Tf, In^{3+} does not closely mimic the general behavior of Fe^{3+} [117].

It has also been suggested that the larger radius of In changes the conformation of In-Tf in such a way as to lower the binding affinity of In-Tf with the transferrin receptor [86]. Spectroscopic data on the effect of the metal ionic radius on the transferrin conformation are inconsistent. Small angle X-ray scattering studies have shown that Fe^{3+} and In^{3+} cause essentially the same degree of closure of the interdomain cleft in the protein in which the metal binding site resides [118]. In contrast, circular dichroism spectra in the aromatic region suggest that indium induces a conformation different from that for Ga^{3+} and Al^{3+} [119].

Despite the apparent problems associated with receptor mediated uptake of In, In still tends to localize in tissues which have large numbers of transferrin receptors [120]. However, In-Tf has not proven to be an effective agent for delivering In to tumors. A number of In-chelates have been tested for selectively delivering In to various tissues [69, 121]. Much of the recent emphasis has been on covalently binding In chelates to monoclonal antibodies for more selective delivery to tumor cells [122]. In these applications one takes advantage of the slower ligand exchange kinetics of In to help prevent dissociation of the label in vivo.

3.3
Aluminum

There is currently a great deal of interest in the biological chemistry of Al^{3+}, and numerous recent reviews and books are available [94, 123–128]. Aluminum was long considered to be relatively non-toxic, primarily because the intestinal barrier to the absorption of Al from the gut is quite effective [94, 123, 125, 129]. However, Al that does enter the blood can cause neurological effects, osteomalacia, and anemia [94, 125]. Aluminum toxicity has been demonstrated most dramatically in the form of dialysis dementia, where neurological impairment is observed in long term dialysis patients [125, 130]. The condition arises from the transfer of toxic amounts of free Al from the dialysis solution into serum due to the relatively high binding affinity of serum ligands, especially transferrin. Aluminum neurotoxicity has also been observed in patients on total parenteral nutrition [131], where once again the normal protections against intestinal aluminum uptake are circumvented. Aluminum-related encephalopathy has even been observed in non-dialysis patients, especially children, who are taking high doses of oral aluminum drugs as phosphate-binders [132–134]. There is even some concern that infants may be exposed to hazardous levels of Al from certain types of infant formula [135].

There has also been much discussion regarding a possible link between the deposition of Al in the brain and Alzheimer's disease [136–141]. This association was initially prompted by the findings that the brains of Alzheimer's patients had unusually high Al concentrations [142] and that the Al appeared to be localized in the specific regions of the brain damaged by Alzheimer's disease [143]. There is even some evidence that treatment of Alzheimer's patients with Al chelating drugs slows the progression of the disease [144]. However, the pathology of Al-induced neurofibrillary tangles is different from the lesions seen in Alzheimer's disease [94, 137], and the importance of Al as a causative factor in Alzheimer's disease has been seriously challenged [94, 125, 137].

The binding of Al to apoTf has been studied by a variety of spectroscopic methods, including ^1H NMR [145–147], ^{27}Al NMR [23, 55, 148–150], ^{13}C NMR [55, 63], circular dichroism [147, 151], fluorescence [151], and difference UV spectroscopy [36, 38–40, 65]. Both difference UV and ^{27}Al NMR indicate that binding levels off prior to saturation of both the Tf binding sites [23, 38]. Investigators often ignore this fact and assume that adding two equivalents of Al^{3+} to apoTf will saturate both the transferrin binding sites.

Harris and Sheldon [38] have conducted difference UV titrations of both forms of monoferric Tf with Al and determined that at pH 7.4 the Al binding is stronger at the C-terminal site. High resolution ^1H and ^{13}C NMR studies have indicated preferential binding to the N-terminal site at pH 8.8 [23, 146]. It has been suggested that the NMR data might reflect a kinetic preference for binding at the N-terminal site, as has been observed for the binding of ferrous ion [37]. However, Al equilibrates relatively quickly with Tf, so kinetic control of the Al distribution is unlikely. Studies on ferric ion indicate that the relative binding affinities of the two Tf binding sites are sensitive to both pH and salt concentrations [152, 153]. The apparent discrepancy between the difference UV and NMR data on Al binding may well be due to variations in experimental conditions.

Al-Tf binding constants have been determined by several groups, and log K_1^* generally falls in the range 10^{12}–10^{14}. Based on difference UV titrations, Harris and Sheldon [38] have reported values of log $K_1^* = 13.5$ and log $K_2^* = 12.5$ for pH 7.4 and 5 mmol/l bicarbonate at 25 °C. Martin et al. [36] have reported somewhat lower values of log $K_1^* = 12.2$ and log $K_2^* = 11.6$. The separation between log K_1^* and log K_2^* in these studies is in the range that has been observed for other metal ions [21, 22]. In contrast,

Cochran et al. [40] reported that the two sites have essentially identical binding constants of 3×10^{15} at 37 °C. In this paper bicarbonate was explicitly included as a reactant in the equilibrium expressions. If these constants are reformulated to conform to Eqs. (1) and (2), then log $K_1^* =$ log $K_2^* = 13.7$ for 18 mmol/l bicarbonate. The variation between the results from 25 and 37 °C raises the possibility that the site selectivity may vary with temperature, but additional studies are needed to confirm this observation.

Fatemi et al. [39] report values of log $K_1^* = 12.23$ and log $K_2^* = 11.76$ for 25 mmol/l bicarbonate. However, the titration curve reported by Fatemi et al. is clearly sigmoidal, in contrast to the hyperbolic curve expected for this system. Furthermore, the data are analyzed by simple Hill plots that tacitly assume different molar absorptivities for the two sites and fail to take into account incomplete saturation of the protein.

Attempts to characterize the protein binding of Al in serum have been inconsistent and confusing. There are several complicating factors. Since there is no convenient radioisotope for Al, serum fractionation studies have been conducted using cold Al. Aluminum contamination is so ubiquitous that it is very difficult to measure low Al concentrations accurately [154]. Aluminum recoveries from size exclusion columns are often highly variable due to the exchange of Al between the column and sample [155–159]. The analysis of chromatographic data is further complicated by the absence of bicarbonate in some of the eluting buffers, which weakens Al-Tf binding and exacerbates the problem of dissociation of the Tf complex on the column [160].

Several groups have now concluded that Tf is the only major Al-binding protein in serum [156–158, 161–164]. The strongest data on this point come from ion exchange HPLC and immunoaffinity chromatography [156, 157, 163, 164]. Early reports of Al-albumin binding in serum [160, 165, 166] based on size exclusion chromatography should be discounted, since there was very poor resolution between Tf and albumin for the columns that were used [36, 125]. Attempts to measure Al-albumin binding directly in vitro have also been somewhat confusing. Fatemi et al. [39] reported that albumin competes with transferrin for Al at physiological concentrations, and weak binding of Al to albumin has been detected by ^{27}Al NMR [148]. However, Cochran et al. [158] detected no Al-albumin binding by equilibrium dialysis, and others have pointed out that, based on stability constants of albumin with other hard metal ions such as Ca^{2+} and Gd^{3+}, one would predict that albumin would bind Al^{3+} very weakly [36, 125]. While one may be able to detect weak, non-specific Al binding under certain conditions, it is highly unlikely that albumin can compete with Tf, citrate, and phosphate in serum for the binding of a significant fraction of Al in vivo.

The infusion into serum of the siderophore desferrioxamine B (DFO), which is also a powerful Al-chelating agent, results in a large increase in the fraction of ultrafilterable or low molecular weight Al. Most investigators attribute this to the formation of the Al^{3+} complex of desferrioxamine [157, 161, 163, 167–171]. Others attribute this new peak to new, relatively small aluminum-binding proteins [172–176]. One group has proposed a new protein, named albindin, with a MW of around 14,000 [172, 174]. Another group has proposed a somewhat smaller (around 8000 MW) Al-binding protein [173, 175]. It has been suggested that these proteins are synthesized as a mechanism to detoxify aluminum [172, 175]. However, neither of these proteins has been isolated and adequately characterized, and it seems more likely that the increase in ultrafilterable Al associated with the administration of DFO results from the simple chelation of Al by DFO. Transferrin remains the only serum protein that is clearly shown to bind Al in vivo.

The most reliable data on the fraction of LMW aluminum in serum come from ultrafiltration studies [156, 167, 168, 177–180]. In normal serum, 15% of the Al is ultrafilterable, and the remaining 85% is presumably protein-bound. The fraction of ultrafilterable Al rises to 25% in uremic serum. The basis for this increase is not firmly established. A computer model for the equilibrium speciation of Al in serum has been developed which calculates that the fraction of LMW Al is 19% for normal serum and 30% for uremic serum, in good agreement with the ultrafiltration data [181]. In this model, the only difference between normal and uremic serum is an increase in the concentration of inorganic phosphate from 1.1 mmol/l for normal serum to 1.8 mmol/l for uremic serum, and phosphate-Al binding accounts for the decrease in the fraction of Al-Tf.

Several other Al speciation models have been reported [182–184]. The major variation among the models is in the set of Al binding constants used for phosphate and citrate. The Al-phosphate binding constants are particularly difficult to measure and represent the greatest uncertainty in the models. In the Harris model [181], the primary LMW species is $[Al(PO_4)(OH)]^-$. Others have suggested the citrate complex $[Al(H_1cta)(OH)]^{2-}$ [124, 178, 183] and the binuclear phosphate complex $[Al_2(PO_4)(OH)_2]^+$ [182] as the major LMW species. More experimental data on Al complexation by phosphate and citrate are needed to resolve this issue.

It appears that Al^{3+} is delivered to cells by Al-Tf via the receptor-mediated iron uptake pathway, and that this has a significant impact on its toxicity. The Al-Tf complex binds to the transferrin receptor, although not as strongly as Fe-Tf [185–187]. The ^{27}Al NMR spectrum indicates that Al^{3+}, like Fe^{3+}, is six-coordinate in the transferrin complex [23]. However, X-ray scattering studies suggest that the binding of Al^{3+} causes a smaller conformational change than does the binding of iron [118], and this change in the protein conformation may adversely affect receptor binding of Al-Tf.

Despite the weaker binding of Al-Tf to the Tf receptor, the addition of Al-Tf rather than simple inorganic Al salts significantly enhances the inhibition of cell growth and division [188–191]. The addition of Al-Tf also interferes with the cellular uptake of iron from Fe-Tf [187, 192, 193]. There is evidence that the interference with iron uptake involves the down regulation of the number of transferrin receptors [187].

Because of the neurotoxicity of Al, it is particularly significant that the passage of Al across the blood-brain barrier appears to be mediated by transferrin receptors [185, 194–196]. There is relatively little difference in the binding of Fe-Tf or Al-Tf to the transferrin receptors in normal or Alzheimer's brain cells [185], and the regional distribution of Al within the brain matches the distribution of Tf receptors [197]. There is also a study, based on the uptake of Ga as a model for Al, which suggests that this receptor mediated movement across the blood brain barrier is unidirectional, leading to the gradual accumulation of Al in the brain at a rate high enough to account for the Al found in the brains of Alzheimer's victims [194].

There have been several studies in which the effects of Al toxicity have been substantially reversed by relatively simple chelation therapy with the siderophore DFO [126, 169]. Treatment with DFO substantially reduces the body burden of Al and is quite effective for reversing both the neurotoxicity and osteomalacia associated with long term dialysis patients. In fact, Al-induced anemia in dialysis patients can sometimes be reversed simply by preparing the dialysate with deionized water [198].

The efficacy of DFO treatment is presumably related to its high Al binding constant. Computer simulations show clearly that DFO should be able to remove essentially all the Al from serum transferrin under in vivo conditions [38]. The aminocarboxylate ligand EDTA also has a higher effective binding constant than Tf, but it is much less effective than DFO in treating Al toxicity [199]. This reflects the difference in the selectivity of these chelating agents for Al over the other metals present in serum. The computer simulations show that Al-chelation in vivo by DFO should be relatively unaffected by interference from serum Ca^{2+} and Zn^{2+} [38]. In contrast, the simulations show that EDTA would be almost completely saturated with zinc and calcium and would be unable to bind significant amounts of Al.

3.4
Plutonium

The serum transport of the actinides has been extensively studied in relation to the health hazards associated with various elements in the nuclear fuel cycle. In general, the early actinides (Th through Pu) are distinguished by a very high biological hazard associated with α-emitting isotopes which have very long physical half-lives [200]. The greatest attention has been focused on plutonium, which is produced in significant amounts during reactor operations by the reaction

$$238_U \xrightarrow{\text{n}, \gamma} 239_U \xrightarrow{\beta^-} 239_{Np} \xrightarrow{\beta^-} 239_{Pu} \qquad (9)$$

The ^{239}Pu, an α-emitter with a physical half-life of 24,400 years, is extremely toxic. Although a few kg of ^{239}Pu can be found in natural uranium ores, the amounts are inconsequential compared to the more than 5000 kg of anthropogenic ^{239}Pu from weapons and other sources that have already been released into the environment [201].

Plutonium can be found in every oxidation state from +3 to +7 [200, 201], but under biological conditions the stable oxidation state is +4. The Pu^{4+} ion has a charge/radius ratio very similar to Fe^{3+}, so it is not surprising that a number of biological studies have shown that Pu is bound in the blood and transported by Tf [49, 202–207]. Bicarbonate is required as a synergistic anion for formation of a stable Pu-Tf complex [204], and Pu binding is blocked by the addition of ferric ion [203, 204]. Thus there is little doubt that Pu is binding at the high-affinity iron binding sites of the protein.

Because there are no stable isotopes of Pu, it is very difficult to conduct the usual chemical studies of protein binding that require macroscopic amounts of the metal ion. Nonetheless, the binding of Pu^{4+} to apoTf has been followed by difference UV spectroscopy [49, 208]. The characteristic spectrum of a metal-Tf complex is observed, which further confirms that Pu^{4+} binds at the iron binding site. A binding constant of log $K_1^* = 21.2$ has been reported for Pu-Tf [209]. This is slightly higher than the Fe-Tf binding constant. Since it is known that Fe^{3+} readily displaces Pu^{4+} from transferrin [203, 204] and that DTPA, citrate, and other chelators rather easily remove Pu from serum transferrin [203, 204, 210], this high Pu-Tf binding constant should be viewed as preliminary.

The percentage of serum Pu reported to be bound to Tf varies from 80 to 100% [49, 204, 206, 207]. A kinetics model for plutonium deposition indicates that the LMW fraction of Pu clears from the blood more rapidly than the Pu which is bound to Tf, so that the fraction of Pu-Tf tends to increase with time and represents >90% of serum Pu anytime after 2 h post-injection [211]. Although there are some early reports of Pu also binding to albumin [205] and to an α-macroglobulin [203, 204], the chromatography peaks in these studies probably reflect the formation of high molecular weight Pu hydroxide polymers, rather than additional Pu-protein complexes. More recent studies using ion exchange and affinity chromatography consistently show that Tf is the only significant protein ligand for Pu in serum [49, 206, 207].

Additional insight into Pu-Tf binding can be gained from the study of the tetravalent metal ions Th^{4+} and Hf^{4+}. The metabolism of Th^{4+} is generally similar to that of Pu^{4+} [212]. This similarity extends to the chromatographic profiles of these two metal ions in serum, both of which show one major band associated with Tf [213]. The in vitro binding of Th^{4+} to apoTf has also been studied by difference UV spectroscopy [214]. The Th^{4+} titration curve was linear over the addition of the first equivalent of metal ion, with a slope corresponding to a molar absorptivity of about 16,000 l/mol cm^{-1} per bound metal ion. The titration curve leveled off at only 24,000 l/mol cm^{-1}, and this was originally interpreted as an indication of nonequivalent binding sites, with two tyrosine involved in the stronger site and only one tyrosine bound to the Th in the weaker site [214]. However, given the number of subsequent studies which have show saturation of the difference UV titrations at less than two metal ions per transferrin molecule [29, 38, 41, 50, 215], it now appears much more likely that two tyrosines bind to Th at both transferrin binding sites, but that hydrolysis of the Th limits the degree of saturation of the weaker site to about 50%. Since Pu^{4+} is even more prone to hydrolysis than is Th^{4+}, the Th data probably represent an upper limit on the binding of Pu^{4+} to apoTf.

The biodistribution of Hf^{4+} also resembles that of Pu^{4+} [216], and >90% of serum Hf is bound in vivo to transferrin [216, 217]. One significant advantage to using Hf^{4+} as a model for Pu is that the binding of $^{181}Hf^{4+}$ to apoTf can be studied by perturbed angular correlation spectroscopy (PAC) [209, 217, 218].

In this method one evaluates the relationship between the direction of two successive γ photons emitted during the decay of ^{181}Hf to ^{181}Ta [217]. A complex analysis of the data provides information related to the local electronic environment around the Hf and the rotational motion of the complex.

Upon addition of small amounts of chelated Hf^{4+} to apoTf, three distinct signals are observed by PAC. One signal, representing about 25% of the Hf, is very broad and is assigned to non-specifically bound Hf [217]. The other two signals were initially assigned to Hf bound to the C- and N-terminal metal binding sites of Tf [217, 218]. This was consistent with a difference UV titration that appeared to show binding of around 2 Hf^{4+} ions per transferrin molecule [208]. However, subsequent PAC studies showed that the addition of Hf^{4+} above about 0.8 equivalent per apoTf molecule leads only to non-specific binding [209]. Based on the Th^{4+} and Hf^{4+} data, it is unlikely that the binding of Pu^{4+} to apoTf will compete effectively with hydrolysis unless there is an excess of apoTf, and that hydrolysis and polymerization of Pu would become dominant reactions near stoichiometric Pu: Tf ratios.

Further PAC studies on Hf bound to isolated C- and N-terminal lobes of ovotransferrin indicate that each binding site exists as a pH-dependent mixture of two conformations [209]. A similar pH-dependent conformational change in the vanadyl-transferrin complex has been detected by ESR spectroscopy [219], but the nature of this change is still unknown. It has been reported that formation of the low pH conformation of Hf-Tf requires phosphate [209]. The requirement for bicarbonate has not been experimentally verified for Hf, but there is no precedent among other metal ions for phosphate acting in place of bicarbonate as a synergistic anion. The involvement of phosphate may indicate the formation of a PO$_4$-Hf-CO$_3$-Tf quaternary complex, since the PAC data on Hf-Tf are consistent with eight-coordinate Hf [217].

The two critical organs for Pu toxicity are liver and bone [220]. In general, monomeric Pu complexes will deposit about 70% of the Pu in the skeleton and about 30% in the liver [221]. There appears to be no significant uptake of Pu from Pu-Tf into liver [211]. In hepatocyte cell cultures, soluble low-molecular-weight complexes, such as Pu-citrate, are taken up more readily than Pu-Tf [222–224]. In fact, the addition of apoTf almost totally inhibits Pu uptake into hepatocytes [222], while the addition of diferric Tf has almost no effect [222, 225].

Even though apoTf inhibits cell uptake of Pu, some Pu does bind to the cell membranes following exposure to Pu-Tf [223, 226, 227]. One possibility is that Pu-Tf binds at the normal Fe-Tf receptor, but cannot be internalized and/or released from Tf for incorporation into the cell [223, 227]. However, the addition of detergent to solubilize membrane proteins from lymphoblasts shows that the Pu is not present either as Pu-Tf or as the Pu-Tf-receptor complex [226]. Instead, the Pu appears to dissociate from Tf and bind to some other membrane protein.

Taylor and co-workers [225] studied Pu uptake in liver cells grown as multicellular spheroids rather than as a monolayer and found much higher uptake of Pu from Pu-Tf. However, there was still no competition from Fe-Tf, so it appears that even under conditions leading to higher cellular uptake of Pu, Pu-Tf does not deliver Pu to the cell via the usual Fe-Tf receptors.

Small angle X-ray scattering studies have shown that the binding of Hf^{4+} to apoTf does not trigger the usual protein conformation change that one observes following the binding of Fe^{3+}, Ga^{3+}, or In^{3+} [118]. Since Hf^{4+} and In^{3+} have essentially the same size, it has been suggested that the failure of Hf^{4+} to promote the normal closure of the apoTf conformation may be due to the preference of this tetravalent cation for a coordination number of eight [118]. Taylor has suggested that the transferrin receptor fails to recognize Pu-Tf because Pu^{4+}, like Hf^{4+}, fails to promote the conformation change to the "closed" conformation characteristic of ferric transferrin [209].

It is well established that soluble, monomeric Pu is preferentially taken up by bone, and that the degree of bone deposition is directly related to bone growth [220] and blood flow [228, 229]. The Pu does not appear to substitute for calcium within the mineral component of the bone, nor is there any indication that the process involves transferrin receptors. Instead, it appears that there is a ligand exchange reaction between transferrin and either phospholipids or proteins at the mineralizing surfaces of the bone [230]. Several bone proteins are reported to have Pu-binding affinities greater than that of apoTf [231]. Soon after the metal is deposited on the bone surface, it is covered over by newly deposited bone matrix and becomes physically inaccessible to serum chelating agents [201].

The role of Tf in Pu metabolism appears to be rather indirect in that the protein does not actively "deliver" Pu to specific sites. Instead, the protein functions in large part by preventing excretion and forming a long-lived Pu buffer in blood [211]. The circulating protein complex is in equilibrium with a small pool of reactive low-molecular-weight Pu complexes, which are more avidly taken up by cells and react with mineralizing bone surfaces.

The chemical composition of this low molecular weight Pu pool in serum is still not clear. Attempts to simulate the speciation of this pool with equilibrium computer models are severely hampered by the lack of reliable stability constants for Pu with important serum components such as citrate (cta) and phosphate. Duffield et al. [232] calculated that this LMW Pu should be present as a mixture of 71% $[Pu(cta)(OH)_2]^-$ and 29% $[Pu(cta)(OH)]$. A subsequent reanalysis suggested that the Pu should be present almost exclusively as $[Pu(cta)_2(OH)_2]^{4-}$ [233]. However, these authors point out that the predominance of a charged species does not account for the efficacy of citrate in mediating the transport of Pu across cell membranes. Thus the computer modeling still needs more reliable data on Pu-citrate complexation.

3.5
Other Actinides

There are biological data available for U, Am, Cm, and to a lesser extent Np and Pa. Their serum chemistry is strongly influenced by the oxidation state of the metal ions. In the case of uranium, the uranyl ion, UO_2^{2+}, is much more stable than U^{4+} under physiological conditions [200]. The serum chemistry of uranium is accordingly quite distinct from Th^{4+} or Pu^{4+}. The binding of the UO_2^{2+} ion to purified serum transferrin has been demonstrated by difference UV spectroscopy [234], but the Tf complex of this dioxo cation is much weaker than the complexes of the tetravalent actinides. As a result, about 60% of serum uranium exists as a uranyl-carbonate complex, while only about 40% is bound to Tf [234]. This weaker protein binding leads to much more rapid clearance of uranium via the kidneys [235], and exposure to high uranium levels leads to renal toxicity. There is relatively little bone deposition compared with Pu.

Serum fractionation studies have shown that essentially all Np and Pa is carried by transferrin [207, 236]. Taylor et al. [207] have suggested that the apparent avidity with which Tf binds these two metal ions in vivo indicates that the protein is able to stabilize the tetravalent state and thus form stable M^{4+}-Tf complexes, which are expected to mimic the chemistry of Th^{4+}-Tf and Pu^{4+}-Tf. The Np^{4+} ion is relatively stable with respect to oxidation to NpO_2^{2+} [200], and one would expect that Tf binding would further stabilize the Np^{4+} ion. However, the tight protein binding of Pa is somewhat surprising, since oxidation of Pa^{4+} to PaO_2^{+} is even more favorable than the oxidation of U^{4+} to UO_2^{2+} [200], and one would have expected to observe the weaker protein binding characteristic of other dioxo cations.

As one progresses across the periodic table past Pu, the metals more closely resemble the lanthanides, and the trivalent oxidation state becomes more stable [74]. Thus one finds only Am^{3+} and Cm^{3+} under physiological conditions. Although this is the same charge as the ferric ion, the much larger ionic radii of Am^{3+} and Cm^{3+} leads to much weaker complexation, and the binding of Am^{3+} and Cm^{3+} to most proteins (including transferrin) is weaker than the binding of Pu^{4+} [231, 237]. Although Tf binding constants have not been measured directly for Am^{3+} or Cm^{3+}, Harris [50] has estimated a K_1^* value of only about $10^{6.5}$ for both metal ions based on linear free energy relationships with the lanthanides Nd^{3+} and Sm^{3+}.

Early studies suggested that Am and Cm were bound in serum to Tf and/or albumin, but the interaction appeared to be weak, possibly reflecting non-specific binding [237]. The issue was complicated by the tendency of the Tf complexes to dissociate during gel filtration chromatography. A more recent study on the fractionation of radiolabeled serum used affinity chromatography to separate the albumin and Tf fractions, and showed that a significant fraction of both Am and Cm is recovered in the Tf fraction [238]. This study appears to rule out binding of Am or Cm to albumin but, because of low recoveries of the metal ions from the columns, it is still difficult to assess quantitatively the distribution of these metal ions between Tf and LMW complexes. Taylor et al. [49] have estimated that only about 20% of serum Am^{3+} and Cm^{3+} is bound to Tf. The large pool of LMW complexes in vivo leads to the rapid clearance of Am^{3+} and Cm^{3+} from the blood [239, 240].

3.6
Lanthanides

Because of their favorable luminescent properties, lanthanide-Tf complexes have been involved in a large number of spectroscopic studies [1–3]. Prior to the successful crystallographic work on Fe-Tf, luminescent energy transfer studies had determined that the two metal binding sites were about 36 Å apart [241] and were buried about 15 Å under the outer surface of the protein [242]. These numbers agree reasonably well with the crystallographic values of around 42 and 10 Å [13].

The binding of the lanthanides to Tf can easily be detected by difference UV spectroscopy [243]. Harris [50] measured the binding constants of Nd^{3+} and Sm^{3+} as models for Am^{3+} and Cm^{3+}, and obtained log K_1^* values of 6.1 for Nd^{3+} and 5.4 for Sm^{3+}. There is a delicate balance regarding the effect of the bicarbonate concentration on lanthanide-Tf binding. Some bicarbonate must be present to enable the formation of the usual $M-CO_3$-Tf ternary complex. However, the lanthanide carbonates ($M_2(CO_3)_3$) are very insoluble, with K_{SP} values in the range of 10^{-30}. Thus the addition of a few mmol/l bicarbonate, as is typically done in most metal-Tf binding studies, actually reduces the extent of lanthanide binding relative to that observed at ambient bicarbonate (around 200 µmol/l). Even at ambient bicarbonate concentrations, the final saturation of Tf with 2 equivalents of the metal is precluded by the formation of the metal-carbonate complexes [47, 50].

Luminescence studies have provided additional information about the binding of lanthanides to Tf. Luminescence lifetime studies indicate that Tf-bound Tb^{3+} retains one water molecule [243], presumably a reflection of the tendency of the lanthanides toward higher coordination numbers. There has even been speculation that an expanded coordination shell of the lanthanides might include a third tyrosine ligand in the transferrin complex [48]. The circularly polarized luminescence spectra have been reported for the Tb^{3+} complexes of transferrin, lactoferrin, and ovotransferrin [244]. The results indicate that the lanthanide coordination geometries are very similar in transferrin and lactoferrin, but that there is a significant difference between these two proteins and ovotransferrin.

Recently there has been great interest in Gd^{3+}-Tf, largely because of the growing interest in Gd complexes as contrast agents in magnetic resonance imaging. The Gd-Tf binding constant for the C-terminal site has been reported as $10^{6.8}$ [48] and $10^{8.0}$ [47]. There appears to be an unusually wide gap between the Gd binding constants of the two Tf sites. One study [48] reported no detectable binding at the N-terminal site, while another study [47] reported a binding constant of only $10^{5.9}$. Binding to the weaker N-terminal site is more difficult to characterize because of the experimental difficulty of coping with the Gd-carbonate solubility problem.

The data base of the K_{SP}'s for lanthanide-carbonates is incomplete [109], but the available data would suggest that these metal ions should form insoluble complexes with carbonate and/or phosphate in vivo. Nevertheless, low concentrations of lanthanides in serum appear to be reasonably stable and to bind to serum proteins, primarily albumin [245]. Since the binding constant of Gd to bovine serum albumin is only about 10^4 [245], its role in Gd binding in vivo presumably reflects its relatively high concentration in serum.

There is some disagreement as to the likelihood that transferrin will bind lanthanides in vivo. The biodistribution of Tm^{3+} is unaffected by the addition of either Fe or Ga, which indicates that Tm is not bound to Tf in serum [246]. Some authors have concluded that Gd-Tf binding is too weak to be of significance in vivo because of the high bicarbonate concentration in serum [47, 48]. However, a computer simulation of the distribution of Gd^{3+} in serum indicated that for low concentrations of Gd (10^{-5} mol/l), the metal ion would be about equally distributed between Tf and citrate [93]. A similar model for Sm^{3+} indicated that 95% of the Sm in serum would be bound to transferrin at equilibrium, although the authors also noted that the formation of Sm-Tf was slow and several hours might be required to reach this equilibrium [247]. In addition, Taylor et al. have reported that Tf binds about 20% of Eu and Yb in vivo [49]. Thus there could be a role for Tf in the transport of lanthanides at low concentrations where the formation of colloidal carbonate and/or phosphate precipitates is suppressed.

Transferrin binding is still unlikely to be important with respect to the use of Gd as an MRI contrast agent. To reduce its toxicity the Gd^{3+} must be introduced as a stable chelate, e.g., $Gd(DTPA)^{2-}$. It is virtually certain that the chelating agent used to bind the Gd will be a much stronger ligand than Tf, so one would expect very little donation of Gd from the low molecular weight ligand to Tf.

3.7
Vanadium

Vanadium is unique in that complexes with both transferrin [248] and lactoferrin [249] have been reported for three different oxidation states: V(III), V(IV), and V(V). The best characterized system involves V(IV) in the form of the vanadyl complex. Because the VO^{2+} ion has only one d-electron and is easily studied by ESR spectroscopy, it has served as a very useful probe of the transferrin metal binding sites [1, 2]. ESR studies have (a) shown the strong binding of two vanadyl ions per transferrin molecule into two spectroscopically distinct sites [250], (b) documented pH dependent conformational changes in the transferrin metal binding sites [219, 250], and (c) provided early evidence that the synergistic anion in the metal-anion-Tf ternary complex was directly coordinated to the metal ion [251].

The vanadyl ion appears to bind rather tightly to apoTf, although no equilibrium constants for serum Tf have been reported. A binding constant of around 10^9 has been reported for the vanadyl complex with ovotransferrin [67]. However, the relevance of this constant to serum transferrin is debatable due to the fact that the VO^{2+}-ovotransferrin complex forms in the absence of bicarbonate [67], whereas a synergistic bicarbonate anion is required for formation of the VO^{2+}-Tf complex [52, 250].

The vanadyl-Tf complex is rather easily oxidized to V(V) by molecular oxygen [52, 252], with a $t_{1/2}$ of only 8 min in air-saturated hepes buffer at

37 °C. This V(V)-Tf species can also be formed directly and reversibly by the addition of vanadate (VO_4^{3-}) to apoTf at physiological pH [52]. Other tetrahedral anions such as arsenate and phosphate also bind to apoTf [58], so it is plausible that vanadate could bind to apoTf as an anion. However, the difference UV spectrum produced by anion-binding to apoTf is quite distinct from that produced by metal-binding [58, 59]. The difference UV spectrum of V(V)-Tf vs apoTf is characteristic of metal binding, which indicates that the vanadium is directly coordinated to the two tyrosines of the metal binding site [52].

V(V) most often reacts with multidentate ligands to form complexes of the VO_2^+ cation, but complexes of V^{5+}, VO^{3+}, and $VO(OH)^{2+}$ are also known [248]. The absence of strong metal-ligand charge transfer bands in the spectrum of the V(V)-Tf complex rules out binding of either V^{5+} or VO^{3+} [253]. There are pH-dependent changes in the difference UV spectrum of V(V)-Tf which indicate that there is a protonation equilibrium with a pK_a of approximately 8 [52]. It has been suggested that V(V) reacts with Tf to form a VO_2^+-Tf complex at high pH, which is then protonated to form $VO(OH)^{2+}$-Tf at neutral pH [52]. This type of protonation equilibrium has been observed for the VO_2^+ complex with the transferrin model compound ethylenebis[(o-hydroxybenzyl)glycine] [253].

The diamagnetic V(V)-Tf system has also been investigated by ^{51}V NMR [248]. The addition of two equivalents of ammonium vanadate to apoTf results in two partially resolved peaks in the ^{51}V NMR spectrum which are assigned to vanadium bound at the C- and N-terminal binding sites. The chemical shifts of these two peaks are consistent with direct coordination of the two tyrosine residues [254].

The ^{51}V NMR spectrum indicates that the V(V) ion is in a trigonal bipyramidal environment [248]. It has been proposed that the VO_2^+ cation binds to the two tyrosines and the aspartic acid of the transferrin binding site, with hydrogen bonds between the two oxo groups and the binding site histidine and arginine residues [248]. Retention of the coordinated oxygen ligands in VO_2^+-Tf and $VO(OH)^+$-Tf presumably accounts for the formation of these V(V) complexes in the absence of a synergistic carbonate anion [52, 62].

A value of log $K_1 = 7.45$ for the binding of the first equivalent of vanadate to apoTf has been determined by competition between vanadate and phosphate [52]. The second equivalent of vanadate binds more weakly, so that one can determine a binding constant of $K_2 \approx 10^{6.5}$ either from the simple titration of apoTf with vanadate [52, 255] or by competition with either phosphate or sulfate [59].

A V(III) complex with Tf has been characterized by UV-vis and circular dichroism spectroscopy [256, 257]. It is most readily distinguished from VO^{2+}-Tf by changes in the d-d bands in the visible region and from VO_2^+-Tf by the presence of intense charge-transfer bands in the UV spectrum of the V^{3+}-Tf complex. The most remarkable attribute of this complex is its purported stability toward oxidation. The V^{3+}-Tf complex, after preparation under an inert atmosphere to avoid oxidation of free V^{3+}, was reported to be stable to oxidation by a ten-fold excess of hydrogen peroxide [256]. We were initially quite skeptical of this report [52] because VO^{2+}-Tf is easily air-oxidized to VO_2^+-Tf. Most low molecular weight phenolate ligands such as ethylenebis[(o-hydroxyphenyl)glycine] also form V(III) complexes which are unstable toward air oxidation [253]. However, there is a recent report of an air-stable V(III) complex with the hexadentate ligand N,N'-bis(o-hydroxybenzyl)-N,N'-bis(2-pyridylmethyl)-ethylenediamine [258].

Very recently Smith et al. [249] prepared the V(III), V(IV), and V(V) complexes of lactoferrin. These workers relied primarily on changes in the ESR spectrum to follow oxidation reactions, since only the V(IV) complex is ESR detectable at room temperature. At neutral pH, the V(III)-lactoferrin complex is air-oxidized to VO^{2+}-lactoferrin in about 5 min. This vanadyl complex air-oxidizes to a vanadate-lactoferrin complex in about 1 h.

Vanadium is an essential element and is found in normal blood at a concentration of about 0.5 μmol/l [259]. The distribution of this vanadium between plasma and red blood cells depends on an intriguing combination of protein binding and vanadium redox chemistry. It was noted early on that the biodistribution of vanadium was essentially the same following the injection of either V(III), V(IV), or V(V) [260]. It appears that vanadium cycles between V(IV) and V(V) within blood. Following injection of VO_4^{3-} or VO^{2+}, there is a very rapid elimination phase, in which about 40% of vanadate and 70% of vanadyl leaves the blood [261]. After this initial phase, the remaining vanadium clears from blood with a $t_{1/2}$ of 24 h regardless of whether the initial form of the injected vanadium was vanadyl or vanadate [261]. In addition, the percentage of total vanadium found in the blood cells remains constant during the second phase of vanadium clearance. Thus the vanadium must be moving between the plasma and cellular components of blood at a rate comparable to the rate of blood clearance.

It has been established that erythrocytes take up vanadate in a biphasic process [262]. The initial phase, which has a $t_{1/2}$ of only 4 min, is attributed to the rapid movement of vanadate across the cell membrane through the phosphate transport system and produces equal intracellular and extracellular vanadium concentrations [263]. The second, slower phase of vanadium uptake occurs at the same rate as the reduction of intracellular vanadate to vanadyl, presumably by glutathione [262].

Chasteen et al. [252] have shown that vanadate is also reduced with a $t_{1/2}$ of only 5 min in fresh, anaerobic serum. For serum treated to remove reductants and then aerated, the half-life for air oxidation of VO^{2+} back to vanadate was 15 min. Essentially the same oxidation half-life was observed for oxidation of vanadyl in hepes buffer in the presence of excess Tf. The rate of oxidation increased with decreasing apoTf concentration, so that it appears that the oxidation reaction actually proceeds via a small pool of free vanadyl ion [252]. Because of these competing oxidation and reduction reactions, it has been proposed that the vanadium in serum will consist of a dynamic mixture of vanadyl and vanadate [252].

When vanadate uptake is carried out with erythrocytes suspended in buffer, 90% of the vanadium accumulates within the cells as vanadyl ion [262]. The metal appears to be bound intracellularly to hemoglobin [262], although neither vanadyl nor vanadate binds to hemoglobin in vitro [264]. In contrast, only about 40% of vanadium in blood accumulates in cells following intravenous injection of vanadate or vanadyl [261], and only 10% of blood vanadium is found in blood cells when the vanadium is absorbed from the gut [255]. Thus it appears that the combination of reductants and vanadium ligands in plasma have a strong effect on the distribution of vanadium between the plasma and red blood cells.

It has been difficult to obtain reliable estimates of the extent of vanadium binding to Tf in vivo. Initially, Sabbioni and Marafante [265] fractionated labeled serum by gel chromatography and reported that the percentage of plasma vanadium bound to Tf started at about 3% 30 min after injection and rose slowly to about 90% after 4 days. Harris et al. [261] analyzed the protein binding of vanadium in serum by agarose gel electrophoresis and found that about 25% of the vanadyl-Tf and essentially all of the vanadate-Tf dissociated during the 2-h electrophoresis experiment. After correcting for this dissociation, they concluded that only about 5% of the vanadium was initially bound (as vanadyl) to Tf regardless of whether vanadyl or vanadate was injected. The percent of vanadium binding to transferrin rose to 70% after 2 days. The percentage was essentially identical following injection of either vanadyl or vanadate. Since only vanadyl-Tf survived the electrophoresis, there was no way to determine to what extent vanadate binds to Tf in vivo.

Chasteen et al. [252] have used ESR to measure the distribution of VO^{2+} between Tf and albumin, both in serum and in hepes buffer. They estimate that the vanadyl-Tf binding constant is about six times larger than the vanadyl-albumin constant. However, since the concentration of albumin in serum is ten times larger than the Tf concentration, they suggest that albumin may also carry a significant fraction of the vanadyl ion in serum. The vanadyl-albumin complex is rapidly air-oxidized and would have likely escaped detection in the electrophoresis studies of Harris et al. [261].

The differences between the ESR and electrophoresis data on the binding of vanadate to transferrin in serum may reflect in part a difference in vanadium concentrations. The ESR studies used vanadium concentrations that were comparable to the serum transferrin concentration. Binding to albumin may become a more significant factor as the stronger transferrin binding site has been saturated and vanadate must bind to the weaker site. Chasteen et al. [252] note that the binding to albumin is higher at higher total vanadium concentrations. In contrast, the serum samples analyzed by Harris et al. [261] never exceed 3 μmol/l, which is only about 10% of the serum transferrin concentration.

The overall model that has emerged for the behavior of vanadium in serum is a combination of extracellular protein binding and oxidation coupled with intracellular reduction. Extracellular vanadium consists of VO^{2+}-Tf, VO^{2+}-albumin, and a mixture of free vanadate and $VO(OH)^{2+}$-Tf. The VO^{2+}-protein complexes in serum are oxidized, presumably by free oxygen. The resulting vanadate anion rapidly equilibrates with an intracellular vanadate pool in the erythrocytes. This intracellular pool is slowly reduced to vanadyl, which leaves the cell and binds to Tf and albumin in the serum compartment.

3.8
Manganese

The binding of manganese to Tf bears some resemblance to the iron-transferrin system. The higher oxidation state manganic ion (Mn^{3+}) is bound very tightly, while the lower oxidation state manganous ion (Mn^{2+}) is bound very weakly. As with iron, the Mn^{3+}-HCO_3-Tf ternary complex can be prepared by air-oxidation of Mn^{2+} in the presence of apoTf and bicarbonate [266]. But while the oxidation of ferrous ion to form ferric transferrin is very rapid, the oxidation of Mn^{2+} is very slow. Indeed, the synthesis of Mn^{3+}-lactoferrin requires peroxide as the oxidant [267].

The visible absorption spectrum of the Mn^{3+}-Tf complex is dominated by intense charge transfer bands at 430 and 330 nm with ε's of about 5000 l/mol cm^{-1} [266]. Studies on model compounds have confirmed that these are associated with the coordinated phenolic groups of the two binding site tyrosine residues [268]. Tyrosine coordination has been confirmed by the enhancement of aromatic ring vibrational modes in the resonance Raman spectrum of Mn^{3+}-Tf [269]. The trivalent oxidation state of the metal has been confirmed by magnetic susceptibility measurements and ESR spectroscopy [266].

No binding constant for Mn^{3+}-Tf has been reported, and, given the poorly behaved aqueous chemistry of the free Mn^{3+} ion, none is likely to be measured in the foreseeable future. The free Mn(III) aquo ion is not very stable in neutral aqueous solution and is subject to disproportionation to Mn(II) and Mn(IV) (MnO_2) [270]. In acidic media, the reduction potential for the reaction Mn^{3+} + $1e^- \rightarrow Mn^{2+}$ is +1.5 V vs NHE, making Mn^{3+} a potent oxidizing agent. Thus many Mn(III) complexes are subject to internal redox reactions in which the ligand is oxidized and the metal is reduced to Mn(II). Even complexation by EDTA only reduces the Mn^{3+} reduction potential to +0.8 V, and the Mn^{3+}-EDTA complex slowly decomposes by oxidation of the ligand [270].

Given the high reduction potential of Mn^{3+}, it might seem surprising that one can form a stable Mn^{3+}-Tf complex by simple air oxidation. This unusual stability of the Mn^{3+}-Tf complex can be attributed to the very high selectivity of the coordinated phenolic groups for Mn^{3+}. For example, the Mn^{3+} complex of the diphenolic ligand EHPG has a reduction potential of 0.0 V vs NHE and is stable to internal redox reactions [268]. In the absence of reliable experimental data on the stability of Mn^{3+}-Tf, one can use an LFER between Mn^{3+} and Fe^{3+} to estimate a binding constant for this complex. An LFER based primarily on aminocarboxylate ligands shows a good linear relationship ($r = 0.987$) between the Mn^{3+} and Fe^{3+} binding constants which is described by the equation

$$\log\ K_{Mn} = 1.09(\log\ K_{Fe}) + 1.09 \tag{10}$$

Based on the K_M value for Fe-Tf [22], this equation predicts a log K_M value of 23.4 for Mn^{3+}-Tf. This is 2 log units larger than the Fe^{3+} binding constant. With such a large binding constant, one would expect that essentially all Mn^{3+} in serum will be bound to Tf, and that the complex will be stable enough to survive most methods which might be used to separate the proteins in serum.

In sharp contrast to Mn^{3+}, the Mn^{2+} ion binds very weakly to apoTf [33]. Since the Mn^{2+} does not hydrolyze at neutral pH, this is one of the rare metal-Tf systems for which one can measure the simple equilibrium between the protein and the free metal ion without the need for competitive chelating agents. An additional experimental advantage is that one can use ESR spectroscopy to measure directly the concentration of free, hexaaquo Mn(II) in equilibrium with the protein. The free ion gives a simple, isotropic six-line ESR spectrum which can be detected at micromolar concentrations, while the slow rotational motion of the protein broadens the Mn^{2+}-protein signal to the point that it is not observed at room temperature [271, 272].

A value of log $K_1^* = 3.98$ for Mn^{2+}-Tf was measured by titration of 50 μmol/l free Mn^{2+} in pH 7.4 Hepes buffer with apoTf [33]. Similar titrations with diferric transferrin were used to correct for non-specific binding. The macroscopic binding constant for the binding of the second Mn^{2+} ion to apoTf was determined by difference UV spectroscopy to be log $K_2^* = 2.96$. The binding of Mn^{2+} to the vacant metal binding sites of apoTf and both C- and N-terminal monoferric transferrins was also measured by difference UV titrations of the proteins with a relatively large excess of Mn^{2+} [33]. Site-specific binding constants of log $k_C = 3.74$ and log $k_N = 3.05$ were deter-mined, where the subscript denotes the site at which the Mn^{2+} is binding. The difference of 1.0 log unit between the macroscopic binding constants can be accounted for by a combination of a statistical separation factor for two binding sites and the difference of 0.7 log units between the intrinsic binding constants for the two sites.

The binding constants for Mn^{3+}-Tf and Mn^{2+}-Tf differ by approximately 19 log units, which would shift the Mn^{3+}-Tf reduction potential 1.1 V negative of the $E°$ for free Mn^{3+}. This contributes to the unusual stability of the Mn^{3+}-Tf toward reduction to Mn^{2+}. There is a similar gap of about 17 log units between the binding constants for Fe^{3+}-Tf and Fe^{2+}-Tf. Since the standard reduction potential for Fe^{3+} is much lower than that of Mn^{3+}, complexation shifts the Fe^{3+}-Tf reduction potential to values sufficiently negative to account for the facile air-oxidation of Fe^{2+}-Tf.

There has been considerable confusion regarding the speciation of Mn in serum. There have been reports that Mn in serum is bound to transferrin [273–279], serum albumin [280, 281], a protein initially identified as transmanganin [282], and a much larger protein which might be α_2-macroglobulin [277–279]. It has also been proposed that a significant portion of serum manganese is present as ionic Mn^{2+} or low molecular weight chelates [273, 277, 283].

Part of the confusion stems from the fact that the extent of transferrin binding in serum depends on the incubation time of Mn^{2+} with apoTf in either serum or buffer [153, 274, 278, 279, 283]. The delay in binding to apoTf is associated with the necessity of oxidizing the Mn^{2+}, which is very weakly bound, to Mn^{3+}, which is very strongly bound. When serum was labeled in vitro with Mn^{2+}, incubated for 1 min, and then fractionated by gel permeation chromatography, the majority of the Mn was associated with low molecular weight species, about 15% was associated with Tf, and a similar percentage appeared at the void volume of the column [273]. After incubation for 1 h, the Tf peak had grown to 40% of the total Mn. A large fraction of ionic Mn was also observed following in vitro labeling by Hancock et al. [277], but the incubation time was not given. Scheuhammer and Cherian [279] reported that it takes about 6 h at 37 °C for Tf binding to maximize in serum labeled in vitro with Mn^{2+}.

Manganese oxidation in serum appears to be faster than oxidation in buffer, and in vitro studies have shown that the oxidation and binding of Mn^{2+} to apoTf can be catalyzed by ceruloplasmin [278, 283]. Whether ceruloplasmin carries out this function in vivo has not been established. The observation of normal Tf loading with Mn in the plasma of copper-deficient rats with low ceruloplasmin levels casts some doubt on the role of this protein as the plasma oxidase for Mn [273].

When Mn^{2+} is injected, there is a very rapid clearance of the metal from the blood, and only about 1% of the injected dose remains in the circulation after 10 min [278, 284]. A large fraction of the dose of free Mn^{2+} is taken up by the liver [284]. Following this rapid clearance of most of the Mn dose, one observes a small but persistent Mn fraction which is primarily associated with Tf [278]. Gibbons et al. [278] injected the pre-formed Mn^{3+}-Tf complex and found that the complex has a plasma clearance half-life of about 3 h, which is similar to that of Fe^{3+}-Tf [278]. These workers proposed that it is Mn^{3+}-Tf which is the species responsible for the longer-term transport and controlled delivery of Mn to extrahepatic tissues. The protein originally identified as transmanganin [282] is now presumed to have been transferrin.

There have been several reports of a very high molecular weight Mn complex in serum, which is often identified as a complex with α_2-macroglobulin [273, 277–279, 285]. Gibbons et al. [278] have shown that Mn^{2+} will bind to purified α_2-macroglobulin. However, when the Mn^{2+} complex of α_2-macroglobulin was injected, the Mn^{2+} was cleared from the blood almost as rapidly as free Mn^{2+}. Thus it does not appear that this complexation by α_2-macroglobulin plays a major role in Mn^{2+} transport.

There have been conflicting reports on the binding of Mn to albumin. The Mn^{2+} ion binds to purified albumin at a single binding site with a binding constant of log $K_1 = 4.3$ [271, 286]. There may be additional low affinity sites, but because of the high concentration of albumin in serum, these should not be important with respect to binding Mn in vivo. Two early reports identified albumin by gel permeation chromatography as the major Mn binding agent in serum [280, 281]. More recently, the protein binding of Mn^{2+} in serum has been studied under conditions designed to limit oxidation to Mn^{3+} [287] Ultrafiltration data showed that 70% of the Mn^{2+} in serum was protein-bound [287]. Attempts to identify the specific protein(s) which bind Mn^{2+} by gel filtration chromatography were thwarted by the dissociation of the weak Mn-protein complexes on the column.

A computer model for the speciation of Mn^{2+} in serum has also been reported [33]. When both albumin and Tf are included in the model, the calculations predict that over 80% of Mn^{2+} would be bound to albumin. About 12% of the Mn would be evenly split between $Mn(H_2O)_6^{2+}$ and $Mn(HCO_3)^+$, and only 1% would be bound to Tf. This is in good agreement with the ultrafiltration data and clearly indicates that albumin, rather than Tf, is the primary transport protein for Mn^{2+}.

The bulk of the data on Mn binding is consistent with the basic serum transport model proposed by Gibbons et al. [278]. Upon entering the blood, Mn^{2+} is loosely bound to albumin and possibly α_2-macroglobulin. This free or loosely bound Mn^{2+} is removed very efficiently (essentially in one pass) by the liver due to the presence of a high affinity, high capacity Mn^{2+} transport system in hepatocytes [288]. Most of this Mn^{2+} is passed quickly into bile for excretion [288]. However, a small fraction of the Mn^{2+} is oxidized to Mn^{3+} and reappears in the serum as Mn^{3+}-Tf. This oxidation may be catalyzed by ceruloplasmin. This Mn^{3+}-Tf has a longer half-life in plasma, and is likely to represent the transport species for the controlled delivery of Mn to extrahepatic tissues [279].

Free Mn(II) in serum is rapidly assimilated into other tissues in addition to liver. Free Mn(II) crosses the blood-brain-barrier and is incorporated into brain in a very rapid, saturable process [287, 289]. It has been proposed that there is a family of metal transport systems with differing selectivities for a series of divalent metal ions, including Mn^{2+} [290]. Other studies have shown that there is receptor mediated cellular uptake of Mn from the Mn^{3+}-Tf complex [283, 289, 291]. One hypothesis is that the uptake of Mn^{2+} is too rapid and difficult to control. To allow for a more controlled delivery of Mn to tissues which need it, the liver rapidly removes Mn^{2+} from the serum and releases low concentrations of Mn^{3+}-transferrin. Uptake of this species is then regulated by the usual transferrin receptor mediated endocytosis.

4
Conclusions

As demonstrated by the data in Table 1, serum transferrin is capable of binding a very wide variety of metal ions in vitro. The stability of these complexes varies widely from around 10^4 for Fe^{2+} and Mn^{2+} to around 10^{22} for Fe^{3+} [21, 22]. Transferrin can also bind oxocations, such as VO^{2+} and UO_2^{2+}, but this binding is relatively weak. The M^{3+} and M^{4+} cations which tend to form strong transferrin complexes also have a strong tendency to hydrolyze, and there is often a close balance between hydrolysis and protein binding. For several metal ions, the stronger C-terminal binding site can compete effectively with hydrolysis, but the weaker N-terminal site cannot. As a result, for many metal ions it is difficult to saturate the N-terminal site.

The extent of transferrin binding in vivo varies significantly from one metal ion to another. Transferrin appears to bind almost all of serum Mn^{3+} [273, 274, 276], Ga^{3+} [71, 85, 87, 88], In^{3+} [88–90, 112], Hf^{4+} [216, 217], Pa^{4+} [236], and Np^{4+} [207, 292]. It also appears to bind around 90% of serum Pu^{4+} [74, 204, 206, 207], around 80% of serum Al^{3+} [167, 168, 177, 179], and around 20% of the trivalent actinides Am^{3+} and Cm^{3+} [49]. Transferrin binds about 40% of

UO_2^{2+} [234] and a significant fraction of serum VO^{2+} [252, 261], but there is little evidence that transferrin binds any other divalent cations in serum.

There are conflicting views on the extent to which Tf binds lanthanide ions. Taylor et al. have reported that Tf binds about 20% of Eu^{3+} and Yb^{3+} [49]. Equilibrium models for serum speciation of the lanthanides tends to give much higher estimates for the degree of Tf binding [93, 247], but others have pointed out that Tf binding is likely to be seriously restricted by the small solubility products for the lanthanide-carbonate complexes [47, 48].

The impact of transferrin binding on the in vivo behavior of a metal ion depends on both the fraction of the metal bound to Tf in serum and the interactions of the metal-Tf complexes with the transferrin receptor. Since LMW metal complexes tend to be cleared from blood rather quickly, blood clearance rates tend to be inversely related to the extent of transferrin binding. Even a relatively small pool of labile LMW complexes may provide a significant source for transferrin-independent metal uptake.

There is also a significant variation in the extent to which metal-transferrin complexes can utilize the transferrin receptor to enhance cellular uptake. Cellular uptake via receptor mediated endocytosis has been demonstrated for Ga^{3+} [82, 99], Mn^{3+} [283, 289, 291] and Al^{3+} [185, 186, 189, 191, 192] although transferrin-independent pathways may also be important [98]. The In-Tf complex can bind to the transferrin receptor, but the In^{3+} is not transported across the membrane [94, 116]. The absence of receptor-mediated uptake or a significant LMW pool of In leads to a much longer residence time in serum for the In-Tf complex. The Pu-Tf complex does not appear to bind to the transferrin receptor [225, 226]. For this metal ion, the impact of transferrin binding is to retard blood clearance, which exposes tissues to a relatively long-lived, pool of more reactive LMW Pu complexes.

The variation in the response of M-Tf complexes to the transferrin receptor may be due to several factors. Conformational differences between M-Tf and Fe^{3+}-Tf, due either to the ionic radius of the metal ion [86, 118, 119] or to a higher coordination number for the metal [118], may reduce the avidity with which the M-Tf complex binds to the transferrin receptor. In addition, the inability to reduce metals like Ga^{3+} and In^{3+} may hinder the release of the metal from the complex between M-Tf and the transferrin receptor [98, 196].

5
References

1. Chasteen ND (1983) Adv Bioinorg Chem 5: 201
2. Harris DC, Aisen P (1989) Physical biochemistry of the transferrins. In: Loehr TM (ed) Iron carriers and iron proteins. VCH, New York, p 239
3. Aisen P (1989) Physical biochemistry of the transferrins, update 1984–1988. In: Loehr TM (ed) Iron carriers and iron proteins. VCH, New York, p 353
4. Brock JH (1985) Transferrins. In: Harrison P (ed) Metalloproteins, pt II. Macmillan, London, p 183
5. Bates GW, Graybill G, Chidambaram MV (1987) Transferrin. In: Boynton AL, Leffert HL (eds) Control of animal cell proliferation. Academic Press, New York, p 153
6. Baker EN (1994) Adv Inorg Chem 41: 389

7. Hider RC, Hall AD (1991) Prog Med Chem 28: 43
8. Aisen P (1982) Chemistry and physiology of the transferrins. In: Dunford HB, Dolphin D, Raymond KN, Sieker L (eds) The biological chemistry of iron. Reidel, Boston, p 63
9. Aisen P, Leibman A (1972) Biochim Biophys Acta 257: 314
10. May WS, Cautrecasas P (1985) J Membr Biol 88: 205
11. Bailey S, Evans RW, Garratt RC, Gorinsky B, Hasnain S, Horsburgh C, Jhoti H, Lindley PF, Mydin A, Sarra R, Watson JL (1988) Biochemistry 27: 5804
12. Sarra R, Garratt R, Gorinsky B, Jhoti H, Lindley P (1990) Acta Cryst B46: 763
13. Anderson BF, Baker HM, Norris GE, Rice DW, Baker EN (1989) J Mol Biol 209: 711
14. Anderson BF, Baker HM, Dodson EJ, Norris GE, Rumball SV, Waters JM, Baker EN (1987) Proc Natl Acad Sci (USA) 84: 1769
15. Baker EN, Lindley PF (1992) J Inorg Biochem 47: 147
16. Dewan JC, Mikami B, Hirose M, Sacchettini JC (1993) Biochemistry 32: 11 963
17. Kurokawa H, Mikami B, Hirose M (1995) J Mol Biol 254: 196
18. Anderson BF, Baker HM, Norris GE, Rumball SV, Baker EN (1990) Nature 344: 784
19. Grossmann JG, Neu M, Pantos E, Schwab FJ, Evans RW, Townes-Andrews E, Lindley PF, Appel H, Theis W-G, Hasnain SS (1992) J Mol Biol 225: 811
20. Schlabach MR, Bates GW (1975) J Biol Chem 250: 2182
21. Harris WR (1989) Equilibrium constants for the complexation of metal ions by serum transferrin. In: Dintzis FR, Laszlo JA (eds) Mineral absorption in the monograstic GI tract. Plenum, New York, p 67
22. Harris WR (1993) Trends Inorg Chem 3: 559
23. Aramini JM, Saponja JA, Vogel HJ (1996) Coord Chem Rev 149: 193
24. L'Eplattenier F, Murase I, Martell AE (1967) J Am Chem Soc 89: 837
25. Motekaitis RJ, Martell AE, Welch MJ (1990) Inorg Chem 29: 1463
26. Motekaitis RJ, Sun Y, Martell AE (1989) Inorg Chim Acta 159: 29
27. Harris WR, Martell AE (1976) Inorg Chem 14: 974
28. Chasteen ND (1977) Coord Chem Rev 22: 1
29. Harris WR, Stenback JZ (1988) J Inorg Biochem 33: 211
30. Harris WR (1983) Biochemistry 22: 3920
31. Hirose J, Fujiwara H, Magarifuchi T, Iguti Y, Iwamoto H, Kominami S, Hiromi K (1996) Biochim Biophys Acta 1296: 103
32. Harris WR, Madsen LJ (1988) Biochemistry 27: 284
33. Harris WR, Chen Y (1994) J Inorg Biochem 54: 1
34. Kojima N, Bates GW (1981) J Biol Chem 256: 12 034
35. Harris WR (1986) J Inorg Biochem 27: 41
36. Martin RB, Savory J, Brown S, Bertholf RL, Wills MR (1987) Clin Chem 33: 405
37. Aisen P, Leibman A, Zweier J (1978) J Biol Chem 253: 1930
38. Harris WR, Sheldon J (1990) Inorg Chem 29: 119
39. Fatemi SJA, Kadir FHA, Moore GR (1991) Biochem J 280: 527
40. Cochran M, Coates JH, Kurucsev T (1987) Life Sciences 40: 2337
41. Harris WR, Pecoraro VL (1983) Biochemistry 22: 292
42. Kulprathipanja S, Hnatowich DJ, Beh R, Elmaleh D (1979) Int J Nucl Med Biol 6: 138
43. Larson SM, Allen DR, Rasey JS, Grunbaum Z (1978) J Nucl Med 19: 1245
44. Clausen J, Edeling C-J, Fogh J (1974) Cancer Res 34: 1931
45. Harris WR, Chen Y, Wein K (1994) Inorg Chem 33: 4991
46. Lurie DJ, Smith FA, Shukri A (1985) Int J Appl Radiat Isot 36: 57
47. Harris WR, Chen Y (1992) Inorg Chem 31: 5001
48. Zak O, Aisen P (1988) Biochemistry 27: 1075
49. Taylor DM, Duffield JR, Williams DR, Yule L, Gaskin PW, Unalkat P (1991) Eur J Solid State Inorg Chem 28: 271
50. Harris WR (1986) Inorg Chem 25: 2041
51. Li HL, Sadler PJ, Sun H (1996) J Biol Chem 271: 9483
52. Harris WR, Carrano CJ (1984) J Inorg Biochem 22: 201
53. Aramini JM, Krygsman PH, Vogel HJ (1994) Biochemistry 33: 3304

54. Bertini I, Luchinat C, Messori L (1983) J Am Chem Soc 105: 1347
55. Aramini JM, Vogel HJ (1993) J Am Chem Soc 115: 245
56. Ichimura K, Kihara H, Yamamura T, Satake K (1989) J Biochem 106: 50
57. Aramini JM, Vogel HJ (1994) J Am Chem Soc 116: 1988
58. Harris WR, Nesset-Tollefson D, Stenback JZ, Mohamed-Hani N (1990) J Inorg Biochem 38: 175
59. Harris WR (1985) Biochemistry 24: 7412
60. Harris WR, Nesset-Tollefson D (1991) Biochemistry 30: 6930
61. Zweier JL, Wooten JB, Cohen JS (1981) Biochemistry 20: 3505
62. Saponja JA, Vogel HJ (1996) J Inorg Biochem 62: 253
63. Bertini I, Luchinat C, Messori L, Scozzafava A, Pellacani G, Sola M (1986) Inorg Chem 25: 1782
64. Dubach J, Gaffney BJ, More K, Eaton GR, Eaton SS (1991) Biophys J 59: 1091
65. Gelb MH, Harris DC (1980) Arch Biochem Biophys 200: 93
66. Shongwe MS, Smith CA, Ainscough EW, Baker HM, Brodie AM, Baker EN (1992) Biochemistry 31: 4451
67. Casey JD, Chasteen ND (1980) J Inorg Biochem 13: 111
68. Green MA, Welch MJ (1989) Nucl Med Biol 5: 435
69. Welch MJ, Moerlein S (1980) Radiolabeled compounds of biomedical interest containing radioisotopes of gallium and indium. In: Martell AE (ed) Inorganic chemistry in biology and medicine. American Chemical Society, Washington, D.C., p 121
70. Gunasekera SW, King LJ, Lavender PJ (1972) Clin Chim Acta 39: 401
71. Vallabhajosula SR, Harwig JF, Siemsen JK, Wolf W (1980) J Nucl Med 21: 650
72. Larson SM, Grunbaum Z, Rasey JS (1981) Int J Nucl Med Biol 8: 257
73. Vallabhajosula SR, Harwig JF, Wolf W (1981) Int J Nucl Med Biol 8: 363
74. Cotton FA, Wilkinson G (1988) Advanced inorganic chemistry, 5th edn. Wiley, New York
75. Harris WR (1986) Biochemistry 25: 803
76. Weiner R, Hoffer PB, Thakur ML (1981) J Nucl Med 22: 32
77. Weiner RE, Schreiber GJ, Hoffer PB, Bushberg JT (1985) J Nucl Med 26: 908
78. Weiner RE, Schreiber GJ, Hoffer PB, Shannon T (1981) Int J Nucl Med Biol 8: 371
79. Weiner RE, Schreiber GJ, Hoffer PB (1983) J Nucl Med 24: 608
80. Weiner RE (1989) J Nucl Med 29: 70
81. Nakamura K, Kawaguchi H, Shimizu K, Orii H (1984) Eur J Nucl Med 9: 237
82. Chitambar CR, Zivkovic Z (1987) Cancer Res 47: 3929
83. Baes CF, Mesmer RE (1976) The hydrolysis of cations. Wiley, New York
84. Clevette DJ, Orvig C (1990) Polyhedron 9: 151
85. Brunetti A, Blasberg G, Finn RD, Larson SM (1988) Nucl Med Biol 15: 665
86. Otsuki H, Brunetti A, Owens ES, Finn RD, Blasberg RG (1989) J Nucl Med 30: 1676
87. Staker BL, Graham MM, Evans ML (1991) J Nucl Med 32: 1439
88. Raijmakers PGHM, Groeneveld ABJ, Den Hollander W, Teule GJJ (1992) Nuc Med Commun 13: 349
89. Tsan M-F, Scheffel U, Tzen K-Y, Camargo EE (1980) Int J Nucl Med Biol 7: 270
90. Kulprathipanja S, Hnatowich DJ, Evans G (1978) Int J Nucl Med Biol 5: 140
91. Raiszadeh M, Harwig JF, Wolf W (1981) J Labelled Comp Radiopharm 18: 167
92. McGregor SJ, Brock JH (1992) Clin Chem 38: 1883
93. Jackson GE, Byrne MJ (1996) J Nucl Med 37: 379
94. Ganrot PO (1986) Env Health Persp 65: 363
95. Nelson B, Hayes RL, Edwards CL, Kniseley RM, Andrews GA (1972) J Nucl Med 13: 92
96. Sohn M-H, Jones BJ, Whiting JH, Datz FL, Lynch RE, Morton KA (1993) J Nucl Med 34: 2135
97. Larson SM, Rasey JS, Allen DR, Nelson NJ, Grunbaum Z, Harp GD, Williams DL (1980) J Nat Canc Inst 64: 41
98. Weiner R (1990) Nucl Med Biol 17: 141

99. Chitambar CR, Zivkovic-Gilgenbach Z (1990) Cancer Res 50: 1484
100. Weiner RE, Spencer RP, Dambro TJ, Klein BE (1992) J Nucl Med 33: 1701
101. Scheffel U, Wagner HN, Frazier JM, Tsan M-F (1984) J Nucl Med 25: 1094
102. Sephton RG, De Abrew S, Hodgson GD (1982) Brit J Radiol 55: 134
103. Chan SM, Hoffer PB, Maric N, Duray P (1987) J Nucl Med 28: 1303
104. Hoffer P (1980) J Nucl Med 21: 282
105. Tsan M-F (1985) J Nucl Med 26: 88
106. Evers A, Hancock RD, Martell AE, Motekaitis RJ (1989) Inorg Chem 28: 2189
107. Emery T (1986) Biochemistry 25: 4629
108. Emery T, Hoffer PB (1980) J Nucl Med 21: 935
109. Martell AE, Smith RM (1974) Critical stability constants. Plenum, New York
110. Wochner RD, Adatepe M, Van Amburg A, Potchen EJ (1970) J Lab Clin Med 75: 711
111. Hulkvist U, Westergren G, Hansson U-B, Lewan L (1987) Res Exp Med 187: 131
112. Hosain F, McIntyre PA, Poulose K, Stern HS, Wagner HN (1969) Clin Chim Acta 24: 69
113. Goodwin DA, Goode R, Brown L, Imbornone J (1971) Radiology 100: 175
114. Anghileri LJ, Ottaviani M, Raynaud C (1983) J Nucl Med Allied Sci 27: 17
115. Evans RW, Ogwang W (1988) Biochem Soc Trans 16: 833
116. Beamish MR, Brown EB (1974) Blood 43: 703
117. Beamish MR, Brown EB (1974) Blood 43: 693
118. Grossmann JG, Neu M, Evans RW, Lindley PF, Appel H, Hasnain SS (1993) J Mol Biol 229: 585
119. Battistuzzi G, Calzolai L, Messori L, Sola M (1995) Biochim Biophys Acta 206: 161
120. Jonsson B-A, Strand S-E, Larson BS (1992) J Nucl Med 33: 1825
121. Jones-Wilson TM, Motekaitis RJ, Sun Y, Anderson CJ, Martell AE, Welch MJ (1995) Nucl Med Biol 22: 859
122. Gansow O (1991) Nucl Med Biol 18: 369
123. Martin RB (1994) Acc Chem Res 27: 204
124. Martin RB (1988) Bioinorganic chemistry of aluminum. In: Sigel H (ed) Metal ions in biological systems. Marcel Dekker, New York, p 1
125. Wills MR, Savory J (1989) Crit Rev Clin Lab Sci 27: 59
126. Yokel RA (1994) J Tox Environ Health 41: 131
127. Chadwick DJ, Whelan J (1992) Aluminum in biology and medicine. Wiley, New York
128. Harris WR (1996) Coord Chem Rev 149: 347
129. Gregor JL (1992) Dietary and other sources of aluminum intake. In: Chadwick DJ, Whelan J (eds) Aluminum in biology and medicine. Wiley, New York, p 26
130. Kerr DNS, Ward MK, Ellis HA, Simpson EW, Parkinson IS (1992) Aluminum intoxication in renal disease. In: Chadwick, DJ, Whelan J (eds) Aluminum in biology and medicine. Wiley, New York, p 123
131. Klein GL, Alfrey AC, Miller NL, Sherrard DJ, Hazlet TK, Ament ME, Coburn JW (1982) J Clin Nutr 35: 1425
132. Sedman AB, Wilkening GN, Warady BA, Lum GM, Alfrey AC (1984) J Pediatrics 105: 836
133. Griswold WR, Reznik V, Mendoza SA, Trauner D, Alfrey AC (1983) Pediatrics 71: 56
134. Russo LS, Beale G, Sandroni S, Ballinber WE (1992) J Neurol Neurosurg Psych 55: 697
135. Hawkins NM, Coffey S, Lawson MS, Delves HT (1994) J Ped Gastroent Nutr 19: 377
136. McLachlan DR, Fraser PE, Dalton AJ (1992) Aluminum and the pathogenesis of Alzheimer's disease: a summary of evidence. In: Chadwick, DJ, Whelan J (eds) Aluminum in biology and medicine. Wiley, New York, p 87
137. Wisniewski HM, Wen GY (1992) Aluminum and Alzheimer's disease. In: Chadwick DJ, Whelan J (eds) Aluminum in biology and medicine. Wiley, New York, p 142
138. McLachlan DRC, Lukiw WJ, Kruck TPA (1989) Can J Neurol Sci 16: 490
139. Martyn CN, Osmond C, Edwardson JA, Barker DJP, Harris EC, Lacey RF (1989) Lancet i: 59
140. Flaten TP (1990) Environ Geochem Health 12: 152
141. Wróbel K, Blanco Gonazález E, Snaz-Medel A (1995) Analyst 120: 809

142. Crapper DR, Krishnan SS, Dalton AJ (1973) Science 180: 511
143. Perl DP, Brody AR (1980) Science 208: 297
144. McLachlan DRC, Dalton AJ, Kruck TPA, Bell MY, Smith WL, Kalow W, Andrews DF (1991) Lancet 337: 1304
145. Kubal G, Sadler PJ (1992) J Am Chem Soc 114: 1117
146. Kubal G, Mason AB, Sadler PJ, Tucker A, Woodworth RC (1992) Biochem J 285: 711
147. Forbes WF, Gentleman JF, Maxwell CJ (1995) Exp Gerontol 30: 23
148. Fatemi SJA, Williamson DJ, Moore GR (1992) J Inorg Biochem 46: 35
149. Aramini JM, Germann MW, Vogel HJ (1993) J Am Chem Soc 115: 9750
150. Aramini JM, Vogel HJ (1996) J Mag Res B 110: 182
151. Tang S, MacColl R, Parsons PJ (1995) J Inorg Biochem 60: 175
152. Chasteen ND, Williams J (1981) Biochem J 193: 717
153. Williams J, Chasteen ND, Moreton K (1982) Biochem J 201: 527
154. Guillard O, Pineau A, Baruthlo F, Arnaud J (1988) Clin Chem 34: 1603
155. Keirsse H, Smeyers-Verbeke J, Verbeelen D, Massart DL (1987) Anal Chim Acta 196: 103
156. Rahman H, Skillen AW, Channon SM, Ward MK, Kerr DNS (1985) Clin Chem 31: 1969
157. Blanco Gonzalez E, Perez Parajon J, Alonso JIG, Sanz-Medel A (1989) J Anal At Spectrom 4: 175
158. Cochran M, Patterson D, Neoh S, Stevens B, Mazzachi R (1985) Clin Chem 31: 1314
159. van Ginkel MF, van der Voet GB, van Eijk HG, de Wolff FA (1990) J Clin Chem Clin Biochem 28: 459
160. Trapp GA (1983) Life Sciences 33: 311
161. Sanz-Medel A, Fairman B (1992) Mikrochim Acta 109: 157
162. Moshtaghie AA, Skillen AW (1991) Indian J Pharmacol 23: 75
163. Garcia Alonso JI, Lopez Garcia A, Perez Parajon J, Blanco Gonzalez E, Sanz Medel A, Cannata JB (1990) Clin Chim Acta 189: 69
164. D'Haese PC, Van Landeghem GF, Lamberts LV, De Broe ME (1995) Mikrochim Acta 120: 83
165. King SW, Wills MR, Savory J (1979) Res Commun Chem Path Pharm 26: 161
166. King SW, Savory J, Wills MR (1982) Ann Clin Lab Sci 12: 143
167. Perez Parajon J, Blanco Gonzalez E, Cannata JB, Sanz-Medel A (1989) Trace Elem Med 6: 41
168. Leung FY, Hodsman AB, Muirhead N, Henderson AR (1985) Clin Chem 31: 20
169. Ackrill P, Day JP (1993) Contr Nephrol 102: 125
170. Leung FY, Niblock AE, Bradley C, Henderson AR (1988) Sci Tot Environ 71: 49
171. Bertholf RL, Savory J, Wills MR (1986) Trace Elem Med 3: 157
172. Favarato M, Mizzen CA, Sutherland MK, Krishnan B, Kruck TPA, McLachlan DRC (1992) Clin Chim Acta 207: 41
173. Khalil-Manesh F, Agness C, Gonick HC (1989) Nephron 52: 329
174. Favarato M, Mizzen CA, McLachlan DR (1992) J Chromatogr 576: 271
175. Khalil-Manesh F, Agness C, Gonick HC (1989) Nephron 52: 323
176. Gonick HC and Khalil-Manesh F (1992) Aluminum-binding protein in plasma and brain of dialysis dementia patients. Proceedings of the 2nd International Conference on Aluminum and Health, p 15
177. Gidden H, Holland FF, Klein E (1980) Trans Amer Soc Artificial Internal Organs 26: 133
178. Elliot HL, MacDougall AI, Haase G, Cumming RLC, Gardiner PHE, Fell GS (1978) Lancet 940
179. Graf H, Stummvoll HK, Meisinger V, Kovarik J, Wolf A, Pinggera WF (1981) Kidney Int 19: 587
180. Galbraith LV, Bradley C, Krijnen PMW, Leung FY (1988) Clin Chem 34: 1304
181. Harris WR (1992) Clin Chem 38: 1809
182. Jackson GE (1990) Polyhedron 9: 163
183. Lundin AP, Caruso C, Sass M, Berlyne GM (1978) Clin Res 26: 636 A
184. Dayde S, Filella M, Berthon G (1990) J Inorg Biochem 38: 241

185. Roskams AJ, Connor JR (1990) Proc Natl Acad Sci (USA) 87: 9024
186. Morris CM, Candy JM, Court JA, Whitford CA, Edwardson JA (1987) Biochem Soc Trans 15: 498
187. McGregor SJ, Naves ML, Oria R, Vass JK, Brock JH (1990) Biochem J 272: 377
188. McGregor SJ, Naves ML, Birly AK, Russell NH, Halls D, Junor BJR, Brock JH (1991) Biochim Biophys Acta 1095: 196
189. Mladenovic J (1988) J Clin Invest 81: 1661
190. Bertholf RL, Wills MR, Savory J (1984) Biochem Biophys Res Commun 125: 1020
191. Kasai K, Hori MT, Goodman WG (1991) Am J Physiol 260: E537
192. Cochran M, Chawtur V, Jones ME, Marshall EA (1991) Blood 77: 2347
193. McGregor SJ, Fernández Menéndez MJ, Naves ML, Elloriaga R, Brock JH, Cannata JB (1994) Trace Elem Electrolytes 11: 187
194. Pullen RGL, Candy JM, Morris CM, Taylor G, Keith AB, Edwardson JA (1990) J Neurochem 55: 251
195. Edwardson JA, Candy JM, Ince PG, McArthur FK, Morris CM, Oakley AE, Taylor GA, Bjertness E (1992) Aluminum accumulation, β-amyloid deposition and neurofibrillary changes in the central nervous system. In: Chadwick DJ, Whelan J (eds) Aluminum in biology and medicine. Wiley, New York, p 165
196. Radunovic A, Bradbury MWB (1994) Life Sci Rep 11: 159
197. Candy JM, Morris CM, Oakley AE, Taylor GA, Mountfort SA, Chalker PR, Bishop HE, Ward MK, Bloxham CA, Edwardson JA (1988) Biochem Soc Trans 17: 669
198. O'Hare JA, Murnaghan DJ (1982) N Eng J Med 11: 654
199. Domingo JL, Llobet JM, Gómez M, Corbella J (1986) Res Commun Chem Path Pharm 53: 93
200. Bulman RA (1980) Coord Chem Rev 31: 221
201. Raymond KN, Smith WL (1981) Struct Bonding 43: 159
202. Boocock G, Popplewell DS (1965) Nature 208: 282
203. Stover BJ, Breunger FW, Stevens W (1968) Rad Res 33: 381
204. Turner GA, Taylor DM (1968) Rad Res 36: 22
205. Stevens W, Breunger FW, Stover BJ (1968) Rad Res 33: 490
206. Lehman M, Culig H, Taylor DM (1983) Int J Radiat Biol 44: 65
207. Taylor DM, Seidel A, Planas-Bohne F, Schuppler U, Neü-Muller M, Wirth RE (1987) Inorg Chim Acta 140: 361
208. Duffield JR, Taylor DM (1987) Inorg Chim Acta 140: 365
209. Taylor DM (1993) In: Hay RW, Dilworth JR, and Nolan KB, (eds) Perspectives in bioinorganic chemistry, vol. 2. JAI Press, London, p 139
210. Taylor DM, Kontoghiorghes GJ (1986) Inorg Chim Acta 125: L35
211. Durbin PW, Horovitz MW, Close ER (1972) Health Phys 22: 731
212. Stover BJ, Atherton DR, Keller N, Buster DS (1960) Rad Res 12: 657
213. Peter E, Lehmann M (1981) Int J Radiat Biol 40: 445
214. Harris WR, Carrano CJ, Pecoraro VL, Raymond KN (1981) J Am Chem Soc 103: 2231
215. Harris WR, Madsen LJ (1988) Biochemistry 27: 282
216. Taylor DM, Lehmann M, Planas-Bohne F, Seidel A (1983) Rad Res 95: 339
217. Then GM, Appel H, Duffield J, Taylor DM, Thies W-G (1986) J Inorg Biochem 27: 255
218. Appel H, Duffield J, Taylor DM, Then GM, Thies W-G (1987) J Inorg Biochem 31: 229
219. Chasteen ND, White LK, Campbell RF (1977) Biochemistry 16: 363
220. Durbin PW (1975) Health Phys 29: 495
221. Thomas RG, Healy JW, McInroy JF (1984) Health Phys 46: 839
222. Planas-Bohne F, Duffield J (1988) Int J Radiat Biol 53: 489
223. Planas-Bohne F, Jung W, Neu-Müller M (1985) Int J Radiat Biol 48: 797
224. Planas-Bohne F, Taylor DM, Duffield JR (1983) Cell Biochem Funct 1: 141
225. Schuler F, Csovcsics C, Taylor DM (1987) Int J Radiat Biol 52: 883
226. Planas-Bohne F, Rau W (1990) Human Exp Tox 9: 17
227. Planas-Bohne F, Taylor DM, Duffield JR (1985) Cell Biochem Funct 3: 217
228. Howells GR, Green D, Sontag W (1984) Radiat Environ Biophys 23: 127

229. Smith JM, Miller SC, Jee WSS (1984) Rad Res 99: 324
230. Priest ND, Giannola SJ (1980) Int J Radiat Biol 37: 281
231. Chipperfield AR, Taylor DM (1968) Nature 219: 609
232. Duffield JR, May PM, Williams DR (1984) J Inorg Biochem 20: 199
233. Duffield JR, Raymond DP, Williams DR (1987) Inorg Chim Acta 140: 369
234. Stevens W, Breunger FW, Atherton DR, Smith JM, Taylor GN (1980) Rad Res 83: 109
235. Dang HS, Pullat VR (1993) Health Phys 65: 303
236. Taylor DM, Farrow LC (1987) Nucl Med Biol 14: 27
237. Taylor DM (1972) Health Phys 22: 575
238. Cooper JR, Gowing HS (1981) Int J Radiat Biol 40: 569
239. Guilmette RA, Cohen N, Wrenn ME (1980) Rad Res 81: 100
240. Lloyd RD, Bruenger FW, Mays CW, Atherton DR, Jones CW, Taylor GN, Stevens W, Durbin PW, Jeung N, Jones ES, Kappel MJ, Raymond KN, Weitl FW (1984) Rad Res 99: 106
241. O'Hara P, Yeh SM, Meares CF, Bersohn R (1981) Biochemistry 20: 4704
242. Yeh SM, Meares CF (1980) Biochemistry 19: 5057
243. Luk CK (1971) Biochemistry 10: 2838
244. Abdollahi S, Harris WR, Riehl JP (1996) J Phys Chem 100: 1950
245. Evans CH (1990) Biochemistry of the lanthanides. Plenum Press, New York
246. Schomäcker K, Franke W-G, Henke E, Fromm W-D, Maka G, Beyer G-J (1986) Eur J Nucl Med 11: 345
247. Jarvis NV, Wagener JM, Jackson GE (1995) J Chem Soc Dalton 1411
248. Butler A, Carrano CJ (1991) Coord Chem Rev 109: 61
249. Smith CA, Ainscough EW, Brodie AM (1995) J Chem Soc Dalton 1121
250. Cannon JC, Chasteen ND (1975) Biochemistry 14: 4573
251. Campbell RF, Chasteen ND (1977) J Biol Chem 252: 5996
252. Chasteen ND, Grady JK, Holloway CE (1986) Inorg Chem 25: 2754
253. Bonadies JA, Carrano CJ (1986) J Am Chem Soc 108: 4088
254. Butler A, Eckert H (1989) J Am Chem Soc 111: 2802
255. Chasteen ND, Lord EM, Thompson HJ, Grady JK (1986) Biochim Biophys Acta 884: 84
256. Bertini I, Luchinat C, Messori L (1985) J Inorg Biochem 25: 57
257. Bertini I, Canti G, Luchinat C (1982) Inorg Chim Acta 67: L21
258. Neves A, Ceccatto AS, Erasmus-Buhr C, Gehring S, Haase W, Paulus H, Nascimento OR, Batista AA (1993) J Chem Soc Chem Comm 1782
259. Chasteen ND (1983) Struct Bonding 53: 107
260. Sabbioni E, Marafante E, Amantini L, Ubertalli L (1978) J Inorg Biochem 8: 503
261. Harris WR, Friedman SB, Silberman D (1984) J Inorg Biochem 20: 157
262. Macara NG, Kustin K, Cantley LC (1980) Biochim Biophys Acta 629: 95
263. Cantley LC, Resh MD, Guidotti G (1978) Nature 272: 552
264. Cantley LC, Aisen P (1979) J Biol Chem 254: 1781
265. Sabbioni E, Marafante E (1978) J Inorg Biochem 9: 389
266. Aisen P, Aasa R, Redfield AG (1969) J Biol Chem 244: 4628
267. Ainscough EW, Brodie AM, Plowman JE (1979) Inorg Chim Acta 33: 149
268. Patch MG, Simolo KP, Carrano CJ (1982) Inorg Chem 21: 2972
269. Tomimatsu Y, Kint S, Scherer JR (1976) Biochemistry 22: 4918
270. Davies G (1969) Coord Chem Rev 4: 199
271. Mildvan AS, Cohn M (1963) Biochemistry 5: 910
272. Reed GH, Ray WJ (1971) Biochemistry 10: 3190
273. Critchfield JW, Keen CL (1992) Metabolism 41: 1087
274. Davidsson L, Lönnerdal B, Sandström B, Kunz C, Keen CL (1989) J Nutr 119: 1461
275. Panić B (1967) Acta Vet Scand 8: 228
276. Keefer RC, Barak AJ, Boyett JD (1970) Biochim Biophys Acta 221: 390
277. Hancock RGV, Evans DJR, Fritze K (1973) Biochim Biophys Acta 320: 486
278. Gibbons RA, Dixon SN, Hallis DK, Russell AM, Sansom BF, Symonds HW (1976) Biochim Biophys Acta 444: 1

279. Scheuhammer AM, Cherian MG (1985) Biochim Biophys Acta 840: 163
280. Friedberg F (1975) FEBS Let 59: 140
281. Nandenkar AKN, Nurse CE, Friedberg F (1973) Int J Peptide Prot Res 5: 279
282. Cotzias GC, Bertinchamps AJ (1960) J Clin Invest 39: 979
283. Aschner M, Aschner JL (1990) Brain Res Bull 24: 857
284. Keen CL, Lönnerdal B, Hurley LS (1984) Manganese. In: Frieden E (ed) Biochemistry of the essential ultratrace elements. Plenum, New York, p 89
285. Lafond JL, Duron C, Favier A (1985) Nutr Res Suppl. I: 121
286. Chapman BE, MacDermott TE, O'Sullivan WJ (1973) J Inorg Biochem 3: 27
287. Rabin O, Hegedus L, Bourre J-M, Smith Q (1993) J Neurochem 61: 509
288. Schramm VL, Brandt M (1986) Federation Proc 45: 2817
289. Aschner M, Gannon M (1994) Brain Res Bull 33: 345
290. Sturrock A, Alexander J, Lamb J, Craven CM, Kaplan J (1990) J Biol Chem 265: 3139
291. Suárez N, Eriksson H (1993) J Neurochem 61: 127
292. Harris WR, Keen C (1989) J Nutr 119: 1677

Photophysics and Photochemistry of Metal Polypyridyl and Related Complexes with Nucleic Acids

Cécile Moucheron[1], Andrée Kirsch-De Mesmaeker[1,#] and John M. Kelly[2,*]

[1] Université Libre de Bruxelles, Chimie Organique Physique, CP 160/08,
50 Avenue F.D.Roosevelt, 1050 Bruxelles, Belgique
Email: AKIRSCH@ULB.AC.BE

[2] Trinity College Dublin, Chemistry Department, Dublin 2, Ireland
Email: JMKELLY@TCD.IE

The effect of DNA or nucleotides on the photophysical properties of metal polypyridyl complexes (especially those of ruthenium, osmium, rhodium, rhenium and copper) is reviewed. On binding to DNA the excited states of certain complexes show reduced non-radiative decay and protection from oxygen or water, while in other cases the excited state is quenched because of electron transfer from the nucleobases. Examples of photochemical induction of strand breaks, light-induced photooxidative damage and the formation of covalent photoadducts between metal complexes and nucleobases are discussed. Consideration is given to approaches being made to improve the affinity and specificity of interaction so as to generate molecular DNA photoprobes for particular DNA sequences or morphologies (e.g. cruciform, Z-DNA, single stranded DNA).

* Corresponding author
# Director of Research at the FNRS (Belgium)

Structure and Bonding, Vol. 92
© Springer Verlag Berlin Heidelberg 1998

1
Introduction

DNA is a biological polyelectrolyte which has been extensively studied in chemistry and biology. Although the knowledge of nucleic acids has progressed tremendously, there are still aspects which are not fully understood. The polymorphism of DNA has been clearly evidenced, but there is an important need in methods and tools in order to determine local topologies or structures of DNA and relate them with the function of nucleic acids. As demonstrated in this review, metal polypyridyl complexes are possible molecular tools for the study of nucleic acids. We first recall very briefly the basic characteristics of polynucleotides and give some examples of their polymorphism. Afterwards we discuss the advantages of designing metal complexes not only to help characterise DNA, but also to use them as anti-cancer or anti-viral drugs.

The primary structure of DNA is composed of a string of nucleosides, each joined to its two neighbours through phosphodiester linkages so that each 5'-hydroxyl group is linked through a phosphate to a 3'-hydroxyl group (Fig. 1).

The regular DNA secondary structures are currently well characterized. Thus A-, B- and Z-DNA constitute the major regular DNA forms (Fig. 2) where

Major Groove

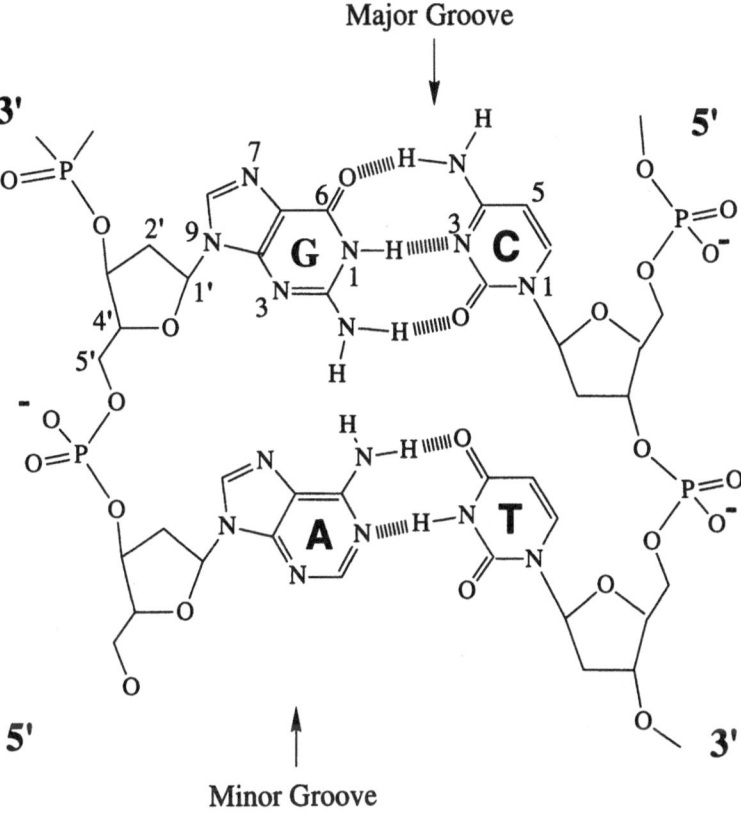

Minor Groove

Fig. 1. Primary structure of DNA [adapted from: Dickerson RE (1983) Scientific American 6: 86]

the base pair formation is based on the Watson Crick model. B-DNA is the most common structure, with a right-handed helix, ten base-pairs per turn, quasi-perpendicular to the main helix axis. The wide major groove and narrow minor groove are both well solvated by water molecules. A-DNA which is also a right-handed helix, has 11 residues per turn, with tilting of the bases of 20°. This helix is fairly stiff, with a deep and narrow major groove and a broad and shallow minor groove.

Z-DNA is a left-handed helix stabilized by high concentrations of $MgCl_2$, NaCl, or ethanol and most favoured for alternating G-C sequences. The phosphate backbone has a zig-zag appearance. The minor groove is narrow and very deep and the major groove is very shallow.

It is known that DNA not only exists in the form of these regular double helix structures described above, but may also exist as single stranded forms or with single stranded portions in double stranded DNA, or as circular form. For example, calf thymus DNA (CT-DNA) when denatured, contains around 60% of double stranded helix and 40% of single stranded portions. Some

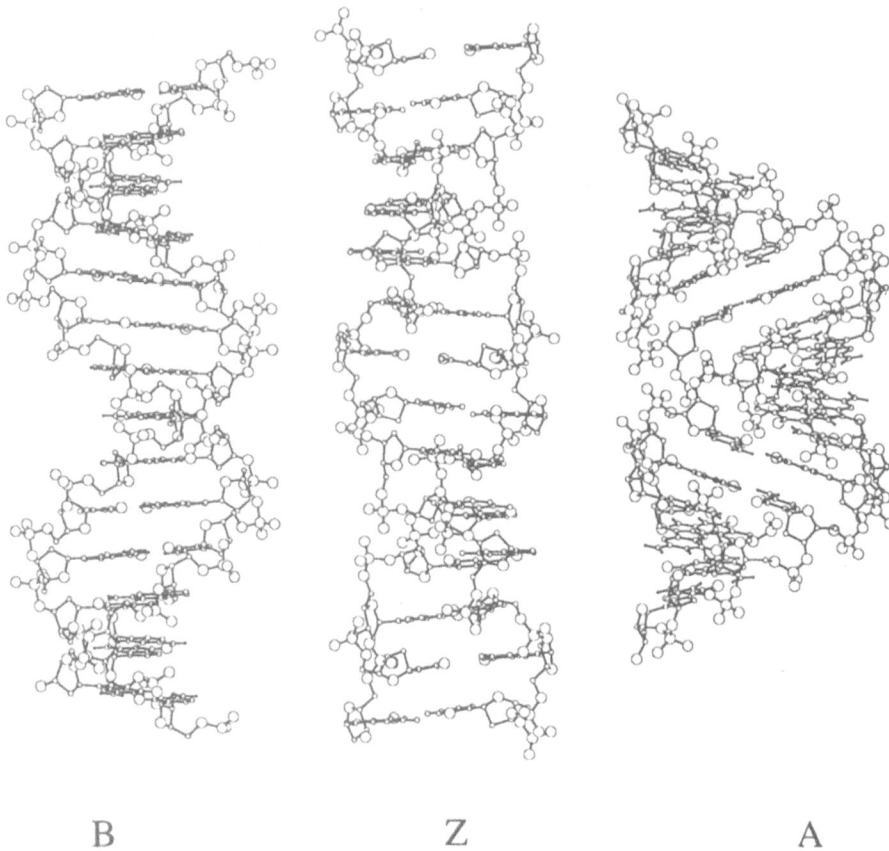

B   Z   A

Fig. 2. Representation of B-, Z-, and A-DNA. Graphics display was performed with QUANTA (MSI) on Silicon Graphics 4D25G Personal Iris workstation, hard copies obtained with Tektronix RGBII. LEDSS, Université J Fourier, Grenoble, France

bacterial DNA exists as a double-stranded closed circle and it has also been shown that the replicative form of bacteriophage ϕX174 DNA is in this form. DNA viruses may have either single- or double-helical circular DNA. Triple helices can also be formed by polynucleotides (e.g. poly(dG) binding to poly(dG).poly(dC) or poly(dT) binding to poly(dA).poly(dT)). The hydrogen bonding interaction involved in triple-helix involve Hoogsteen as well as Watson-Crick hydrogen bonds [1].

DNA can also be present as irregular forms which play important roles for the DNA functions. For example B-DNA, in contrast to the A- and Z-forms, is rather flexible and thus capable of making small adjustments in local helix structure in response to particular base sequences. It is probable that these local variations of structure help proteins to read and recognize one base sequence in preference to another. The particular changes in local bases sequences are translated by conformational responses in the backbone. For example some particular repeats of sequences give rise to bent DNA such as the repeat of tracts in homopolymers which contain $(dA)_5.(dT)_5$ sections occurring in phase in each turn of 10-fold or 11-fold helix.

Other types of DNA structures are also formed. The reader should refer to standard textbooks [2, 3] for a description of these different possible structures. As an example, we mention *cruciform structures* as they are the best explored species of unusual DNA structure. They have been characterized for many inverted repeat sequences in plasmids and in phages:

$$\underline{\text{TGTGGATCCGGTACCAGAATTCTGGTACCGGATCCTCT}}$$
$$\underline{\text{ACACCTAGGCCATGGTCTTAAGACCATGGCCTAGGAGA}}$$

In such a DNA structure (when torsionally underwound) stem-loop structures are formed by intrastrand pairing (Fig. 3). The loops are sensitive to single-strand nucleases. Figure 3 shows different pathways for formation of cruciform DNA [2].

It is clear nowadays that these different forms and structures of DNA play important biological roles. Therefore it would be particularly attractive to have at one's disposal a battery of molecular tools so as to detect, in vitro and in

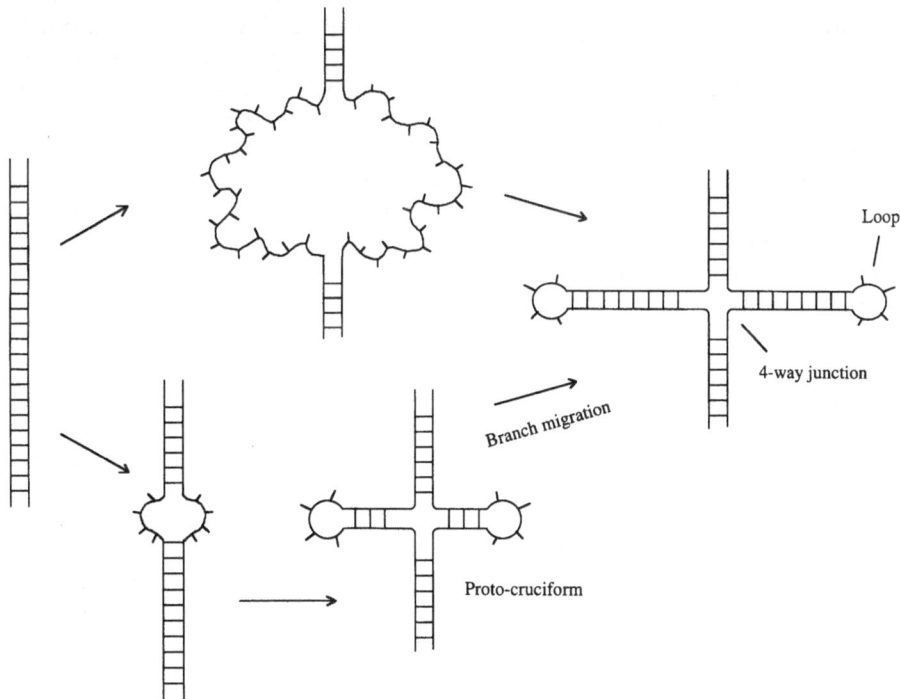

Fig. 3. Structures of a cruciform DNA and alternative pathways for its formation (base-paired sections are helical throughout) [adapted from: Blackburn GM and Gait MJ (1992) Nucleic acids in chemistry and biology. Oxford University Press, Oxford]

vivo, appearance of one of these particular structures at a certain stage of development of the system. This would allow the correlation of a particular structure with a particular function. For example the design of luminescent probes or photoreagents of particular sites of DNA, such as single-stranded portions characteristic of the numerous steps preceding formation of cruciform DNA (Fig. 3), or any other site, would be particularly useful. Indeed these probes might recognize these sites at a particular moment during the function of DNA such as the replication or transcription mechanism.

Some of the molecular tools developed at present in this area of DNA study, which will be discussed in this review, are based on metal complexes. The useful properties of these metal compounds are mainly their luminescence and photoreactivity in the presence of DNA, as outlined below. The development of such molecular tools could also lead to a new generation of potential anti-cancer agents, as has been the case of the Pt compounds. In contrast to these Pt drugs, which are active in the dark, the transition metal complexes which are presented here would be active only under illumination. Formation of some photoproducts or photoadducts of these metal complexes with DNA could indeed inhibit or interfere with the normal functions of DNA, for example by inhibiting the enzymes operating on the DNA or RNA.

It has been shown that organic molecules and complexes can bind to DNA in a number of ways [4]. As the interacting molecules are usually positively charged, they can of course interact with the negatively charged phosphate backbone and this can lead in some cases to aggregation of the molecules through stacking. Molecules may also enter the minor or major DNA grooves (see Fig. 2). These include antitumour agents such as netropsin and various dyes such as Hoechst 33,258. In many cases there are specific hydrogen bonding and van der Waals interactions between the molecules and DNA. Another possibility is the intercalation of a planar portion of the molecule into the stacking of bases. This causes unwinding and lengthening of the helix. Complex molecules often bind to DNA by combination of these modes of bindings.

2
Why Metal Complexes with DNA?

For many years, the interaction and (photo)reactions of polypyridyl transition metal complexes with biomolecules such as DNA have been the focus of intensive research. The interest in such coordination compounds originates from their particular structures and properties.

The DNA polyanion interacts strongly with positively charged transition metal complexes that are stable in aqueous medium. The binding constants reported in the literature for metal complexes with CT-DNA range from 1×10^3 to 6×10^7 M^{-1} [5].

The three dimensional structure of these rigid complexes, which are coordinatively saturated, makes them very attractive as spatial probes for the nucleic acids. These complexes are able to recognise a complementary shape in the different DNA binding sites, which results in a self-assembling. As nucleic acids are chiral, chirality in the design of a complex may be used advantageously for making the interaction enantioselective. For example, the Δ-[Ru(phen)$_3$]$^{2+}$ enantiomer is preferred with B-DNA relatively to the Λ-enantiomer (a report on enantioselectivity in binding of metal complexes to DNA is presented in [5]).

By varying the metal centre, it is possible to change the geometry of the complex (square planar, tetrahedral, octahedral, ...) and also, to modify its photophysical properties (see below). Changes in the chelating ligands (Fig. 4) may also modify the photophysical properties [6–8] and the interaction with nucleic acids.

For example, [Ru(phen)$_2$(DPPZ)]$^{2+}$ does not emit in water whereas the parent complex, [Ru(phen)$_3$]$^{2+}$, displays luminescence in the same conditions. By binding to nucleic acids, the luminescence of [Ru(phen)$_2$(DPPZ)]$^{2+}$ is switched-on, reflecting an efficient protection of the DPPZ ligand from the solvent quenching. For this complex, an intercalative mode of interaction has been unambiguously demonstrated (see further), in contrast to [Ru(phen)$_3$]$^{2+}$. Thus, the addition of a quinoxaline subunit to a phenanthroline motif clearly results in different photophysical properties in solution and with DNA.

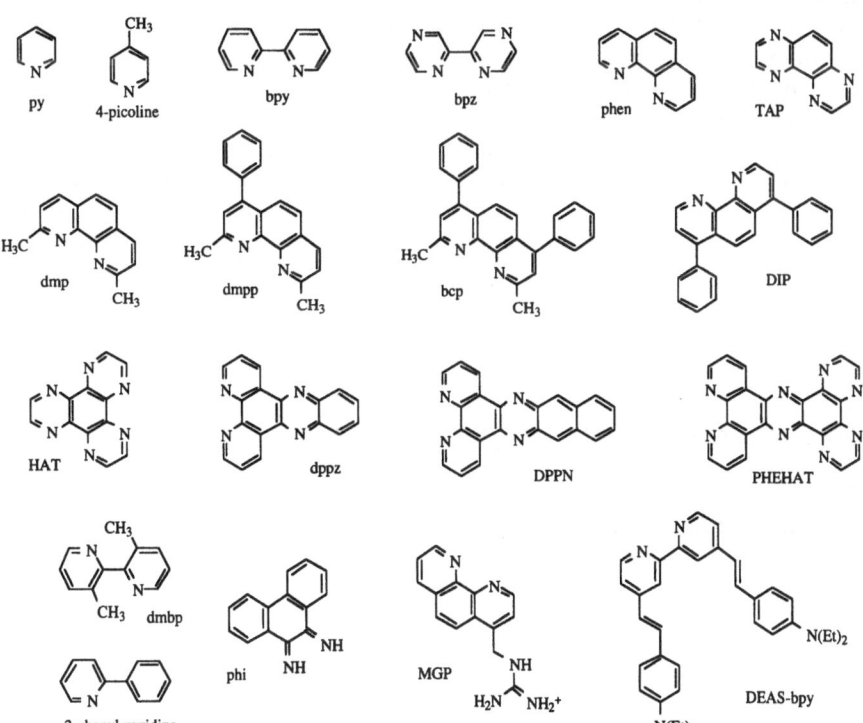

Fig. 4. Ligands (and their abbreviation) corresponding to the complexes cited in the review

Thanks to the interesting photophysical properties of the metal complexes, these compounds may also behave as sensitive spectroscopic probes to examine different DNAs; the intensive research on metal complexes with phenanthroline ligands for example is the best illustration of this application ($[Ru(phen)_3]^{2+}$ [9, 10], $[Os(phen)_3]^{2+}$ [11], $[Rh(phen)_3]^{3+}$ [12], $[Fe(phen)_3]^{2+}$ [13, 14], $[Co(phen)_3]^{2+}$ [15, 16], $[Cu(phen)_2]^+$ [17, 18], ...).

Metal complexes generally exhibit intense optical transitions in the visible and most of them are able to luminesce at room temperature. These transitions are very much affected by the microenvironment surrounding the complex. Binding to the DNA double helix (without photoreaction) generally results in an absorption hypochromicity and a luminescence enhancement. These changes may be used for monitoring the interaction with the polynucleotides.

Finally, the transition metal complexes are of course better oxidizing and reducing agents in the excited state than in the ground state. Changes of ligand will also modulate the photoredox properties of the complex.

The ensemble of these properties of the metal complexes make them extremely useful in the context of DNA biochemistry and biomedicine. Thus, they can be used as luminescent reporters for DNA instead of using radioactive markers [19, 20]. Moreover when they induce photo-cleavages of the DNA backbone, they can be applied for footprinting or mapping experiments. Metal complexes which produce DNA-photoadducts may also mark irreversibly a specific site of DNA; those which photocleave DNA may also be regarded as artificial endonucleases [10]. Finally, some of the metal complexes could find applications in photochemotherapy [21, 22] or in the field of new chemotherapeutic agents. Indeed, as much work in this area concerns Pt(II) complexes [23–26], there is a need to find new anti-tumour drugs that are less toxic than the Pt compounds. In addition, a drug active exclusively under illumination might also be used advantageously for localized tumour treatments. Obviously, all these aspects cannot be treated in this review and only the interactions and photoreactions of polypyridyl complexes (especially those of Ru(II), Os(II), Re(I), Rh(III) and Cu(I)) with DNA will be discussed.

3
Photophysics of the Different Metal Complexes

3.1
The Ru(II) and Os(II) Complexes

The photophysical behaviour of Ru(II) tris(polypyridyl) complexes has been extensively studied during the last two decades and several reviews cover this subject [8, 27]. It has been mentioned that more than 200 chelating polypyridine ligands (LL) have been used in the Ru(II) chemistry and that numerous homoleptic ($[Ru(LL)_3]^{2+}$) and heteroleptic ($[Ru(LL)_2(LL')]^{2+}$ or $[Ru(LL)(LL')(LL'')]^{2+}$) complexes have been synthesized; they exhibit a wide range of redox potentials (E_{ox} extending from 1.0 to 2.1 V/SCE and E_{red} from

−1.5 to −0.6 V/SCE) and show quite different excited state properties (λ_{max} of emission ranging from 610 to 750 nm, excited state lifetimes varying from <10 ns to >2 μs and ϕ_{em} from 10^{-3} to 0.3) [6, 28, 29].

Irradiation of these d^6 octahedral Ru(II) and Os(II) complexes gives rise to the dπ(Ru) → π*(ligand) ^{1}MLCT (Metal to Ligand Charge Transfer) excited state which, by intersystem crossing ($\phi_{ISC} = 1$), populates the triplet MLCT state [30, 31] (Fig. 5).

From that state, the complex may deactivate by radiative or radiationless processes to the ground state, or reach the upper triplet MC (Metal Centred) state by thermal activation [32, 33]. This latter process depends obviously on the energy difference between the two triplet states (^{3}MC and ^{3}MLCT). Normally for the Os(II) complexes, the MC states are sufficiently high in energy so that they are not populated [34, 35]. Two different photochemical behaviours are governed by these two states. Generally, from the ^{3}MLCT state, the complex reacts by redox processes whereas from the ^{3}MC state, dechelation or ligand dissociation takes place, which results in the rupture of a Ru-nitrogen bond. Indeed, the distortion of the complex in the ^{3}MC state [36, 37] facilitates the break of the Ru-N bond and leads to a five-coordinate system. This unstable species can fill its empty coordination site either by regenerating the initial complex or by adding a monodentate ligand present in the medium (substitution process). The resulting compound can regenerate the initial product or give rise to a second Ru-N break, with the loss of the bidentate ligand (photodechelation process). The reaction pathway depends on the presence of reductants or oxidants in the medium and on the relative energy of the MLCT and MC states which is controlled to a certain degree by the choice of the ligands. This is illustrated with a whole series of Ru(II) complexes based on the HAT (or TAP) ligand, which, because of its very important π-acceptor character, corresponds to the electron-accepting ligand of the lowest MLCT excited state (Fig. 6).

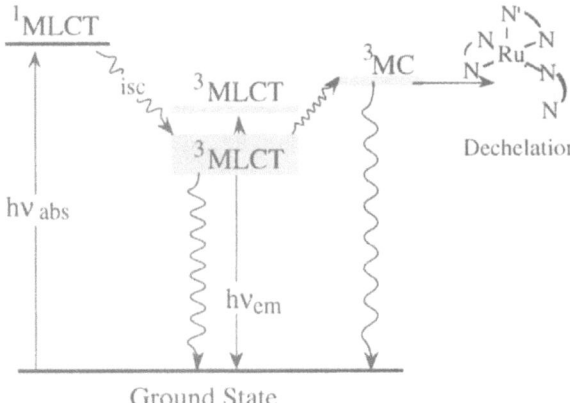

Fig. 5. Diagram of the excited states levels of Ru(II) polypyridyl complexes. $h\nu_{abs}$ = absorption, $h\nu_{em}$ = luminescence, ISC = intersystem crossing

Detailed studies of the photophysical mechanisms controlling the luminescence lifetimes and intensities [29, 38] show that the energy level of the ^{3}MLCT state increases by increasing the number of HAT (or TAP) in the complex, whereas the level of ^{3}MC remains at more or less the same energy. For $Ru(HAT)_3^{2+}$ (or $Ru(TAP)_3^{2+}$), for example, containing the greatest number of HAT (TAP) ligands, both excited states (MLCT and MC) are populated and, consequently, both photoredox and photodissociation reactions take place. Moreover, according to their high oxidation power in the excited state, $Ru(HAT)_3^{2+}$ (or $Ru(TAP)_3^{2+}$) can abstract an electron from rather poor reductants. The relative contribution of the reductive quenching and dechelation depends on the oxidation potential and concentration of the reducing agent. In contrast, for complexes containing only one HAT ligand (or one TAP), the ^{3}MC is not accessible at room temperature and, consequently, only redox processes take place from the ^{3}MLCT state.

The tuning of the excited state reduction potentials of the Ru(II) complexes allows the control of the photophysical and photochemical behaviours of these compounds in the presence of the four DNA bases, as we will discuss further.

As mentioned above, the photophysics of the Os(II) complexes is similar to that of the Ru(II) compounds [30] except that the energy difference between the ^{3}MLCT and ^{3}MC states is much higher so that the ^{3}MC state is not populated at room temperature. The consequence of this slight change in the

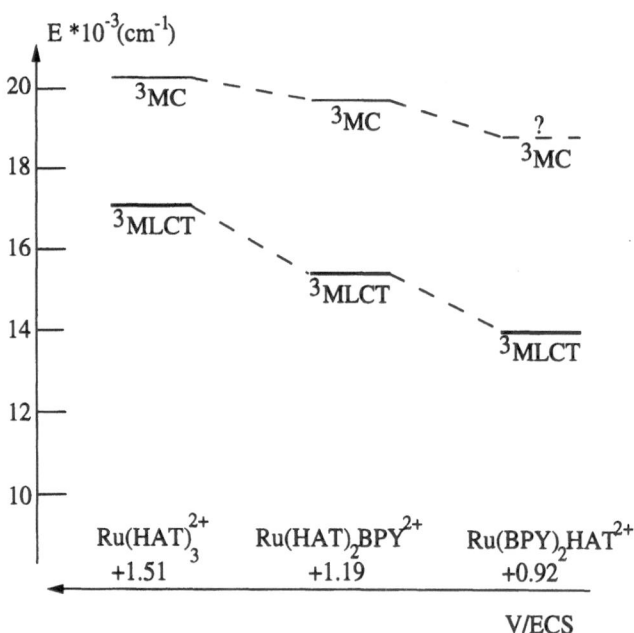

Fig. 6. Energies of the ^{3}MLCT and ^{3}MC states in the series of HAT complexes and corresponding E^0 (Ru^{2+*}/Ru^{1+}) [adapted from: Jacquet L, Kirsch-De Mesmaeker A (1992) J Chem Soc Faraday Trans 88: 2471]

photophysics is that the tris-chelated Os(II) complexes are photostable. Moreover, as Os is much heavier than Ru, the spin orbit coupling is more important and consequently the decay from the triplet to the ground state singlet is favoured. This, of course, shortens the luminescence lifetimes of the Os(II) complexes. This tendency is illustrated when the Ru centre is replaced by an Os centre in complexes such as $Ru/Os(bpy)_3^{2+}$, $Ru/Os(phen)_3^{2+}$ [30, 32, 33, 39]. However, when Ru is replaced by Os in $Ru/Os(TAP)_3^{2+}$ [39], unexpectedly, an increase of emission lifetime is observed. This is attributed to the fact that in $Ru(TAP)_3^{2+}$ the emission lifetime is totally controlled by the activation to the ^3MC, which results in a very short luminescence lifetime. In contrast, as in $Os(TAP)_3^{2+*}$ the thermal activation to the ^3MC has disappeared, the lifetime of this excited state is controlled by k_r and k_{nr}, which results in a longer excited state lifetime than for $Ru(TAP)_3^{2+*}$.

Finally, it should be mentioned that the ^3MLCT state of Ru(II) or Os(II) complexes containing specific extended ligands, such as DPPZ or PHEHAT, is quenched by water to such an extent that no luminescence is detectable. This particular property is very important and is, in part, responsible for the numerous works published on these complexes in the presence of polynucle-otides in aqueous medium [40–45]. It may be noted that $Ru(TAP)_2(DPPZ)^{2+}$, in contrast, displays luminescence in water; this is due to the fact that the lowest lying excited state corresponds to the $d\pi(Ru) \rightarrow \pi^*(TAP)$ transition in $Ru(TAP)_2(DPPZ)^{2+}$ and does not involve the DPPZ ligand as in $Ru(phen)_2$-$(DPPZ)^{2+}$ [46].

3.2
The Re(I) Complexes

An investigation of the photophysical properties of several d^6 Re(I) complexes of the type $[fac\text{-}(LL')Re(I)(CO)_3L]^{1+}$ (where LL' is a bidentate diimine ligand such as 2,2'-bipyridine and L is a monodentate one such as pyridine) has been the focus of recent research [47–52].

Except for a few cases (see below), the absorption spectra are characterized by a broad structureless absorption band in the near UV with a less intense absorption band (or tail) in the visible [48, 49, 53–57], this latter being attributed to the $d\pi(Re) \rightarrow \pi^*(LL')$ MLCT absorption. The emission observed at room temperature originates from the energetically lowest lying $d\pi(Re) \rightarrow \pi^*(LL')$ ^3MLCT excited state [58–61]. These states are characterized by broad featureless luminescence (ϕ_{em} ranging from 0.01 to 0.3) with λ_{max} ranging from 500 to 650 nm and excited state lifetimes in the range 20 ns—2 μs (Fig. 7).

However, examples are also mentioned in the literature, where the lowest lying excited state corresponds to an intraligand $^3\pi,\pi^*$ transition (^3IL) centred on the diimine ligand ($^3IL_{LL'}$) [49, 62–65] (such as for example in (DPPZ)-Re or (DEAS-bpy)-Re complexes). In those cases, the populations of both MLCT and intraligand ($IL_{LL'}$) excited singlet states are followed by an intersystem crossing to the triplet MLCT and triplet IL states. After internal conversion it is exclusively the lowest lying ^3IL state which governs the luminescence.

Therefore, the emission appears as a structured band with well-resolved vibronic components and with low deactivation rate constants [64, 65, 49].

For $[fac\text{-}(DPPZ)Re(I)(CO)_3L]^{1+}$ (with L = 4-picoline), Schanze et al. have demonstrated the existence of such a lowest lying $^3IL_{DPPZ}$ excited state and observed a quenching by water. This is in contrast to the behaviour of the ^3IL state of the DPPZ free ligand which exhibits the same lifetime in MeOH and MeOH/H$_2$O (1/1), demonstrating that H$_2$O does not quench the DPPZ triplet state. Therefore, according to Schanze and coworkers, in water the energy of the ^3MLCT state in this complex would be slightly above that of the $^3IL_{DPPZ}$ state which would become depopulated by thermal activation to this ^3MLCT state. Thus this energetically close lying ^3MLCT state would provide a faster deactivation channel of the ^3IL state in H$_2$O than in an organic solvent (Fig. 8).

The photophysical behaviour observed for another Re(I) complex with a DPPN ligand, is also in agreement with this model [66].

In conclusion, the ^3MLCT state in these Re(I) complexes would influence the photophysics of the $^3IL_{DPPZ}$ state, exactly as the ^3MC state influences the photophysics of the ^3MLCT state in the Ru(II) polypyridyl complexes [38, 67–69].

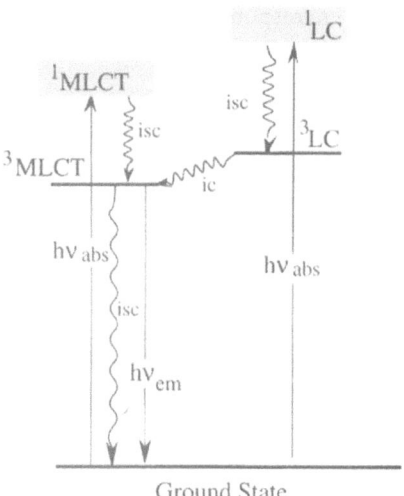

Fig. 7. Photophysical scheme admitted for most of Re(I) polypyridyl complexes. $h\nu_{abs}$ = absorption, $h\nu_{em}$ = luminescence, ISC = intersystem crossing, IC = internal conversion

3.3
The Cu(I) Complexes

The Cu(I) complexes, which are d^{10} systems, have been studied for many years. An MLCT transition is observed in the visible (absorption maxima ranging from 440 to 460 nm) for complexes involving ligands with low-lying

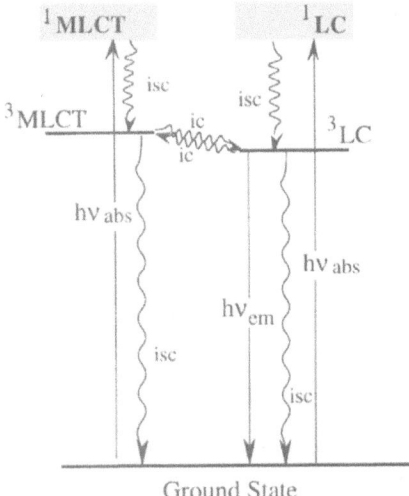

Fig. 8. Photophysical scheme for (DPPZ)- type Re(I) polypyridyl complexes [adapted from: Stoeffler HD, Thornton NB, Temkin SL, Schanze KS (1995) J Am Chem Soc 117: 7119]. $h\nu_{abs}$ = absorption, $h\nu_{em}$ = luminescence, ISC = intersystem crossing, IC=internal conversion

unoccupied π^* orbitals such as 2,9-dimethyl-1,10-phenanthroline (dmp), or 2,2'-bipyridine derivatives [70–72].

Emission is observed in dichloromethane or chloroform (λ_{max} ranging from 670 to 750 nm) with a very small luminescence quantum yield, $\phi_{em} \approx 2 \times 10^{-4}$ [73, 74], and with lifetimes ranging from 10 to 100 ns. The luminescence originates from at least two thermally equilibrated excited states (separated by ≈ 1800 cm^{-1}). The lowest energy of these two excited states corresponds to a triplet state associated with the configuration $d^9\pi^{*1}$ [74, 75], whereas the other state has a substantially greater radiative rate constant and is assumed to correspond to a singlet state [74, 76] (Fig. 9). Therefore, the emission intensity decreases when the temperature is lowered, due to a less efficient population of the singlet state.

In solvents such as water, methanol, ethanol, acetonitrile or acetone, the emission is extremely weak or even undetectable at room temperature [73, 77]. This quenching by basic solvents occurs via exciplex formation favored by a change of geometry from the Cu(I) into a Cu(II) centre. In the excited state indeed, the copper centre can be regarded as a Cu(II). Although immediately after excitation, the coordination geometry corresponds to the ground state Cu(I) system (pseudo-tetrahedral), it transforms rapidly into the geometry which is characteristic of a Cu(II), with a coordination number of 5 or 6. In terms of potential energy surfaces, whereas the ground state surface (for Cu(I) + a solvent molecule) is assumed to be purely dissociative, in the excited state, when the solvent molecule is close to the complex, the charge transfer interaction (resulting from the overlap between the donor orbital of the quencher and an unfilled orbital of the complex) lowers the total energy

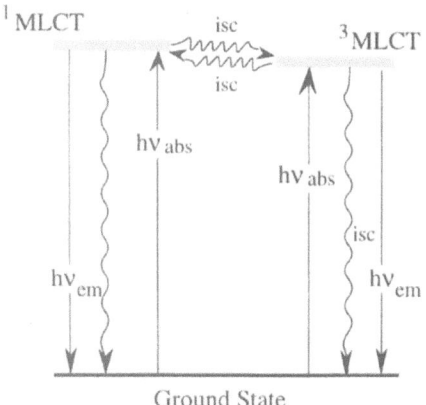

Fig. 9. Photophysical scheme for bis(polypyridyl) Cu(I) complexes. $h\nu_{abs}$ = absorption, $h\nu_{em}$ = luminescence, ISC = intersystem crossing, IC = internal conversion

(Fig. 10). Consequently, radiationless deactivation to the ground state is strongly facilitated, in accord with the energy gap law [78–80].

[*Cu...Q] corresponds to a five coordinate adduct with a new coordinating bond between the copper centre and the solvent. Studies of the temperature [80] and pressure [81] dependence of this quenching, also support this mechanism. Similarly, McMillin reported luminescence quenching by exciplex formation from the ion pair state [78, 82, 83]. Other studies have evidenced the dramatic influence of substituents in the 2,9-positions of the phenanthroline on the photophysical properties of the corresponding complexes [84, 85]. Thus the introduction of bulky substituents in these positions results in an extension of the MLCT absorption band into the red accompanied by a substantial absorptivity decrease [71, 86]. In emission, replacing methyl groups in the 2,9-positions by phenyl groups results in lifetime and luminescence enhancements, and in the appearance of luminescence in polar media such as ethanol or water [84, 86–88]. This emission originates from an efficient protection of the central metal from associative quenching interactions involving the polar solvent.

Studies performed with phenanthroline and tetraazaphenanthrene complexes have shown that substitution in the α-position of the chelating nitrogens of the ligands also influences the electrochemical behaviour of the compounds. Indeed, these complexes, because of the constraint introduced by the ligands, are forced to adopt a tetrahedral-like geometry characteristic of Cu(I), which prevents a square planar structure characteristic of Cu(II). Consequently, ortho-substituted phenanthroline or tetraazaphenanthrene complexes are less easily oxidized than the corresponding unsubstituted compounds, which are in contrast good reducing agents (Cu(phen)$_2^{2+/1+}$: +0.08 V/SCE and Cu(dmp)$_2^{2+/1+}$: +0.58 V/SCE [89]; Cu(TAP)$_2^{2+/1+}$: +0.76 V/SCE and Cu(dmTAP)$_2^{2+/1+}$: +0.97 V/SCE [85]). These differences of redox properties

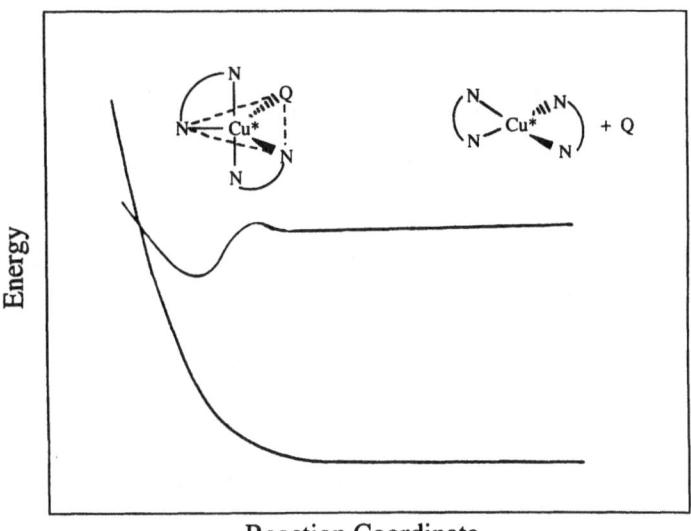

Fig. 10. Potential energy profiles for exciplex formation [adapted from: Stacy EM, McMillin DR (1990) Inorg Chem 29: 393].The ground-state surface is assumed to be repulsive, but the excited state forms a five-coordinate adduct. The minimum in the excited state surface corresponds to the exciplex

will be reflected in the reactivity of these complexes vs the nucleotides, as discussed below.

3.4
The Rh(III) Complexes

Two types of polypyridyl rhodium(III) complexes are generally studied in the literature: the tris(polypyridyl) and the bis-chelated complexes.

3.4.1
Tris-Polypyridyl Complexes

The absorption spectra of these octahedral species are dominated by intraligand (IL) transitions in the UV region, while no absorption is observed in the visible part of the spectra, making tris(polypyridyl) complexes practically colorless [8].

In emission, the low-temperature spectra of these complexes are well-structured and indicate that the luminescence originates from $\pi \to \pi^*$ ligand centred transitions [90–92]. Recent low-temperature investigations [93] of $Rh(bpy)_3^{3+*}$ and deuterated derivatives by high resolution spectroscopy lead to the conclusion that the lowest triplet state also contains weak MC and MLCT contributions.

For $Rh(dmbp)_3^{3+}$ (dmbp = 3,3'-dimethyl-2,2'-bipyridine) at low temperature, an emission from a second excited state, slightly higher in energy than the LC one, has also been detected and assigned to a metal centred triplet state [94]. At 77 K, the emission is dominated by this MC state while in equilibrated fluid solution, both LC and MC states emit.

In the case of Rh(III) terpyridine complexes [95], the only emitting state at low temperature is assigned to the lowest triplet metal centred state. It has to be noted that Rh(III) complexes with orthometallation (presence of two 2-phenyl-pyridine ligands and one polypyridyl such as TAP or HAT) exhibit emissions at room temperature typical of the ^{3}MLCT and/or ^{3}SBLCT (σ Bond to Ligand Charge Transfer) states [96, 97].

As tris(polypyridyl) Rh(III) complexes are weak emitters, their photophysical behaviour at room temperature is not very clear: some publications mention no emission [98, 99], while others report emission only from the ligand centred or metal centred state [94, 100].

The photochemistry of Rh(III) complexes based on $Rh(phen)_3^{3+}$, dictated by the lowest excited state, ^{3}LC, involves electron transfer because the excited state is strongly oxidant. Indeed, according to the reduction potential ($E_0 = -0.75$ V/SCE) and the ^{3}LC energy (2.75 eV), $Rh(phen)_3^{3+}$ acts as a very oxidizing compound ($E_0 \approx +2.0$ V for $Rh(phen)_3^{3+*/2+}$) [8, 101–103]. The compounds with phi ligands such as $Rh(phi)_2(bpy)^{3+}$ have not been photophysically characterized as they do not emit at room temperature [98]. However it has been suggested that the ability of such Rh(III) compounds to function as photo-oxidant originates from the ^{3}LC state and LMCT state resulting from it.

3.4.2
Bis-Polypyridyl Complexes

These complexes such as cis-$Rh(bpy)_2Cl_2^+$ and cis-$Rh(phen)_2Cl_2^+$ contain two photo-labile ligands. The longest wavelength transitions are assigned to metal-centred d-d transitions, overlapped by other absorption bands assigned to intraligand transitions [104, 105]. In emission, the lowest lying excited state corresponds to a ^{3}MC state [94, 106].

Upon irradiation, these complexes undergo photoaquation from the ^{3}MC excited state (substitution of Cl$^-$ by H_2O) [107, 108]:

$$\left[Rh^{III}(L)_2(X_2)\right]^+ + H_2O \xrightarrow{h\nu} \left[Rh^{III}(L)_2(X)(H_2O)\right]^{2+} + X^-$$
$$(L = bpy, phen, trpy \qquad X = Cl, Br, I)$$

The efficiency of this photosubstitution process constitutes a potential mechanism by which other nucleophiles (such as electron-rich DNA sites) can substitute Cl$^-$ and bind to the metal. An intermediate containing one chloride ion and one water molecule can also react with nucleophiles [109].

The electrochemical data show the presence of two overlapping one-electron waves in reduction for cis-$Rh(III)(phen)_2Cl_2^+$ ($E_0 = -0.57$ V/NHE

[110]), accompanied by the loss of one chloride [111]. The reduction potential of the corresponding excited state $Rh^{III*/II}$ is estimated to be 1.18 V/NHE [112]. Photoredox processes have been reported for *cis*-Rh(III)(phen)$_2$Cl$_2^+$ and *cis*-Rh(III)(dpphen)$_2$Cl$_2^+$ (dpphen=4,7-diphenyl-1,10-phenanthroline): the reader should refer to the review devoted to this subject [113].

4
Photophysics with Mono- and Polynucleotides

In the following sections, the photophysics in the presence of mono- or polynucleotides is examined successively for complexes with different metal ions.

4.1
The Ru(II) and Os(II) Polyazaromatic Complexes

This section will be divided into two parts depending on the behaviour of the ^{3}MLCT state of the complex towards the guanine and adenine bases. In the first (Sect. 4.1.1.), we will consider the couples: complex/DNA bases where there is no luminescence quenching. In the second (Sect. 4.1.2), we will examine the couples: complex/DNA bases, where the emission intensity is reduced.

4.1.1
Couples Ru(II) or Os(II) Complexes/DNA Bases,
Without Emission Quenching by the Bases

For many Ru(II) and Os(II) complexes, the luminescence intensity is not affected by the presence of a mononucleotide, for example Ru(phen)$_3^{2+}$ (or Os(phen)$_3^{2+}$), Ru(bpy)$_3^{2+}$ (or Os(bpy)$_3^{2+}$) and their derivatives, as well as Ru(phen/bpy)$_2$(TAP/HAT)$^{2+}$, or Ru(phen/bpy)$_2$(DPPZ)$^{2+}$ (or Os(phen)$_2$(DPPZ)$^{2+}$). For these complexes indeed, the lowest-lying excited states are not sufficiently oxidizing to be quenched by abstracting an electron from a nucleotidic base. The existence or not of such an electron transfer process can of course be checked from a comparison of the reduction potentials of the excited complexes (Table 1) with the oxidation potential determined for the G of GMP (the most reductive of the 4 bases).

As can be concluded from inspection of Table 1, the complexes mentioned above exhibit values for the reduction potentials of their excited states which are not positive enough as compared to the oxidation potential of GMP, i.e. +1.09 V/SCE, in order to induce an electron transfer from GMP to the excited complex. Therefore as the guanine is the most reducing base among the four DNA bases, no luminescence quenching for this series will take place in the presence of a mononucleotide or polynucleotide. We shall also discuss the cases of the couples "complex/nucleotidic base" where the photo-electron transfer is not possible, not because of the choice of the complex but because of the base. Thus the charge process can become inefficient if, instead of a G, we consider an A. For example for the couple "Ru(TAP)$_2$(HAT)$^{2+}$/[poly (dA-dT)]$_2$", no photoelectron transfer process occurs although this is not the case when a guanine is present.

Thus in the following sections we will discuss the effect of the addition of polynucleotides on the photophysics of the Ru(II) (and Os(II)) complexes in the absence of charge transfer processes.

4.1.1.1
Comparison Between Ru(bpy)$_2$(HAT)$^{2+}$, Ru(phen)$_3^{2+}$, Ru(TAP)$_3^{2+}$ and Ru(TAP)$_2$(HAT)$^{2+}$

Generally in these systems, the interaction of a complex with a polynucleotide is accompanied by an increase of the luminescence intensity and lifetime. The importance of this enhancement depends not only on the degree of protection of the complex vs the aqueous phase, that is on the mode of interaction of the complex with DNA, but also on the type of luminophore or optical transition which is involved in emission. Moreover other processes such as the thermal activation from the ^{3}MLCT state to the ^{3}MC state and the quenching by oxygen are also affected by the interaction with DNA. The DNA effects on these different

Table 1. Oxidation (E$_{ox}$) and reduction potentials (E$_{red}$) of a series of complexes, as well as the corresponding reduction potentials in the excited ^{3}MLCT state (E$_{red}$*). e.t.: electron transfer evidenced between GMP and the complex in the excited state. a: from Orellana G, Quiroga M, Braun A (1987) Helv. Chim. Acta 70: 2073

	E$_{ox}$	E$_{red}$	E$_{red}^*$	e.t. with GMP	Ref
Ru(HAT)$_3^{2+}$	+2.07	−0.62	+1.46	e.t.	29
Ru(HAT)$_2$(TAP)$^{2+}$	+2.03	−0.64	+1.43	e.t.	29
Ru(TAP)$_2$(HAT)$^{2+}$	+2.02	−0.68	+1.36	e.t.	29
Ru(TAP)$_3^{2+}$	+1.94	−0.75	+1.32	e.t.	38
Ru(BPZ)$_3^{2+}$	+1.86	−0.86	+1.27	e.t.	130
Ru(Me$_2$ TAP)$_3^{2+}$	+1.80	−0.84	+1.24	e.t.	130
Ru(HAT)$_2$(phen)$^{2+}$	+1.86	−0.66	+1.23	e.t.	130
Ru(BPZ)$_2$(DPPZ)$^{2+}$	+1.77	−0.78	+1.21		196
Ru(HAT)$_2$(bpy)$^{2+}$	+1.79	−0.76	+1.12	e.t.	29
Ru(bpy)(TAP)(HAT)$^{2+}$	+1.78	−0.75	+1.11	e.t.	29
Ru(pzth)$_2$(HAT)$^{2+}$	+1.83	−0.67	+1.18	e.t.	a
Ru(phen)$_2$(PHEHAT)$^{2+}$	+1.35	−0.84	+1.03	e.t.	45
Ru(TAP)$_2$(DPPZ)$^{2+}$	+1.77	−0.80	+1.15	e.t.	46
Ru(TAP)$_2$(bpy)$^{2+}$	+1.70	−0.83	+1.06	e.t.	130
Ru(TAP)$_2$(phen)$^{2+}$	+1.73	−0.83	+1.06	e.t.	130
Ru(bpy)$_2$(DPPZ)$^{2+}$	+1.24	−1.02	+0.97		196
Ru(phen)$_2$(DPPZ)$^{2+}$	+1.30	−1.00	+0.96		45
Ru(phen)$_2$(HAT)$^{2+}$	+1.53	−0.86	+0.87		45
Ru(bpy)$_2$(HAT)$^{2+}$	+1.56	−0.84	+0.83		29
Ru(bpy)$_2$(TAP)$^{2+}$	+1.51	−0.88	+0.86		38
Os(phen)$_3^{2+}$	+0.85	−1.30	+0.42		39
Os(TAP)$_3^{2+}$	+1.55	−0.67	+1.11	e.t.	39

processes have been demonstrated by comparing, in the presence of different polynucleotides, the behaviour of complexes which exhibit three typically different photophysical mechanisms of deactivation of the excited state.

As shown in Table 1, although excited $Ru(TAP)_3^{2+}$ and $Ru(TAP)_2(HAT)^{2+}$ oxidize guanine, no emission quenching is present with [poly (dA-dT)]$_2$. Therefore for these two complexes, [poly (dA-dT)]$_2$ corresponds to the polynucleotide which is added, whereas for the other complexes which are discussed, it is CT-DNA. The addition of CT-DNA or [poly (dA-dT)]$_2$ to this series of complexes increases the luminescence intensity and lifetime. However the origin of this enhancement is quite different for each complex because their emission is not controlled by the same photophysical parameters [114, 115]. For $Ru(bpy)_2(HAT)^{2+}$, the emission is controlled exclusively by the radiative (k_r) and non-radiative (k_{nr}) rate constants of the excited ^{3}MLCT state to the ground state. Therefore in this case, as the luminescence is not oxygen sensitive, it has been shown, from an examination of the dependence of the luminescence lifetime with temperature, that the main effect of the polynucleotide is on k_{nr}. A decrease of k_{nr} is attributed to the intercalation of the HAT ligand between the stacking of bases, which diminishes vibrations of the complex inside the DNA. In contrast, for $Ru(TAP)_2(HAT)^{2+}$ or $Ru(TAP)_3^{2+}$, for neither of which luminescence is quenched by oxygen, the ^{3}MLCT lifetimes are entirely controlled by the thermal activation to the ^{3}MC state and the effect of [poly (dA-dT)]$_2$ originates from the imprisonment or encaging of the complex by the [poly (dA-dT)]$_2$ matrix. Indeed the DNA scaffolding prevents, or at least diminishes, the activation to the ^{3}MC state, because the complex in the ^{3}MC state cannot distort in the DNA grooves as easily as in solution; the net result is thus an increase in luminescence intensity and lifetime. This mechanism has been demonstrated from a study of the emission lifetimes as a function of temperature in the presence and absence of [poly (dA-dT)]$_2$. Finally, with $Ru(phen)_3^{2+}$, the case is very complicated because the photophysics is intermediate between that of $Ru(bpy)_2(HAT)^{2+}$ and that of $Ru(TAP)_2(HAT)^{2+}$ (or $Ru(TAP)_3^{2+}$); thus there is a control on the luminescence lifetime by k_r, k_{nr} and by the thermal activation to the ^{3}MC state. Moreover the emission of $Ru(phen)_3^{2+}$ is very sensitive to oxygen, in contrast to the other complexes. The study shows that, for this complex, the protection of the excited state by the DNA double helix from the oxygen in solution is the most important factor responsible for the luminescence enhancement.

These analyses underline clearly the different origins of the luminescence enhancement by the polynucleotide, i.e. the different photophysical mechanisms, and this without consideration of the geometries of interaction of the complex with the polynucleotide.

4.1.1.2
Photophysics of Ru(phen/bpy)$_2$DPPZ^{2+} in the Presence of Polynucleotides

Even if the Ru(II) complexes discussed in the previous section could be regarded as non-radioactive probes for nucleic acids, two drawbacks at least prevent their broader application in this area: the background luminescence of the free compounds in aqueous solution and their relatively weak binding to DNA. Therefore, several works have been focused on transition metal complexes containing the extended dipyridophenazine ligand [116–118]. One of the first complexes to be investigated was $Ru(bpy)_2(DPPZ)^{2+}$, acting as a true molecular "light switch" for DNA because it shows no photoluminescence in aqueous medium but displays intense luminescence in the presence of DNA [40]. Unwinding experiments suggested for this complex an intercalative mode of binding. Emission data have shown that both $Ru(bpy)_2(DPPZ)^{2+}$ and $Ru(phen)_2(DPPZ)^{2+}$ exhibit biexponential luminescence decays in the presence of B-form DNA, and two separate binding modes were proposed to interpret the results [41, 119]. Emission has also been studied in the presence of various types of DNA. The luminescence quantum yield was found to decrease from triplex DNA to Z form, to B form and finally to A form, reflecting the level of protection of the Ru-(DPPZ) luminophore from the water quenching [119]. Recently, Turro et al. evidenced the quenching by water of the MLCT excited state of $Ru(phen)_2(DPPZ)^{2+}$ in the presence of DNA (for racemic mixture as well as Δ and Λ-$Ru(phen)_2(DPPZ)^{2+}$) attributed to proton transfer [120]. In that study, the existence of two binding geometries for each enantiomer was also proposed to explain their findings.

Several studies from Norden et al. have been published for enantiomerically pure Δ and Λ-$Ru(phen)_2(DPPZ)^{2+}$. Equilibrium binding constants were calculated to be higher than 10^6 M^{-1} for both enantiomers (with 50 mmol/l NaCl in solution), considering a covering site size, n, equal to 3 [43]. Luminescence titrations and isothermal titrations calorimetry experiments on homochiral $Ru(phen)_2(DPPZ)^{2+}$ revealed that the free energy of binding to DNA varies little between the enantiomers. It is largely governed by nonelectrostatic forces for an important part, and is comparable to the free energy of interaction of organic intercalators with DNA. More surprisingly, the complete thermodynamic data showed that the binding to DNA is totally entropically driven, in contrast to organic intercalators.

The same authors reported that each of the enantiomers, when bound to DNA, has two distinct lifetimes [42]. For both enantiomers the longer lifetime contribution was found to increase with increasing degree of occupancy, suggesting that the two lifetimes do not result from two separate intercalative binding modes but from the distribution of metal complexes on DNA. The longer lived excited state originates from an effect of aggregation of the complex, resulting in a better protection from the quenching by water. The relative luminescence quantum yield of the Δ enantiomer bound to DNA was found to be six times greater than that of the Λ enantiomer, the Δ form being therefore primarily responsible for the luminescence increase reported for the racemate in the presence of DNA.

Linear dichroism combined with emission anisotropy excitation spectra provided evidence that the DPPZ-Ru complexes are intercalated, with a small clockwise rotation of the molecular DPPZ plane from the plane perpendicular to the helix axis [121]. This study allowed the determination of the directions of four principal transition moments in Δ and Λ-$Ru(phen)_2(DPPZ)^{2+}$, thus characterizing the DNA binding mode. Whether the complexes intercalate from the major or minor grooves is still a matter of controversy. NMR data led Barton et al. to suggest an intercalative binding within the major groove [122] at least for the Δ isomer [123] whereas comparisons between the binding of Ru(phen)$_2$(DPPZ)$^{2+}$ to T4 DNA (glycosylated in the major groove), poly(dA).[poly(dT)$_2$] triplex and CT-DNA, suggest a minor groove location [121, 124].

4.1.1.3
Difference Between the Ru(II) and Os(II) Complexes in the Presence of Polynucleotides

The different behaviour of these two types of metallic complexes in the presence of polynucleotides results from their slightly different photophysics. As the ^{3}MLCT states of some of the Ru(II) complexes can easily be thermally activated to the ^{3}MC states, these compounds may lose a ligand under illumination. Thus a reactive species is generated on the polynucleotide and is thus likely to react with monodentate ligands such as a guanine or adenine base (see further). The formation of this kind of photoproduct will compete with the processes characteristic of the ^{3}MLCT state (see below). In contrast, as thermal activation to the ^{3}MC does not take place with the Os(II) complexes, no dechelation occurs, and the photophysical processes are entirely controlled by the ^{3}MLCT state.

The addition of increasing amounts of CT-DNA to Os(phen)$_3^{2+}$ induces an enhancement of the complex luminescence, in the same fashion as observed with Ru(phen)$_3^{2+}$ although the emission increase is less important than with Ru(phen)$_3^{2+}$ [39]. The DPPZ ligand has also been tested with the Os(II). As observed with Ru(phen)$_2$(DPPZ)$^{2+}$, the addition of CT-DNA to Os(phen)$_2$ DPPZ^{2+} also switches the luminescence on. Therefore, the complex may also play the role of a DNA probe, with, however, the characteristics of a red-shifted emitter which probes the DNA on a much shorter time-scale (from 1 to 10 ns) [125].

4.1.2
Couples Ru(II) or Os(II) Complexes/DNA Bases, with Emission Quenching Originating from a Photoinduced Electron Transfer

4.1.2.1
With Mononucleotides

Some Ru(II) and Os(II) complexes of Table 1 should be sufficiently oxidizing in the ^{3}MLCT state to abstract an electron from GMP (Ru(TAP/HAT)$_2$(bpy/phen)$^{2+}$, Ru(TAP)$_3^{2+}$, Ru(HAT)$_3^{2+}$, Ru(TAP)$_2$(HAT)$^{2+}$, Ru(HAT)$_2$(TAP)$^{2+}$, Ru(TAP)$_2$-(DPPZ)$^{2+}$, Ru(bpz)$_2$(DPPZ)$^{2+}$, Ru(bpz)$_3^{2+}$, Ru(Me$_2$TAP)$_3^{2+}$, Os(TAP)$_3^{2+}$) and some of them could even photooxidize AMP (Ru(HAT)$_3^{2+}$, Ru(HAT)$_2$(TAP)$^{2+}$). However, on the basis of these electrochemical data, it is rather difficult to predict with certainty whether the electron transfer will actually take place. Indeed the oxidation potentials of the bases in mononucleotides are not known with much precision because the oxidation waves are irreversible and thus no thermodynamic data are available [126–129]. Therefore, luminescence quenching experiments by GMP and AMP have been carried out for most of these complexes [130], in order to determine the quenching rate constants. The plot of the logarithms of these constants (obtained with GMP) as a function of the reduction potential values of the excited complexes, leads to a curve typical for a quenching by an electron transfer process, where a plateau is reached for the most exergonic charge transfers, when they become diffusion controlled [130]. From a mathematical treatment of this curve, according to the Marcus and Rehm-Weller's equation, a value for the oxidation potential of GMP has been obtained. This value (+ 1.16 V/NHE) is in relatively good agreement with some of the values of the oxidation potentials of guanine determined electrochemically or obtained from pulsed radiolysis experiments [131, 132]. There are indeed differences in oxidation potentials obtained by different authors, from different techniques, attributed to the irreversible character of this oxidation. This mathematical treatment furnishes another valuable method for the determination of the GMP oxidation potential. However it is interesting to note that this value corresponds to that obtained by pulsed radiolysis for the deprotonated radical cation.

Further evidence in favour of a photo-electron transfer, has been gained from laser flash photolysis experiments. After a pulsed excitation of the complex in the presence of the mononucleotide under argon (GMP and AMP for the most oxidizing complexes), the transient absorptions of the mono-reduced complex and monooxidized mononucleotide (probably deprotonated) are detected. They disappear according to a bimolecular process in the hundreds of µs time domain. When the experiments are performed with the complex plus mononucleotide in the presence of oxygen, the monoreduced complex disappears according to a monomolecular process, by reoxidation by oxygen, leaving the unreacted radical cation of GMP accumulated in solution; the absorption spectrum of (deprotonated) $GMP^{\cdot+}$ can then be recorded.

In Table 1 the complexes for which the electron transfer has been clearly demonstrated with GMP, or with both GMP and AMP, are denoted by e.t. Complexes such as $Ru(TAP)_2(DPPZ)^{2+}$ [46] and $Os(TAP)_3^{2+}$ [39] have also been examined with GMP and produce an electron transfer. The quantum yield of production of the reduced complex formed from the ion pairs when they separate, has also been determined for some compounds and is rather efficient (of the order of 0.2–0.3).

It should also be mentioned that for complexes such as $Ru(TAP)_3^{2+}$, in addition to the photoinduced electron transfer with a G base, other processes take place. Indeed, as the energy difference between the ^3MLCT and the ^3MC states is small, the loss of ligand is important from the ^3MC state, so that the bidentate ligand may be substituted by at least one GMP or one AMP where one nitrogen of the base forms a coordination bond with the Ru(II) [133]. Some indications have been obtained in favour of this process for $Ru(TAP)_3^{2+}$ and GMP or AMP.

4.1.2.2
With Polynucleotides

When the luminescence intensity of the complexes of the above series is measured at constant complex concentration as a function of increasing amounts of CT-DNA or, for the complexes photo-oxidizing the adenine, in the presence of [poly (dA-dT)]$_2$, instead of observing an emission increase, the ratio I/Io (Io = the intensity in the absence of polynucleotide; I = the intensity in the presence of polynucleotide) decreases until a plateau value is reached (see Table 2).

These results have of course to be correlated with the emission quenching by the mononucleotides; with CT-DNA the luminescence inhibition should operate by electron transfer via the guanines (also via the adenines for the most oxidizing complexes).

Flash photolysis experiments have been carried out with the complexes in the presence of DNA. However, in those cases the transients which are formed are rather weak, as compared to the amount of transients formed with the mononucleotides. This indicates an important back electron transfer from the reduced complex to the oxidized base when the process takes place on the polynucleotide; such a low efficiency of detected reduced complex could also originate from any other irreversible reaction which occurs once the ion pair is formed on the DNA after the charge transfer (see further, formation of photo-adducts).

The case of Ru(Me$_2$TAP)$_3^{2+}$ [130] deserves some comments. Indeed, although a photo-induced electron transfer is observed with GMP for this complex, neither a luminescence quenching nor a reduced complex is detected in the presence of CT-DNA. This is attributed to the fact that this compound does not interact with DNA or at least involves minimal overlap with the DNA

Table 2. Effect of increasing concentrations of various polynucleotides on the luminescence intensity at a fixed wavelength

Complexes	DNA	poly(dGC)	poly(dAT)	References
Ru(bpy)$_2$(TAP)$^{2+}$	↑		↑	11, 200
Ru(bpy)(TAP)$_2^{2+}$	↓		↑	200
Ru(TAP)$_3^{2+}$	↓	↓	↑	200, 202
Ru(bpy)$_2$(HAT)$^{2+}$	↑		↑	115, 130
Ru(bpy)(HAT)$_2^{2+}$	↓		↑	11, 130
Ru(HAT)$_3^{2+}$	↓		↓	11, 130
Ru(bpy)(TAP)(HAT)$^{2+}$	↓		↑	11
Ru(TAP)$_2$(HAT)$^{2+}$	↓	↓	↑	11
Ru(TAP)(HAT)$_2^{2+}$	↓		↓	11
Ru(phen)$_2$(PHEHAT)$^{2+}$	↑		↑	45
Ru(phen)$_2$(DPPZ)$^{2+}$	↑		↑	45
Ru(TAP)$_2$(DPPZ)$^{2+}$	↓		↑	46
Os(phen)$_3^{2+}$	↑		↑	39
Os(TAP)$_3^{2+}$	↓		↑	39

bases because of steric hindrances of the methyl groups with the DNA double helix backbone.

In contrast, more recently, it has been shown that $Ru(TAP)_2(DPPZ)^{2+}$ [46] binds to the polynucleotide with a high affinity, like the other DPPZ complexes, and spectroscopic studies suggested the intercalation of the DPPZ ligand between the base pairs of DNA. However this complex does not act as a light-switch for DNA as the other DPPZ complexes. Indeed, in this case, the lowest lying excited state corresponds to an MLCT transition from the Ru centre to the TAP ligand ($d\pi \rightarrow \pi^*(TAP)$) and not to the DPPZ moiety ($d\pi \rightarrow \pi^*(DPPZ)$). Therefore the complex does emit in water as does $Ru(TAP)_3^{2+}$ or $Ru(TAP)_2(phen)^{2+}$. Moreover, the $d\pi \rightarrow \pi^*(TAP)$ excited state of $Ru(TAP)_2(DPPZ)^{2+}$, like the other bis-TAP Ru(II) complexes, is also quenched by CT-DNA according to a photo-electron transfer with the G bases. This complex thus combines the advantage of binding strongly to DNA with that of photoreacting with the guanine bases of DNA.

For the other metal ions, we have not separated the systems where luminescence quenching occurs upon addition of polynucleotides from those where an emission enhancement is induced.

4.2
The Re(I) Complexes

As mentioned before in Sect. 4.1.1., the luminescence intensity and lifetimes often increase when the complex interacts with DNA. In the system [*fac*-$(DPPZ)Re(I)(CO)_3L]^{1+}$ (L = 4-picoline), spectroscopic information indicates that the complex binds to CT-DNA [49]. Addition of DNA induces hypochromism and a red shift of the DPPZ absorption bands, evidencing a strong interaction between the DPPZ ligand and the polynucleotide, such as intercalation of the planar structure between the base pairs of the double helix.

The DPPZ-Re(I) complex does not emit in water but displays moderate luminescence upon binding to DNA, which is very similar to that observed for the complex in MeOH. Schanze and co-workers suggest that the emission of the complex bound to DNA originates from the $^3IL_{DPPZ}$ state. The enhancement in the lifetimes of the $^3IL_{DPPZ}$ state is explained from the photophysics of the complex. Indeed the $^3IL_{DPPZ}$ state is quenched in water via internal conversion to the energetically close ^3MLCT state. The lower polarity of the DNA-binding sites compared to H_2O increases the energy gap between the ^3MLCT and $^3IL_{DPPZ}$ states, thus decreasing the internal conversion process and increasing the luminescence lifetime in DNA as the DPPZ ligand is less accessible to H_2O.

The emission and transient absorption decays of the DPPZ-Re(I) complex with CT-DNA follow biexponential kinetics, where both the short-lived and the long-lived components originate from $^3IL_{DPPZ}$. As for the Ru(II) complexes probing the DNA [41, 42, 125], these two decay components are attributed to the presence of more than one type of binding site of the complex, one resulting in a less efficient protection of the complex from the polar environment.

The binding constants of $(DPPZ)Re(I)(CO)_3(4\text{-picoline})^{1+}$ to DNA from emission measurements corresponds to $K = 6.0 \times 10^5$ M^{-1} with a binding site of n = 3.2 (using the Scatchard analysis, 25 mmol/l Tris-buffer). This binding constant is an order of magnitude lower than the values for the DPPZ-Ru(II) complexes (for example, $K = 4.9 \times 10^6$ M^{-1} for $[Ru(bpy)_2(DPPZ)]^{2+}$, using 50 mmol/l phosphate (pH 7), with a binding site of n = 1.7 [134]); this is not surprising as the Re(I) is a monocation and Ru(II) is a dication. The determination of the "polyelectrolyte" and "non electrostatic" contributions to the free energy of binding, suggests that the non-electrostatic contribution to the binding constant for the DPPZ-Re(I) and DPPZ-Ru(II) complexes, differs only by a factor of 2 [12]. This is in agreement with an intercalation for both complexes inside the DNA [43].

Yam and co-workers reported similar trends in absorption and emission for $(DPPZ)Re(I)(CO)_3(\text{pyridine})^+$, which was studied in 5% methanol/aqueous buffered solution in the presence of DNA [66]. In the same conditions, the emission intensity of $(DPPN)Re(I)(CO)_3(\text{pyridine})^+$ drops at low [DNA]/[C] ratio before gradually increasing to a plateau-value. This observation is similar to that described for the DPPN-Ru(II) complex [41]. It is suggested that the complex could aggregate on DNA, resulting in the suppression of its luminescence. This observation is different from that for similar experiments reported for $Ru(phen)_2(DPPZ)^{2+}$, $Ru(phen)_2(PHEHAT)^{2+}$ or $Cu(bcp)_2^+$ where first, with addition of DNA, an increase of the luminescence intensity is observed, even when aggregation is considered at low [DNA]/[C] ratios [42, 45, 135]. It is also interesting to note that Yam et al. observe an aggregation behaviour only for DPPN-Re(I) complex and not for DPPZ-Re(I) complex.

4.3
The Cu(I) Complexes

The studies reported for Cu(I) complexes in the presence of polynucleotides are dominated by the work published on 1,10-phenanthroline Cu(I) compounds. Thus, even if the primary processes are not yet clearly known, these results are briefly summarized in order to point out the goals of the works performed with these systems.

Sigman and co-workers have shown that some Cu(I) complexes containing phenanthroline act as chemical nucleases [136–139]. The nucleolytic activity of these complexes results from the binding of the tetrahedral species to DNA, but also from their ability to induce redox processes. Ortho-substituted phenanthroline Cu(I) complexes bind to the double helix (see further) but have no nuclease activity, due to their stability to oxidation. Indeed, these substituents constrain the complex to a tetrahedral geometry characteristic of Cu(I) and prevent the reaching of a coordination number of 5 or 6, characteristic of Cu(II). The mechanism of the nucleolytic activity, not yet clearly known, involves a non-covalently bound form of the complex to DNA, and requires hydrogen peroxide, which oxidizes the cuprous complex on the surface of the helix to provide the species responsible for the cleavage [140,

141]. Sequence-dependent cleavages are due to the site-specific binding of the tetrahedral cuprous complex in the minor groove [142]. Cleavages studies and comparative reactivities of cuprous chelates of a series of 5-substituted phenanthroline [138, 143] suggest that the copper complex binds within the minor groove, whereas absorption and viscosity measurements would indicate an intercalative binding [17, 144]. As the nuclease activity is not induced by light, we will not discuss these works further in this photophysical part, but the reader should refer to recent reviews written by Sigman on this subject [145, 146].

In order to understand clearly the interaction of Cu(I) complexes based on the phenanthroline motif with DNA, McMillin and co-workers have studied copper complexes with different 2,9-dimethyl-1,10-phenanthrolines, structurally analogous to the naked 1,10-phenanthroline but with much weaker reducing power, and therefore, without nuclease activity [147]. In the presence of DNA, the CT absorption bands of $Cu(dmp)_2^+$, $Cu(dmpp)_2^+$ (dmpp = 2,9-dimethyl-4-phenyl-1,10-phenanthroline)   and $Cu(bcp)_2^+$ (bcp = bathocuproine = 2,9-dimethyl-4,7-diphenyl-1,10-phenanthroline) exhibit bathochromic shifts (2–3 nm) and hypochromicity, which indicates binding to DNA [135, 148, 149]. However these effects are small for $Cu(dmp)_2^+$ and, under different conditions, Veal and Rill observed no absorption changes for this complex [17]. Equilibrium dialysis experiments suggest that $Cu(dmp)_2^+$ binds to DNA [150] and studies performed in the presence of ethidium bromide indicate that it binds competitively [151]. On the other hand, DNA has no significant effect on the luminescence intensity of $Cu(dmp)_2^{+*}$ which is still strongly quenched by the solvent. This indicates that the complex in interaction with the DNA is still quite accessible to the aqueous phase, suggesting an external mode of binding to DNA rather than intercalation. That the bound complex has no effect on the relative viscosity of DNA also supports this model [148].

In contrast, the introduction of a phenyl substituent at the 4-position of the dmp [149] leads to an important enhancement in the luminescence intensity of the resulting complex in the presence of DNA or RNA. Spectroscopy studies, as well as viscosity measurements, indicate aggregation processes at high loading levels of the polynucleotide. At low degrees of occupancy, when no aggregation occurs, no change in the viscosity of DNA is observed, suggesting that binding of $Cu(dmpp)_2^+$ (dmpp = 2,9-dimethyl-4-phenyl-1,10-phenanthroline) to DNA is not intercalative.

Still more important are the effects of DNA on the photophysics of $Cu(bcp)_2^+$ (bcp=bathocuproine = 2,9-dimethyl-4,7-diphenyl-1,10-phenanthroline), containing two phenyl groups in the positions 4 and 7 [135, 151]. For this complex, the luminescence intensity, barely detectable in the absence of DNA, strongly increases in the presence of an excess of polynucleotide (DNA or RNA). Similarly, the excited state lifetime, of about 2 ns in nucleophilic solvent [80, 152], increases about 30-fold in the presence of a DNA excess. These data suggest that binding $Cu(bcp)_2^+$ to DNA results in an efficient protection of the complex from the solvent environment, as solvent quenching of the MLCT state, very efficient in homogeneous solution, is suppressed in the presence of the polynucleotide. Absorption and emission results suggest that $Cu(bcp)_2^+$ interacts with DNA by two different modes. In emission, for example, the luminescence intensity measured at 700 nm increases by increasing the ratio [DNA]/[Cu], reaches a maximum and decreases till a plateau value is reached. At low [DNA]/[Cu] ratios, the behaviour is attributed to closely bound metal complexes, resulting in an emissive aggregated form of $Cu(bcp)_2^+$. This interpretation is supported by studies in the presence of poly(styrenesulfonate) (PSS) [153]. Indeed, polyelectrolytes are known to be able to induce aggregation/colloid formation, and, in agreement with this, small amounts of PSS strongly increases the emission of $Cu(bcp)_2^+$ in 33% MeOH. At higher PSS concentrations, when the complex dissolves/disperses in the polymer solution, the emission is quenched again. At high [DNA]/[Cu] ratios, the luminescence intensity reaches a plateau value that is ascribed to the $Cu(bcp)_2^+$ monomer bound to DNA.

The spectral results obtained for this complex, as well as emission polarization data [135], led McMillin and co-workers to propose an intercalative mode of binding for $Cu(bcp)_2^+$. Indeed, this geometry protects the complex from solvent quenching, prevents a rotation independent of the double helix, and induces strong hypochromicity. Viscosity measurements performed afterwards [154] seem to be inconsistent with classical intercalation. Indeed, the same authors observed a large viscosity enhancement ($\eta/\eta_0 = 1.5$) at [base]/[Cu] ≈ 20 (in 33% methanolic solution), when only a fraction of the binding sites are occupied (for comparison, $\eta/\eta_0 = 1.5$ at [base]/[dye] ≈ 3 and 1.1 at [base]/[dye] ≈ 20 for ethidium bromide in ethanolic solution [17]). These results are explained in terms of bridged adducts (Fig. 11) that increase the effective length of the polyelectrolyte. This bridged structure could be stabilized by the presence of several complexes between the DNA molecules.

The complex would be protected from solvent attack by the two helices, which is consistent with the luminescence observed. However, taking into account that the authors took as reference the viscosity of ethidium bromide in alcoholic solutions and that, in those conditions, this dye causes an unusually low extension of the polynucleotide (a 10% EtOH ethidium bromide solution results in an extension which is 50% smaller than the viscosity measured in water) [17, 155], this bridged adduct model is not required to explain the viscosity results [5]. Moreover, viscosity increases corresponding to helix extensions of even more than 3.4 Å per complex have already been observed

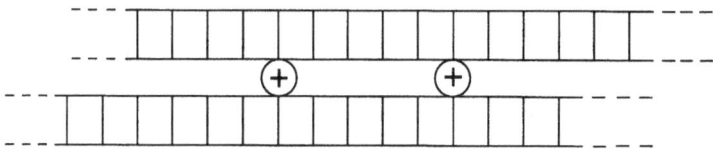

Fig. 11. Schematic picture of two B-DNA helices bridged by $Cu(bcp)_2^+$ complexes [adapted from: Liu F, Meadows KA, McMillin DR (1993) J Am Chem Soc 115: 6699]. The *circumscribed crosses* represent the Cu(I) complexes

for pure intercalators (for example ihremycin [156] and 9-aminoacridine [157], but also, in those cases, there is evidence for pure helix extension, enhanced by tilting of the intercalator in the pocket.

It is noted that those large viscosity enhancements are observed only with DNA rich in adenine-thymine base pairs (salmon testes DNA and [poly (dA-dT)]$_2$), which is relatively flexible due to the presence of only two hydrogen bonds between the base pairs.

4.4
The Rh(III) Complexes

The Rh(III) complexes examined in the presence of DNA are based on DIP and phi ligands [158, 159]. The mode of interaction of $Rh(DIP)_3^{3+}$ with DNA is not yet well established [160, 161]. Nevertheless, this complex has been shown to probe DNA cruciforms through photocleavage [9, 162]. In contrast, mechanistic studies [163, 164], NMR [165, 166], DNA-binding studies or unwinding of supercoiled DNA studies [163] have shown that Rh(III)-phi complexes bind to DNA via the intercalation of the phi ligand between the nucleobases. Indeed, the phi centred $\pi \rightarrow \pi^*$ transitions of $Rh(phen)_2phi^{3+}$ or $Rh(phi)_2bpy^{3+}$ for example, are strongly affected in the presence of DNA, resulting in an important hypochromicity and a red-shift of these transitions [163]. It has been shown by photocleavage studies that these complexes are able to act as shape-selective and enantioselective recognition agents. These results will be discussed further in the next section.

The influence of DNA on the absorption spectrum of cis-Rh(III)(phen)$_2$Cl$_2^+$ under low ionic strength conditions consists of hyperchromicity at long wavelength, furnishing evidence for interaction. This interaction appears however to be weak: the binding constants determined by equilibrium dialysis are on the order of $10^2 \times M^{-1}$ [167]. Nevertheless, because of the interest in the development of new photochemotherapeutic drugs, the ability of this type of complexes to covalently bind to polynucleotide upon irradiation has been studied in some detail (see below).

5
Photochemical Reactions with DNA

While many studies have concentrated on the use of photophysical techniques to exploit the properties of metal complexes and nucleic acids, much can also be learnt by using photochemical methods. Potential and actual uses of these compounds could be as probes for particular DNA conformers and structures (e.g. cruciforms, single-stranded sections, single-strand: double strand junctions), as photo-footprinting reagents, as components in sequence-specific anti-sense or anti-gene therapeutic reagents, or as photochemical anti-tumour agents.

There is an extensive literature on the photochemical induction of lesions in DNA using organic sensitisers [168, 169]. A wide range of sensitisers has been employed including ketones (such as acetone [170], benzophenone [171]), quinones [172] and dyes such as methylene blue [173]. The excited states of these agents may react either directly with DNA (Type I) or may sensitise the formation of singlet oxygen (Type II) [169]. This latter species is known to attack DNA bases (especially guanine) and may also attack the DNA backbone leading to strand breaks. The energy, redox potentials and chemical reactivity are factors which determine which reactions of the sensitiser excited state dominate. If the energy of the sensitiser triplet state is higher than that of the DNA bases then triplet-triplet energy transfer may take place, which will typically lead to photoproducts such as thymine photodimers. The excited states of many sensitisers are strong oxidising agents and the resultant transfer of an electron to the sensitiser from the nucleo-base may lead to oxidative damage. As the oxidation potentials for the nucleobases lie in the order $G<A<C<T$ [174], the oxidative damage to DNA is often located at guanine. The radical cation of the nucleobase appears to be able to mediate strand cleavage, probably by forming the sugar radical [175]. Alternatively the sensitiser excited state may directly extract a H-atom from the nucleotide (particularly from the sugar), leading eventually to breaks in the DNA backbone. Some excited states will directly add to a DNA base, for example psoralens and other furocoumarins which form cycloadducts with the pyrimidine bases, a feature which is exploited in the use of these compounds as agents for the PUVA procedure for the control of psoriasis [169].

For metal polypyridyl complexes the excited state is, in general, too low in energy to allow direct production of the triplet state of the bases via energy transfer. However there are examples of other forms of DNA damage, resulting from singlet oxygen attack, electron- or H-atom- transfer, or photo-adduct formation. Below we consider three important classes of reaction – DNA strand breaks, chemical modification of the base (which may lead to the bases being sensitive to treatment with alkali or nucleophiles) and adduct formation.

5.1
Induction of Strand Breaks

Photochemically-induced strand breaks in DNA have been extensively investigated as they can be readily monitored using super-coiled covalently circular plasmid DNA or radioactively-labelled oligonucleotides. In most cases the strand break involves only the cleavage of one strand. Double strand breaks are much less common, although, of course, reagents which could produce such lesions could be useful as therapeutic agents as biological repair of double strand breaks is difficult.

There is an extensive and growing literature on the induction of strand breaks by ground state oxidising agents and radical species which are able to abstract an H atom from the deoxyribose of DNA. This mode of reaction appears to be responsible for the cleavage of DNA by species such as manganese complexes, and antitumour drugs such as bleomycin or neo-carcinostatin and other enediyne drugs [176]. Which hydrogen atom is abstracted depends on the type of reactive species and the manner in which it is non-covalently bound to DNA. The 1'-, 2'-, 4'- and 5'-H atoms can be accessed from the minor groove but the 3'-atom can only be approached from the major groove. The radical once formed can be scavenged by oxygen or form a hydroxylated deoxyribose. These species are themselves unstable and will further decompose. Scheme 1 shows the reactions proposed following 3'-H atom abstraction.

It will be noticed that under anaerobic conditions, the radical 1 forms 2 which subsequently decomposes leading to the cleavage of the DNA strand and yielding DNA molecules with 3'- and 5'-phosphate termini, as well as the 2-methylene-3-furanone 3 and the free nucleobase. The oxygen dependent route also induces strand breaks, although in this case a DNA molecule with a 3'-phosphoglycolaldehyde terminus and a nucleo-base propenoic acid are formed. As, in general, different products are formed after abstraction of other H atoms in the deoxyribose ring, the nature of the products can be used to diagnose the nature of the site of the initial attack of the reagent.

In a series of papers Barton and coworkers have studied the UV-induced photochemical reaction of various 9,10-phenanthrenequinone diimine (phi) Rh(III) complexes with DNA. NMR, DNA-unwinding and other experiments indicate that the phi-ligand can intercalate between the base pairs of DNA [165, 177]. By appropriate design of the ancillary ligands it is possible to tune the binding properties of the complexes so as to preferentially target particular sequences.

Marked differences were observed upon UV irradiation of [Rh(phen)$_2$-(phi)]$^{3+}$ and [Rh(phi)$_2$(phen)]$^{3+}$with ^{32}P-labelled oligonucleotides [163, 178]. The reaction, which is quite efficient (at 313 nm $\Phi = 1.2 \times 10^{-3}$ for

Decomposition of radical formed after abstraction of 3'-H by (a) oxygen independent and (b) oxygen dependent routes.

$[Rh(phen)_2\text{-}(phi)]^{3+}$ and 3×10^{-4} for $[Rh(phi)_2(bpy)]^{3+}$), has been shown to be brought about by 3'-H atom abstraction as evidenced by the formation of the products shown in Scheme 1. The reactive species appears to be the phi-localised radical centre of the LMCT state [98]. The predominance of 3'-H atom abstraction is consistent with the complex being located in the major groove on the DNA. For $[Rh(phen)_2(phi)]^{3+}$ the reaction, which is dominated by the oxygen-independent pathway, shows significant sequence specificity with a preference for 5'-py.py.pur-3' sequences as shown by the two preferred cleavage sites in double-stranded 5'-CTGGCATGCCAG-3'. On the other hand the reaction of $[Rh(phi)_2(bpy)]^{3+}$ is essentially sequence-neutral, and the oxygen-dependent cleavage route is much more significant.

The factors controlling the recognition sites for $[Rh(phen)_2(phi)]^{3+}$ have been further examined by considering the cleavage patterns induced in three sets of oligonucleotides 4, 5 and 6 which have been characterised in the B-form of DNA by X-ray crystallography [179, 180]. It has been found possible to correlate the preferential cleavage sites (and hence the binding site) with sequence-dependent features of

5'-CGCGAATTCGCG-3'
3'-GCGCTTAAGCGC-5'
4

5'-ACCGGCGCCACA-3'
3'-TGGCCGCGGTGT-5'
5

5'-CCAACGTTGG-3'
3'-GGTTGCAACC-5'
6

individual DNA structures. For the Δ- and Λ-enantiomers, pronounced differences were found in the selectivity of cleavage they induced and also in the extent of O_2-dependent and O_2-independent pathways. The Δ-enantiomer showed a much greater site selectivity, especially for 5'-py-*py*-pur-3' sites (e.g. 5'-T-*C*-G-3' in 4 and 5'-C-*C*-A-3' or 5'-C-*C*-G-3' in 5), but also (though less so) for 5'-pur-*py*-pur-3' (e.g. 5'-A-*C*-G-3' in 6). Significantly less site selectivity was exhibited by the Λ-enantiomer and this compound also produced a greater percentage of oxygen-dependent photo-products (phosphoglycaldehydes). This can be taken to indicate that the better the fit of the complex for its binding site, the less oxygen can access the C-3' radical site. The Δ- enantiomer appears therefore to intercalate preferentially into a 5'-Nu-py-pur-Nu-3' binding site (Nu = any nucleotide), with the excited state abstracting the C-3' H-atom from the sugar of the pyrimidine base. Comparison with the X-ray structural data points to a strong correlation between the efficiency of cleavage by Δ-$[Rh(phen)_2(phi)]^{3+}$ and the openness in the major groove as a result of the sequence-dependent base pair propeller twisting.

The photochemical reactions of $[Rh(phen)_2(phi)]^{3+}$ with various tRNA have also been studied, and its use as a new reagent for characterising the folded structures of RNAs proposed [181]. It cleaves at neither double-helical regions (which are A-form) nor unstructured single-stranded regions, but only where the bases are triply bonded. At these sites a third base H-bonds to a Watson-Crick base pair widening the groove and permitting the $[Rh(phen)_2(phi)]^{3+}$ to bind strongly.

The photocleavage reactions of a series of derivatives of $[Rh(R_2bpy)_2-(phi)]^{3+}$ and $[Rh(R_2bpy)(phi)_2]^{3+}$ ($R_2bpy=5,5'$-dimethyl-2,2'-bipyridyl (Me_2-bpy); 4,4'-diphenyl-2,2'-bipyridyl; 4,4'-amido-2,2'-bipyridyl) have been compared [164]. The introduction of the hydrophobic phenyl and methyl groups significantly restricts the DNA sites targetted. Enantiomeric specificity is observed with the 4,4'-diphenyl complex, where the Δ-isomer binds strongly to DNA and selectively cleaves the sequence 5'-CTCTAGAG-3' [182] whereas, by contrast, the Λ-isomer binds very weakly. Marked differences in selectivity are also found for Δ- and Λ-$Rh(Me_2bpy)(phi)_2]^{3+}$. Thus the Δ-isomer targets 5'-CTTG-3' site while the Λ-isomer prefers 5'-ACTG-3' and 5'-AGT-3' sites. This contrasting affinity has been attributed primarily to stabilising attractive interactions of the methyl groups of the ligand with the thymine methyl groups in DNA.

Remarkable affinity and specificity for DNA has been achieved by using Λ-$Rh(MGP)_2(phi)]^{3+}$ (7) (Fig. 12) which binds at nanomolar concentrations preferentially to a 5'-CATATG-3' sequence [183].

The specificity of binding is attributed to recognition of sequence-dependent twistability of the DNA. Thus while the binding relies on interaction of the guanidinium side-groups with guanine, this is only possible if the 6 base-pair section unwinds. This unwinding also opens up both the nucleotides at the AT intercalation site for cleavage.

Excellent site specificity has also been observed in a series of Rh(III)-phi complexes containing amine ligands by exploiting particular hydrogen-bonding and van der Waals interactions of the complexes with the nucleobases of DNA [184]. Initially complexes including $[Rh(NH_3)_4(phi)]^{3+}$ (8), $[Rh(en)_2-(phi)]^{3+}$ (9), $[Rh([12]-aneN_4)(phi)]^{3+}$ (10), and $[Rh([12]-aneS_4)(phi)]^{3+}$ (11) (en = 1,2-diaminoethane; [12]-aneN$_4$ = 1,4,7,10-tetraazacyclododecane; [12]-aneS$_4$ = 1,4,7,10-tetrathiacyclododecane) were examined. The quantum yields are generally about an order of magnitude less for these compounds than for $[Rh(phen)_2(phi)]^{3+}$ (e.g. 1.0×10^{-4} for rac-$[Rh(en)_2(phi)]^{3+}$ 9). In photocleavage experiments carried out on a 140-base pair restriction fragment complexes 8, Δ-9 and 10 cut the DNA at a 5'-GC-3' step, a feature attributed to H-bonding between the amine ligand and the guanine O6 located in the major groove. 11, which cannot H-bond shows a preference for a 5'-ATG-3' site, whereas Λ-9 showed a strong preference for 5'-TNu-3' steps – a probable consequence of stabilising van der Waals' interactions between the CH_2 groups of the en ligand and the methyl group of the thymine. The specific role of the thymine methyl group was confirmed by demonstrating that the specificity was eliminated when thymine was replaced by uracil. These binding interactions have been examined in more detail for the enantiomers of $[Rh(en)_2(phi)]^{3+}$ 9 [185]. Thus the effect of replacing the guanine in the DNA by 6-methylguanine or 7-deazaguanine supports the essential role of the H-bonding to the guanine O6,

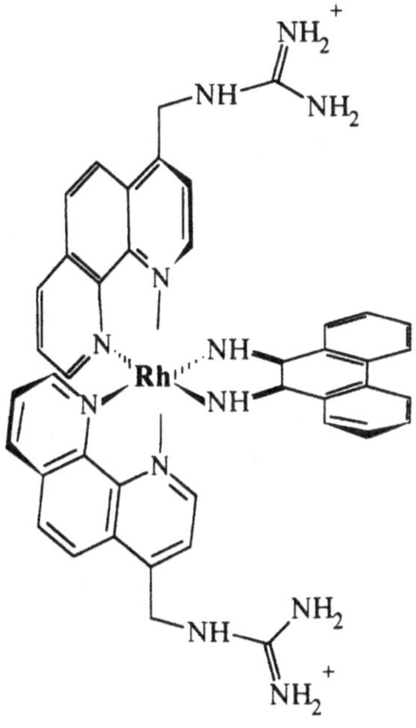

(7)

Fig. 12. Structure of Λ-Rh(MGP)$_2$(phi)$^{3+}$

while replacement of thymine by uracil confirms the importance of van der Waals interactions with the methyl group of this nucleobase.

In order to exploit further the properties of this class of rhodium complexes, four optical isomers of [Rh(2,9-diamino-4,7-diazadecane)(phi)]$^{3+}$ **12** were synthesised and the photo-induced cleavage of a 140 base-pair restriction fragment studied [186]. The isomer shown in Fig. 13 {Δ,α-(2R,9R)-**12**} cleaved preferentially at 5′-TGCA-3′ compared to other 5′-GC-3′ sites.

This can be attributed to the H-bonding with the guanine on both strands and to attractive van der Waals contacts with the thymines on both strands. The behaviour of the Λ-isomers of **12** contrasts markedly with the behaviour of Λ-[Rh(en)$_2$(phi)]$^{3+}$ (Λ-**9**) in that they do not cleave at the 5′-TG-3′ step favoured by Λ-**9** but rather at 5′-GG-3′ or 5′-GC-3′. This disfavouring of sites containing thymine can be attributed to steric clashes of the methyl groups in the thymine with those in the Λ-**12**.

The cleavage properties of [Rh(phi)$_2$(phen)]$^{3+}$ have been exploited to show the effects of binding of a 13-residue oligopeptide to the complex [187, 188]. The non-specific binding affinity is dominated by the metal complex but the side-chains of the peptide allow site specificity of a 180 base-pair DNA restriction fragment. The protein-conjugate chosen cleaved preferentially at a 5′-CCA-3′ site. Replacement or modification of one of the glutamates in the peptide chain removed this specificity. It is suggested that this glutamate is

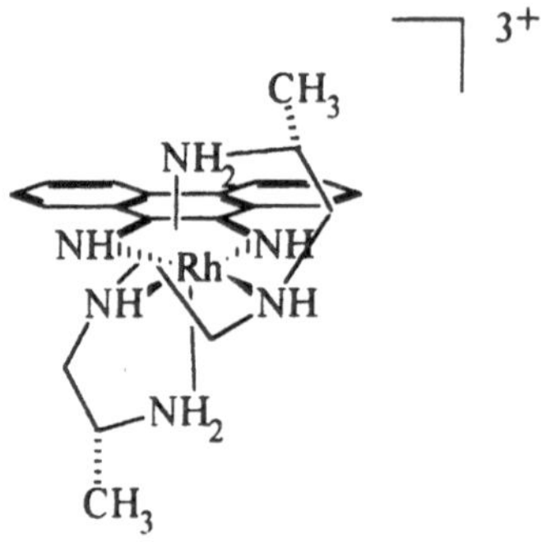

$$3+$$

(12)

Fig. 13. Structure of Δ-α-(2R,9R)Rh(2,9-diamino-4,7-diazadecane)(phi)$^{3+}$

required both to form the unique conformation of the peptide and to interact with DNA by a specific interaction of its carboxy group with cytosine.

In earlier studies Barton and coworkers demonstrated the use of 4,7-diphenyl-1,10-phenanthroline (DIP) complexes of Co(III) and Rh(III) as photochemical probes for particular conformations in DNA. For example the pattern of single strand breaks induced by Λ-Co(DIP)$_3^{3+}$ in pBR322 plasmid and pBR322 plasmid with a (C-G)$_{16}$ insert were compared [189]. It was shown that the insert area was particularly targetted in the modified plasmid, consistent with reaction at a well-defined Z-DNA region. Similarly with pBR322, cleavage was found at segments of alternating purine-pyrimidines. Λ-Co(DIP)$_3^{3+}$ was also shown to cut SV40 DNA at sites important as control elements along the genome [190]. Rh(DIP)$_3^{3+}$ has been proposed as a useful reagent for recognising cruciform DNA. Upon UV irradiation (e.g. 315 or 332 nm) this reagent induces single and double strand breaks in plasmid pBR322 DNA [191]. The double strand breaks occur at one particular site (position 3238 on the 3'-strand; 3250 on the 5'-strand), which is at an AT-rich site close to the stem of a cruciform cleavage occurring at a particular specific site on the plasmid (Fig. 14).

Preferential cleavage at this site is not observed with the linearised plasmid. Interestingly both the Λ- and Δ-enantiomers are effective at inducing the damage. There is also a striking contrast with the reactions of Co(DIP)$_3^{3+}$ which does not induce double strand breaks, and this was ascribed to better recognition and longer "residence time" of the rhodium complex at the site. The fact that the cleavage sites on the two strands are separated by 12 base-pairs is also particularly noteworthy.

Single strand cleavage formed directly upon excitation have also been observed for Ru(phen)$_3^{2+}$ and Ru(bpy)$_3^{2+}$ using plasmid DNA [192–194]. In these cases, however, the quantum yield is low (1.2×10^{-6} in degassed solution; 6.6×10^{-6} in the presence of air for Ru(bpy)$_3^{2+}$) [195]. The nature of the initial processes are uncertain, although it is probable that both Type I and Type II processes are involved, as the involvement of singlet oxygen is

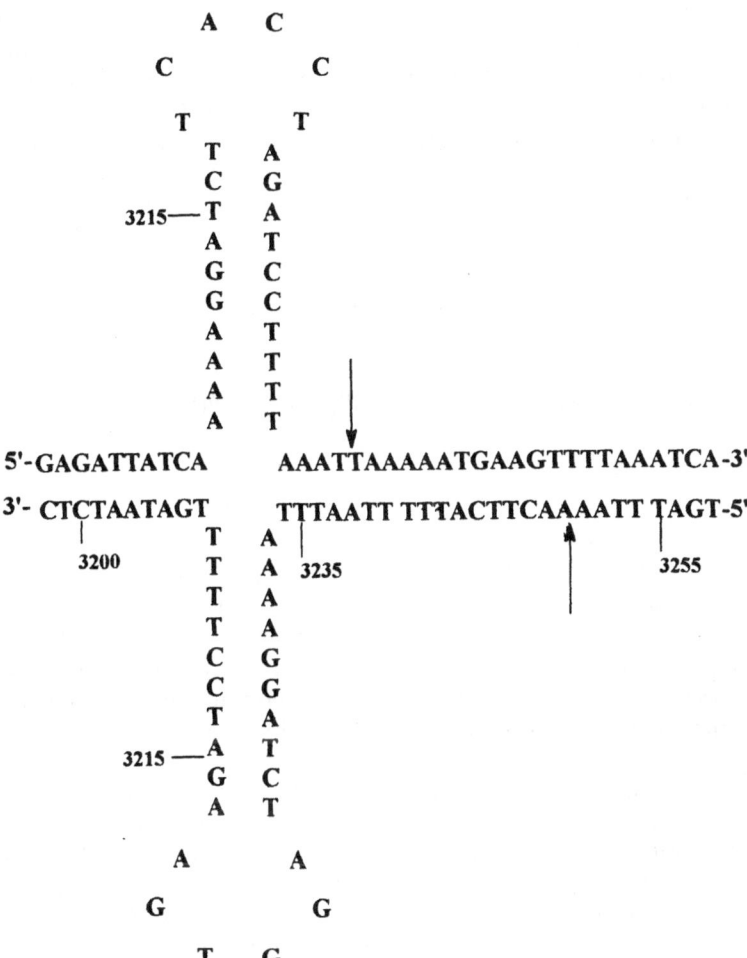

Fig. 14. Principal cleavage sites induced by Rh(DIP)$_3^{2+}$ at a cruciform site in plasmid pBR322 DNA

supported by studies with azide ion and D_2O. It has been shown that the efficiency of DNA cleavage induced by Ru(bpy)$_2$(DPPZ)$^{2+}$ is about eight times less than that caused by Ru(bpy)$_3^{2+}$ [196]. This has been attributed to less efficient sensitisation of singlet oxygen by the intercalating DPPZ complex. An opposite behaviour has recently been reported for (η-C$_5$Me$_5$)Ru(NO)(DPPZ)$^{2+}$ and (η-C$_5$Me$_5$)Ru(NO)(bpy)$^{2+}$ where the DPPZ was found to cleave DNA while the bpy complex was inactive [197]. Co(phen)$_2$(DPPZ)$^{2+}$, but not Ni-(phen)$_2$(DPPZ)$^{2+}$, has also been shown to photosensitise the formation of single strand breaks in plasmid DNA [198].

Higher quantum yields are observed for cleavage of plasmid DNA sensitised by ruthenium complexes of TAP, HAT and bpz whose excited states are sufficiently oxidising to abstract an electron from guanine in DNA. This is clearly shown for the series of complexes $Ru(L)_n(bpy)_{3-n}^{2+}$ (L=TAP or HAT) [130, 199–201] and for $Ru(bpy)_2(DPPZ)^{2+}$ and $Ru(bpz)_2(DPPZ)^{2+}$ [196]. As discussed in Sect. 4.1.2, flash photolysis studies with 5′-GMP have provided good evidence for the initial electron transfer and also for subsequent proton transfers, depending on the medium pH [130, 202]. The deprotonated form of the radical cation (14) can exist as various canonical forms and tautomers (Scheme 2).

$$Ru(TAP)_3^{2+} + G \rightarrow Ru^{II}(TAP)_2(TAP^{\cdot-})^+ + G^{\cdot+} \tag{1}$$
$$(13)$$

$$Ru^{II}(TAP)_2(TAP^{\cdot-})^+ + H^+ \rightarrow Ru^{II}(TAP)_2(TAPH)^{2+} \tag{2}$$

$$G^{\cdot+} \rightarrow G(-H)^{\cdot} + H^+ \tag{3}$$
$$(14)$$

With 5′-GMP most of the reduced ruthenium reacts with the oxidised guanine species to reform the starting materials, but a small proportion form adducts. In double-stranded DNA these reactions are very rapid [200] and will require picosecond techniques to elucidate them fully. The species leading to strand breaks in the DNA is not known with certainty but has been supposed to be the guanine radical cation. This could abstract a H-atom from a neighbouring ribose [175]. Indeed experiments with ^{32}P-labelled oligonucleotides (either single- or double-stranded) indicate that the cleavage process is much less important than adduct formation [130, 200, 201].

High quantum yields (e.g. 7.6×10^{-3}) of strand breaks have also been been observed when $Ru(bpy)_3^{2+}$ is excited in the presence of persulphate [195, 203–205]. The active strand-cleaving species are generated by reaction (4). The main agent for causing strand breaks is probably the sulphate radical, although it is possible that the Ru(III) complex may also react with the DNA nucleobases.

$$Ru(bpy)_3^{2+*} + S_2O_8^{2-} \rightarrow Ru(bpy)_3^{3+} + SO_4^{2-} + SO_4^{\cdot-} \tag{4}$$

The irradiation also causes efficient biological deactivation of the DNA, which was determined by monitoring the transformation efficiency of the plasmid in *E. coli* bacteria. The yield of strand breaks in the presence of persulphate is significantly lower for $Ru(phen)_3^{2+}$, probably because of side-reactions between the phenanthroline complex and the sulphate radical.

5.2
Base Damage

Strand breaks in greater yield can be induced on B-DNA following sensitisation by $Ru(phen)_3^{2+}$ and related complexes in aerated solution, if it is treated with piperidine after photolysis. Studies with ^{32}P-end labelled oligonucleotides indicate that such lesions occur primarily at guanine bases [206, 207]. This is consistent with chemical modification of the guanine, as it is known that treatment of such species with piperidine can yield apurinic sites and produce strand breaks at this point. One route for the formation of such lesions is the addition of singlet oxygen to the guanine base. With other sensitisers such as methylene blue it has been shown that $^1O_2$ leads to 7,8-dihydro-8-oxo-2'-deoxyguanosine (15) [208]. Interestingly the same product is also formed from the radical cation of guanine in DNA (Scheme 2) [209] (but not in the free nucleoside 2'-deoxyguanosine) although accompanied by the oxazolone derivative (16) which is characteristic of such Type 1 processes [210]. Unlike the oxazolone derivative, which is alkali-labile and may therefore be responsible for piperidine-induced strand breaks in DNA, 7,8-dihydro-8-oxo-2'-deoxyguanine does not react rapidly with piperidine [211]. Another striking finding is that, while singlet oxygen forms 4-hydroxy-4,8-dihydro-8-oxo-2'-deoxyguanine with 2'-deoxyguanosine [212], this product is not obtained in DNA, probably because of conformational restrictions in the polynucleotide.

While $Ru(phen)_3^{2+}$ appears to bind preferentially to B-DNA, it has been demonstrated from a study of direct strand breaks and alkali-labile sites formed by its tetramethyl analogue Λ-$Ru(TMP)_3^{2+}$ that this complex primarily binds at A-DNA conformations [206]. In experiments carried out with a linear DNA sample, excised from plasmid DNA, it was found that as well as damage at guanine, likely caused by $^1O_2$, there was also a cleavage at homopyrimidine stretches which are specifically recognised by Λ-$Ru(TMP)_3^{2+}$. The recognition of these homopyrimidine-homopurine regions of DNA may be because they present a non-B-form DNA, in which there is a shallow surface feature to which the complex binds preferentially.

5.3
Formation of Covalent Adducts

In certain cases the metal polypyridyl sensitiser can covalently bind to DNA. This has been shown for ruthenium, rhodium and chromium polypyridyl complexes.

As discussed above, certain ruthenium complexes are sufficiently oxidising that they can oxidise guanine and/or adenine [11, 130]. While studies with plasmid DNA had indicated that these complexes were more efficient at inducing strand breaks in DNA, experiments with ^{32}P-labelled oligonucleotides revealed that the dominant reaction with these small DNA molecules was the formation of adducts [200]. Adducts are formed with both double-stranded and single-stranded oligonucleotides and the formation of adducts with calf thymus DNA can be conveniently monitored spectroscopically [130, 201]. It is proposed that the adducts are formed by reaction of the reduced ruthenium complex with the guanine radical (see Eqs. 1–3 above):

$$Ru^{II}(TAP)_2(TAPH\cdot)^{2+} + G(-H)\cdot \rightarrow Adduct \tag{5}$$

It was also observed that the formation of adducts with CT-DNA is insensitive to medium pH or to the presence of O_2. While species with similar absorption spectra to the DNA-adducts are formed with 5′-GMP, they are only the dominant species in the absence of oxygen and with pH <7. Under other conditions with 5′-GMP, side-reactions such as photodechelation of the complex and oxidation of the guanine become important, whereas with DNA these side reactions are suppressed. In the presence of [poly(dA-dT)]$_2$ or AMP the absorption spectra show that $Ru(TAP)_3^{2+}$ photodechelates [133]. Adducts are formed in lower yield and it is proposed that they are formed by direct coordination of the Ru to the nucleobase.

The $Ru(TAP)_3^{2+}$/5′-GMP adduct has been isolated, the ribose-phosphate removed by treatment with acid, and the resulting complex structurally characterised by NMR and electrospray mass spectrometry [213]. The structure (17) (Fig. 15) shows that, as predicted from the UV/vis spectrum, the guanine is bonded to the complex via one of the TAP ligands.

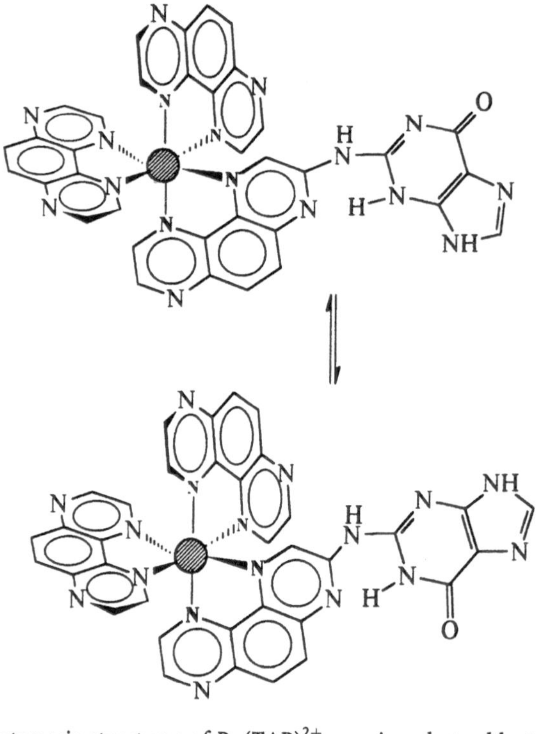

(17a)

(17b)

Fig. 15. Tautomeric structures of $Ru(TAP)_3^{2+}$-guanine photoadducts

The coordination is through the exocyclic NH_2 group of the guanine, consistent with the combination of the radicals proposed in Eq. (5), and subsequent dehydrogenation. The NMR spectra show that the compound exists as two rapidly interconverting tautomeric forms. More recently adducts formed from $Ru(TAP)_2(bpy)^{2+}$ and calf thymus DNA have been excised by enzymatic and acid hydrolytic procedures and isolated by HPLC [214]. Two isomeric guanine adducts (**18a** and **18b**) (Fig. 16) are formed and in both of

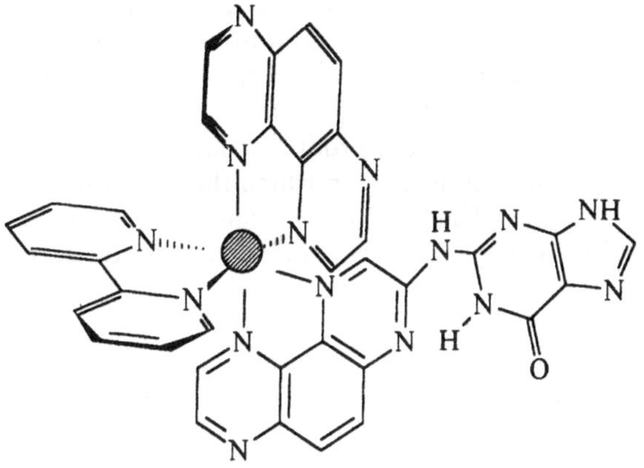

(18a)

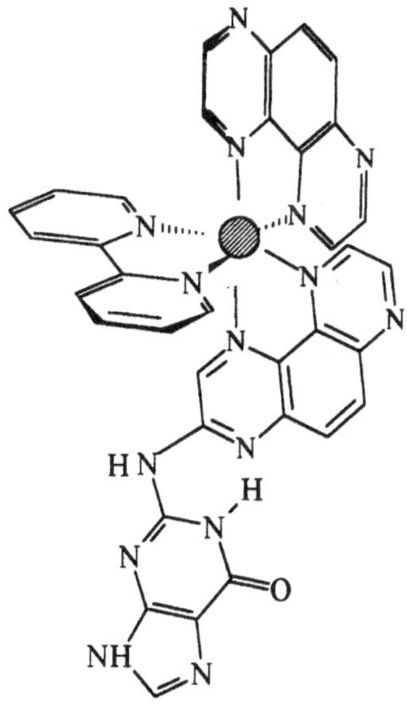

(18b)

Fig. 16. Structures of isomeric photoadducts excised from DNA after photolysis in the presence of $Ru(TAP)_2(bpy)^{2+}$

these the guanine is again bonded via its exocyclic amino group to one of the C atoms β to the coordinating N of the TAP ligand.

The linking of the complex via the guanine 2-NH$_2$ group is consistent with Ru(TAP)$_2$(bpy)$^{2+}$ being located in the minor groove of the DNA.

Adducts have also been isolated following photolysis of *cis*-Rh(phen)$_2$Cl$_2^+$ in the presence of calf thymus DNA [215]. The first studies were carried out with calf thymus DNA, nucleotides and nucleosides [167]. It was found that the reaction proceeded with a quantum yield of around 1×10^{-3} for double-stranded DNA and 6×10^{-3} for denatured DNA and was not significantly dependent on oxygen. After enzymatic excision three deoxyguanosine adducts could be isolated. Spectroscopic data for one of these suggested that the metal had bound to the N1 of guanine (19) (Fig. 17).

The other adduct, which is the main product from DNA or dGMP, could be a diastereoisomer or possibly an isomer in which the rhodium is bound to an exocyclic atom of guanine. These products can be interconverted either photochemically or by heating. The ratio of adducts is different when *cis*-

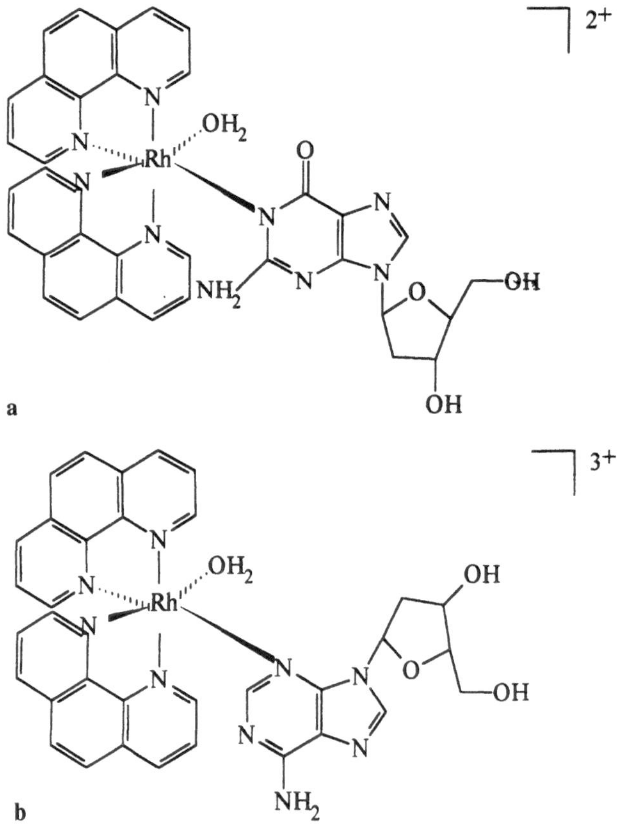

Fig. 17. a Guanosine-adduct obtained by photolysis of Rh(phen)$_2$Cl$_2^{3+}$ in the presence of DNA. **b** Adenosine-adduct obtained by photolysis of Rh(phen)$_2$Cl$_2^{3+}$ in the presence of AMP

$Rh(phen)_2Cl_{2+}$ is photolysed in the presence of deoxyguanosine, as here product (19) is the dominant product. With deoxyadenosine, the N3-bonded species (20) (Fig. 17) is the main product.

Further studies showed that under anaerobic conditions the rate of photoreaction of cis-$Rh(phen)_2Cl_2^+$ in the presence of uric acid or deoxyguanosine (but not deoxyadenosine) was greatly increased, while photoaquation of the complex yielding cis-$Rh(phen)_2Cl(H_2O)^{2+}$ decreased [111]. These results indicated that the reaction proceeded not by a simple photosubstitution (to be expected for a low-lying MC state) but rather by electron transfer (e.g. Eq. 6):

$$cis\text{-}Rh(phen)_2Cl_2^+ + G \rightarrow cis\text{-}Rh(phen)_2Cl_2 + G^{\cdot+} \qquad (6)$$

In deaerated solution photolysis in the presence of deoxyguanosine gave two diastereoisomeric products $Rh(phen)_2(N7\text{-}dG)Cl^{2+}$ (21), in which the chloride has been replaced and the metal coordinates to the N7 of guanine [216]. The quantum yield for the formation of the two products is very high [1.82 at 308 nm (π-π^*excited state) and 1.87 at 355 nm (MC excited state)], and the reaction is proposed to proceed by a chain reaction, which is initiated by electron transfer to the rhodium complex excited state (Eq. 6) and propagated by loss of Cl^- from the Rh(II) complex (Eq. 7) (possibly accompanied by hydration), subsequent reaction with deoxyguanosine (Eq. 8) and oxidation of the resultant Rh(II)-deoxyguanosine complex by the starting material (Eq. 9):

$$cis\text{-}Rh(phen)_2Cl_2 \rightarrow Rh(phen)_2Cl^+ + Cl^- \qquad (7)$$

$$Rh(phen)_2Cl^+ + dG \rightarrow Rh(phen)_2(N7\text{-}dG)Cl^+ \qquad (8)$$

$$Rh(phen)_2(N7\text{-}dG)Cl^+ + cis\text{-}Rh(phen)_2Cl_2^+ \rightarrow$$
$$Rh(phen)_2(N7\text{-}dG)Cl^{2+} + cis\text{-}Rh(phen)_2Cl_2$$
$$(9)$$

Consistent with the formation of the guanine radical cation, small quantities of 7,8-dihydro-8-oxo-2'-deoxyguanosine (15) were detected. In the presence of air the reaction is much less efficient and the principal product is $Rh(phen)_2(N1\text{-}dG)Cl^{2+}$ (19) (Fig. 17), which is proposed to arise from the reaction of the deprotonated radical cation of guanine $[G(\text{-}H)^{\cdot}]$ with cis-$Rh(phen)_2Cl(H_2O)^{2+}$.

Covalent binding of cis-$Cr(phen)_2Cl_2^{2+}$ to DNA has also been reported [215, 217]. Although studies with polynucleotides showed a preference for purines, it has not so far been possible to isolate photoproducts. As with the rhodium analogue, deoxyguanosine reductively quenches the excited state of cis-$Cr(phen)_2Cl_2^{2+}$.

6
Tuning the Interactions

As discussed in the previous sections, when the complexes interact with DNA, their luminescence intensities and lifetimes change, and this furnishes preliminary diagnostics for their mode of interaction. As outlined above, the oxidizing character of some of the Ru complexes in the excited state, which induces electron transfers and generates reactions of the DNA, makes them very attractive. However, with most such complexes, many problems remain to be solved. For example many reactive complexes show a rather low affinity constant for DNA, while other complexes which have high affinity constants (such as $Ru(phen)_2(DPPZ)^{2+}$) are not photoreactive as there is no DNA photosensitization induced by electron transfer from a base. Moreover, so far the DPPZ-complexes discussed above show low specificity of interaction with special DNA structures or with a specific DNA sequence. Although the Δ and λ enantiomers of $Ru(phen)_2(DPPZ)^{2+}$ exhibit some selectivity for a right-handed or left-handed DNA, the degree of selectivity is rather poor. In the future, research should thus be directed toward an improvement of selectivity in the interaction, if, as outlined above, efficient and specific molecular tools for the study of DNA remain one of the main targets of this research area. With this in mind, some examples of Ru(II) and Re(I) complexes have been developed and these are discussed below. These should have increased affinity while retaining appropriate photoreactivity, or they may interact specifically with targeted DNA structures.

As discussed in Sect. 5.1, Rh(III) phi complexes have been designed to have remarkable specificity for particular sequences of DNA and to be able to carry out site specific photocleavage with reasonable quantum efficiency. It may be possible to develop this work further to prepare synthetic endonucleases. One disadvantage of these complexes, compared to the Ru and Os complexes, is that they require illumination in the UV.

6.1
Complexes with Extended Aromatic Ligands

An attractive possibility to increase at the same time the affinity and the photoreactivity, is to design a special ligand which would keep the characteristics for the intercalation similar to those of the DPPZ ligand, and which moreover would confer oxidizing properties to the resulting complex in the ^3MLCT state. This has been achieved with PHEHAT (Fig. 18) which contains a phen and a HAT motif.

It has been shown that $Ru(phen)_2PHEHAT^{2+}$ interacts with DNA [45] with an affinity as high as that of $Ru(phen)_2(DPPZ)^{2+}$ but, by contrast, is able to photooxidize the guanines of GMP. As observed with the DPPZ complexes, the PHEHAT complex does not luminesce in water whereas its emission is switched on by interaction with DNA.

The other possibility, in order to combine the efficiency of interaction by intercalation with the photo-reactivity, is to keep the DPPZ ligand, and

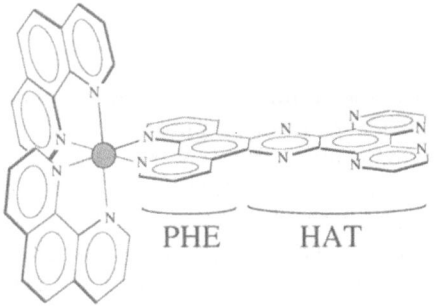

Ru(phen)$_2$(PHEHAT)$^{2+}$

Fig. 18. Bis(1,10-phenanthroline)(1,10-phenanthrolino[5,6-b]1,4,5,8,9,12-hexaazatriphenyl-ene)ruthenium(II) cation [Ru(phen)$_2$(PHEHAT)$^{2+}$]

combine it with two other oxidizing ligands such as two TAP or two bpz, in Ru(TAP)$_2$(DPPZ)$^{2+}$ or in Ru(bpz)$_2$(DPPZ)$^{2+}$ already discussed above. However, in those two cases, the luminophore has changed from Ru-DPPZ in Ru(phen)$_2$DPPZ^2 into Ru-TAP and Ru-bpz; this change not only makes the complexes luminescent in water, but also modifies their photophysical and probably photochemical behaviour with DNA.

6.2
Bifunctional Complexes

These complexes are composed of two units, each of which play a particular role and which may also increase the binding affinity of the resulting compound for DNA.

Bifunctional compounds not based on the metal ions considered in this review, have already been used successfully to target chemical reactions on DNA. For example a cisplatin analogue containing a tethered fluorescent dansyl group and mimicking properties of cisplatin allows one to monitor by a nondestructive technique much lower levels of [Pt(dansen)Cl$_2$] than cisplatin or its ethylenediamine analogue [218]. Organic bifunctional compounds composed of an acridine intercalator for DNA binding, a nucleic base for abasic site recognition and a linker with a cleavage function, have also been prepared to recognise and cleave DNA at apurinic sites [219].

For bifunctional complexes, one unit can play the role of luminescent probe, while the other behaves as a quencher (by electron or energy transfer), or as developed above, the metallic unit behaves as DNA photoreagent, while the other unit may target specific sites. The bifunctional complexes can thus be designed for different purposes.

6.2.1
Ru(II) Complexes

In order to increase the affinity of the TAP complexes for DNA (these complexes being already sufficiently photoreactive), they have been derivatized by an organic molecule such as an aminoquinoline which also interacts with DNA [220]. Thus the $Ru(TAP)_2(POQ)^{2+}$ (Fig. 19) contains a metallic unit, $Ru(TAP)_2^{2+}$, which interacts with DNA according to its own geometry and is also the photoreactive part. This unit is linked via a phen ligand derivatized by an aliphatic chain to the aminoquinoline unit. Interestingly this bifunctional compound is present in solution under a folded and unfolded form. In the folded conformer, an intramolecular electron transfer from the aminoquinoline unit to the excited complex unit inhibits the luminescence of the complex

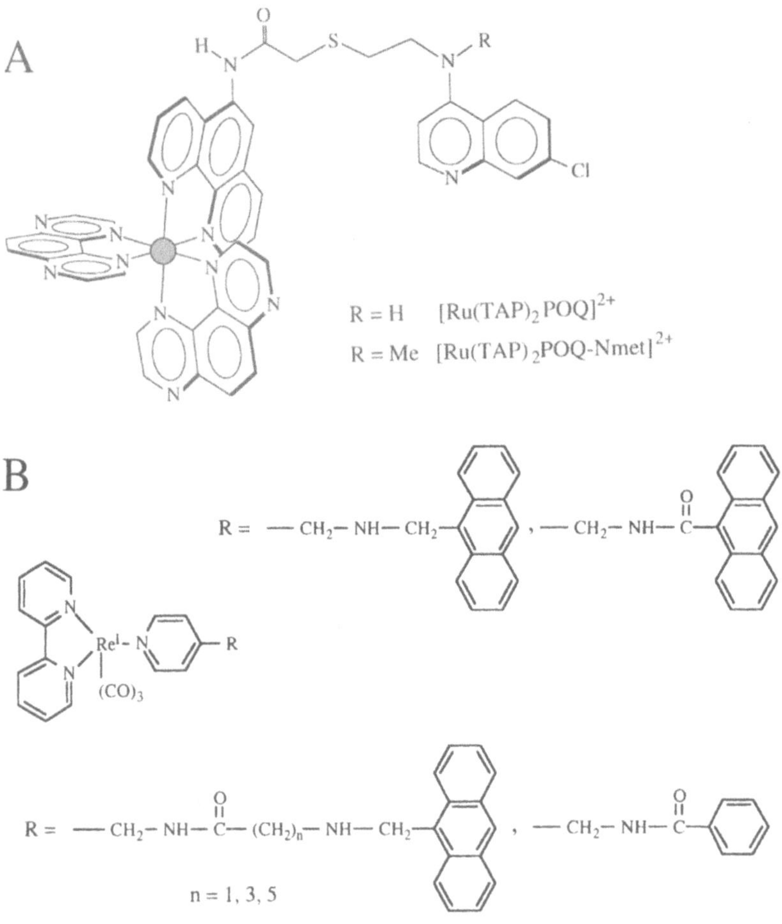

Fig. 19 A,B. Structures of the bifunctional complexes: A $Ru(TAP)_2(POQ)^{2+}$ and $Ru(TAP)_2$-$(POQ$-Nmet$)^{2+}$; B chromophore-quencher Re(I) complexes

in solution, whereas the complex emits in the unfolded conformer. When the aminoquinoline unit is protonated, the luminescence of the metallic complex unit is also restored [221]. In neutral medium, when the weakly emitting complex (due to the intramolecular electron transfer quenching) interacts with [poly (dA-dT)]$_2$ (no guanine content, thus no photoreaction), its luminescence is enhanced, due to the unfolding process of the compound and/or protonation of the quinoline unit. It was also shown that the chemical attachment of the organic unit to the metal complex unit does indeed increase the affinity of the resulting bifunctional compound for the polynucleotide. The design of this type of bifunctional complex, which exhibits good photoprobing properties for DNA and also a good photoreactivity towards DNA, thus constitutes an attractive possibility towards photoreactive compounds which bind more avidly to DNA. Depending on the interaction characteristics of the complex moiety and organic unit, these bifunctional species could exhibit in the future, some particular specificity of interaction for special DNA structures.

6.2.2
Re(I) Complexes

This strategy of bifunctional complexes has also been followed by Schanze for Re(I) complexes [47, 50]. In that case, the addition of a second unit binding strongly to DNA is particularly useful because Re(I) complexes do not bind avidly to DNA (see Sect. 4.2). These bifunctional compounds contain a luminescent metal complex (a bpy-Re(I) derivative) covalently linked, via a flexible tether, to a quencher, operating as an energy acceptor or as an electron donor (Fig. 19). Anthracene derivatives were selected as quenchers because the anthracene moiety binds strongly to double stranded DNA via intercalation [222–224] and quenches the ^{3}MLCT state of the Re(I)-bpy chromophore [225, 226].

In order to understand the photophysical behaviour of these bifunctional complexes in the presence of DNA, it is necessary to know their photophysics in homogeneous solution. Population of the ^{1}MLCT excited state of the Re(I) complex is followed by crossing to the ^{3}MLCT state [48, 227, 228] which decays by luminescence, non-radiative processes and intramolecular energy transfer to the anthracene chromophore to yield the ^{3}An* state [50]. Alternatively, excitation of the anthracene derivatives populates the ^{1}An* state which decays by luminescence, or inter-crosses to the ^{3}An* state [50] or transfers its energy intramolecularly to the Re(I)-bpy chromophore to yield the ^{3}MLCT excited state (via a Förster dipole-dipole mechanism) [50]. The intramolecular energy transfer from the Re(I)-bpy chromophore to the anthracene chromophore (Re $\rightarrow$ An E_nT) [229, 230], is supposed to occur via a Dexter spin exchange mechanism, which increases when the length of the flexible linker is decreased [50]. However, this dependence is smaller than that expected for an extended chain, suggesting that the tether can easily adopt conformations that allow the two chromophores to come into close contact even if the length of the spacer is different.

Photophysical studies of the same compounds in the presence of calf thymus-DNA reveal that the addition of an anthracene subunit to the Re(I) complex results in behaviour in sharp contrast to that of the mono-chromophoric Re(I)-bpy complex which interacts very weakly with the double helix. The binding of the bifunctional compound via the intercalation of the anthracene subunit is accompanied by changes in the absorption of the anthracene chromophore and by an important enhancement in the yield of the ^{3}MLCT emission of the Re(I) complex. The increase in MLCT emission quantum yield probably originates from a decrease in the rate of intramo-lecular energy transfer from the Re(I)-bpy chromophore to the anthracene chromophore. Thus binding of the bifunctional compound to DNA makes the Re(I)-bpy chromophore unable to approach the anthracene subunit closely which is protected by the surrounding bases of the double helix. This study evidences the advantages of designing new supramolecular systems that allow a synergism of different properties.

6.3
Polymetallic Complexes

A study of bimetallic compounds of the type $[Ru(bpy)_2(bpy')(CH_2)_n$ $(bpy')(bpy)_2Ru]^{4+}$ (Fig. 20) [231] has shown that the affinity of such complexes for polynucleotides is markedly improved (binding constant 100

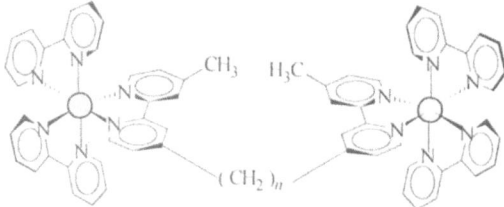

$$[Ru(bpy)_2(bpy')(CH_2)_n(bpy')(bpy)_2Ru]^{4+}$$

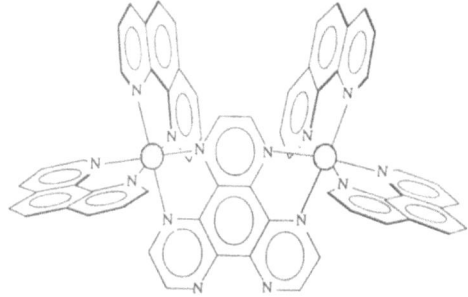

$$[(phen)_2Ru(HAT)Ru(phen)_2]^{4+}$$

Fig. 20. Structure of the bimetallic complexes

times greater) as compared to the affinity of the monometallic species $Ru(bpy)_2(Me_2bpy)^{2+}$ which is a rather poor DNA binding agent. Quite interestingly, the efficiency of photocleavage in aerated solution has also increased for this compound, particularly at high ionic strengths where the binuclear complex remains bound but the mononuclear complex is displaced.

Another bimetallic complex, based on the bridging hat ligand, $[(phen)_{2-}Ru(hat)Ru(phen)_2]^{4+}$ (Fig. 20) also appears as a novel attractive photoreagent of DNA [29, 232, 233]. Indeed this complex has been shown to interact exclusively with denatured CT-DNA. Because of its bulk, it cannot penetrate inside the DNA grooves of a double helix, so that only the portions of important DNA deformations, such as single stranded portions (Fig. 3), are accessible to this complex. Moreover, quite interestingly, because of its high charge of 4+, this compound binds rather avidly to denatured DNA and gives rise to photoadducts exclusively with denatured DNA where the bases are accessible to the complex. Formation of these photoadducts also originates from a photo-induced electron transfer from the guanine bases. These bimetallic complexes could thus be regarded in the future as interesting tools in order to target or detect the presence of single stranded DNA portions as irregular DNA structure illustrated in Sect. 1.

6.4
Complexes Anchored to Oligonucleotides

Another approach, which has been exploited to introduce some specificity of interaction or reaction with a particular sequence of bases, consists of the antigene or antisense strategy. This strategy allows the targeting of a specific sequence of bases. Such an approach has already been performed with $Cu(phen)_2^+$ attached to synthetic oligonucleotides, and shown to induce cleavage at specific DNA sequences [1, 234].

For this purpose, TAP complexes have also been derivatized in order to be chemically attached to an oligonucleotide sequence complementary to the targeted sequence [235, 236]. The Ru(TAP) complex anchored to the 5-position of a thymidine photoreacts, by electron transfer, with guanines of the targeted complementary strand, in the vicinity of the attached complex. This could offer a means to photo-crosslink irreversibly the derivatized conjugate to the targeted sequence, thus allowing one to direct the photoreactions discussed in the previous sections, on particular sequences of bases.

Oligonucleotides derivatized by metallic complexes have also been used to examine the possibility of long range photo-electron transfer between a donor and an acceptor chemically attached to the oligonucleotide, and mediated by the DNA bases (see for example [237]).

7
Concluding Remarks

As illustrated above, research during the the last decade or so has shown that the varied excited state properties of metal polypyridyl complexes make them excellent candidates as photochemical and photophysical probes for DNA. In particular, by modifying the ligands it is possible to change the photochemical properties dramatically. In this way photooxidation of DNA can be favoured (e.g. for the TAP or HAT complexes of ruthenium), the generation of singlet oxygen enhanced or suppressed (c.f. $Ru(phen)_3^{2+}$ or $Ru(TAP)_3^{2+}$), or the quenching of the excited state by water controlled (RuL_2DPPZ^{2+}). By appropriate design of the ligands it is possible to facilitate specific interations between the metal complex and functional groups on the nucleobases and hence achieve remarkably high selectivity in binding. This has been most effectively demonstrated with the series of rhodium (III)-phenanthrolinedii-mine complexes, where specificities approaching those of endonucleases have been achieved. Various photo-chemical reactions have been characterised, including strand cleavage, base oxidation and photadduct formation. While the quantum yield of some of these reactions is rather low, for others the efficiency is such that the complexes could be considered for practical applications including DNA-targeted phototherapeutic procedures. Studies with bifunctional compounds or oligonucleotides show that it will be possible both to increase the binding affinity of the complexes and hence enhance the specificity of photochemical reaction. It is to be expected therefore that there will be even more significant progress in the use of these compounds as molecular DNA probes in the next few years.

Acknowledgements. AKD and CM are grateful to the SSTC (program "Pôles d'attraction interuniversitaires" 4/11) for financial support. The EU Human Capital and Mobility (CHRX-CT92–0016) and COST D1 programmes are also gratefully acknowledged.

8
References

1. Thuong NT, Hélène C (1993) Angew Chem Int Ed Engl 32: 666
2. Blackburn GM, Gait MJ (1992) Nucleic acids in chemistry and biology. Oxford University Press, Oxford
3. Saenger W (1983) Principles of nucleic acid structure, C.R. Cantor edn. Springer, Berlin Heidelberg New-York
4. Neidle S, Waring MJ (1993) Molecular aspects of anti-cancer drug-DNA interactions. MacMillan, London
5. Norden B, Lincoln P, Akerman B, Tuite E (1996) DNA interactions with substitution-inert transition metal ion complexes. In: Sigel A and Sigel H (eds) Metal ions in biological systems, vol 33. Dekker, New York, p 177
6. Juris A, Balzani V, Barigelletti F, Campagna S, Belser P, von Zelewsky A (1988) Coord Chem Rev 84: 85
7. Balzani V, Ballardini R (1990) Photochem Photobiol 52: 409
8. Kalyanasundaram K (1992) Photochemistry of polypyridine and porphyrin complexes. Academic Press, San Diego
9. Pyle AM, Barton JK (1990) Progr Inorg Chem 38: 413
10. Basile LA, Barton JK (1989) Metallonucleases: real and artificial. In: Sigel H, Sigel A (eds) Metal ions in biological systems, vol 25. Dekker, New York, p 31
11. Kirsch-De Mesmaeker A, Orellana G, Barton JK, Turro NJ (1990) Photochem Photobiol 52: 461

12. Kalsbeck WA, Thorp HH (1993) J Am Chem Soc 115: 7146
13. Härd T, Norden B (1986) Biopolymers 25: 1209
14. Yamagishi A (1984) J Phys Chem 88: 5709
15. Carter MT, Bard AJ (1987) J Am Chem Soc 109: 7528
16. Carter MT, Rodriguez M, Bard AJ (1989) J Am Chem Soc 111: 8901
17. Veal JM, Rill RL (1991) Biochemistry 30: 1132
18. Sigman DS, Landgraf R, Perrin DM, Pearson L (1996) Nucleic acid chemistry of the cuprous complexes of 1,10-phenanthroline and derivatives. In: Sigel A, Sigel H (eds) Metal ions in biological systems, vol 33. Dekker, New York, p 485
19. Davidson RS, Hilchenbach MM (1990) Photochem Photobiol 52: 431
20. Glazer AN, Hays SR (1992) Nature 359: 859
21. Henderson BW, Dougherty TJ (1992) Photochem Photobiol 55: 145
22. Anderson-Engels S, Johansson J, Svanberg S, SvanbergK (1989) Anal Chem 61: 1367 A
23. Bruhn S, Toney J, Lippard SJ (1990) Progress in Inorganic Chemistry, Bioinorganic Chemistry 38: 1477
24. Sip M, Leng M (1993) Nucleic acids and molecular biology, vol 7, Eckstein F and Lilley DMJ edn. Springer, Berlin Heidelberg New York, p 1
25. Boudvillain M, Dalbiès R, Leng M (1996) Evidences for a catalytic activity of the DNA double helix in the reaction between DNA, platinum(II), and intercalators. In: Sigel A, Sigel H (eds) Metal ions in biological systems, vol 33. Dekker, New York, p 87
26. Lippert B (1996) *trans*-Diammineplatinum(II): what makes it different from *cis*-DDP? Coordination chemistry of a neglected relative of cisplatin and its interaction with nucleic acids. In: Sigel A, Sigel H (eds) Metal ions in biological systems, vol 33. Dekker, New York, p 105
27. Kalyanasundaram K (1992) Coord Chem Rev 46: 159
28. Krause RA (1987) Synthesis of ruthenium(II) complexes of aromatic chelating heterocycles: towards the design of luminescent compounds. In: Clarke MJ et al. (eds) Structure and bonding. Springer, Berlin Heidelberg New York, 67: 1
29. Jacquet L, Kirsch-De Mesmaeker A (1992) J Chem Soc Faraday Trans 88: 2471
30. Kober EM, Meyer TJ (1982) Inorg Chem 21: 3967
31. Demas JN, Crosby GA (1971) J Am Chem Soc 93: 2841
32. Kober EM, Caspar JV, Lumpkin RS, Meyer TJ (1986) J Phys Chem 90: 3722
33. Durham B, Caspar JV, Nagle JK, Meyer TJ (1982) J Am Chem Soc 104: 4803
34. Meyer TJ (1986) Pure & Appl Chem 58: 1193
35. Demas JN, DeGraff BA (1991) Anal Chem 63: 829
36. Wilson RB, Solomon EI (1980) J Am Chem Soc 102: 4085
37. Miskowski VM, Gray HB, Wilson RB, Solomon EI (1979) Inorg Chem 18: 1410
38. Masschelein A, Jacquet L, Kirsch-De Mesmaeker A, Nasielski J (1990) Inorg Chem 29: 855
39. Content S, Kirsch-De Mesmaeker A (1997) J Chem Soc Faraday Trans 93: 1089
40. Friedman AE, Chambron JC, Sauvage JP, Turro NJ, Barton JK (1990) J Am Chem Soc 112: 4960
41. Hartshorn RM, Barton JK (1992) J Am Chem Soc 114: 5919
42. Hiort C, Lincoln P, Norden B (1993) J Am Chem Soc 115: 3448
43. Haq I, Lincoln P, Suh D, Norden B, Chowdhry BZ, Chaires JB (1995) J Am Chem Soc 117: 4788
44. Bolger J, Gourdon A, Ishow E, Launay JP (1996) Inorg Chem 35: 2937
45. Moucheron C, Kirsch-De Mesmaeker A, Choua S (1997) Inorg Chem 36: 584
46. Ortmans I, Kirsch-De Mesmaeker A, Chambron JC, Sauvage JP (to be submitted)
47. Thornton NB, Schanze KS (1993) Inorg Chem 32: 4994
48. Schanze KS, MacQueen DB, Perkins TA, Cabana LA (1993) Coord Chem Rev 122: 63 and references therein
49. Stoeffler HD, Thornton NB, Temkin SL, Schanze KS (1995) J Am Chem Soc 117: 7119
50. Thornton NB, Schanze KS (1996) New J Chem 20: 791
51. van Outersterp JWM, Stufkens DJ, Vlcek A Jr (1995) Inorg Chem 34: 5183

52. van Outersterp JWM, Stufkens DJ, Fraanje J, Goubitz K, Vlcek A Jr (1995) Inorg Chem 34: 4756
53. Ruminski R, Cambron RT (1990) Inorg Chem 29: 1575
54. Wrighton M, Morse DL (1974) J Am Chem Soc 96: 998
55. Giordano PJ, Wrighton HS (1979) J Am Chem Soc 101: 2888
56. Worl LA, Duesing R, Chem P, Della Ciana L, Meyer TJ (1991) J Chem Soc Dalton Trans 849
57. Striplin DR, Crosby GA (1994) Chem Phys Lett 221: 426
58. Van Wallendael S, Shaver RJ, Rillema DP, Yoblinski BJ, Stathis M, Guarr TF (1990) Inorg Chem 29: 1761
59. MacQueen DB, Petersen JD (1990) Inorg Chem 29: 2313
60. Baiano JA, Carlson DL, Wolosh GM, DeJesus DE, Knowles CF, Szabo EG, Murphy WR Jr (1990) Inorg Chem 29: 2327
61. Yoblinski BJ, Stathis M, Guarr TF (1992) Inorg Chem 31: 5
62. Juris A, Campagna S, Bidd I, Lehn JM, Ziessel R (1988) Inorg Chem 27: 4007
63. Sacksteder LA, Lee M, Demas JN, DeGraff BA (1993) J Am Chem Soc 115: 8230
64. Shaw JR, Schmehl RH (1991) J Am Chem Soc 113: 389
65. Shaw JR, Webb RT, Schmehl RH (1990) J Am Chem Soc 112: 1117
66. Yam VWW, Lo KKW, Cheung KK, Kong RYC (1995) J Chem Soc Chem Commun 1191
67. Caspar JV, Meyer TJ (1983) J Am Chem Soc 105: 5583
68. Caspar JV, Meyer TJ (1983) Inorg Chem 22: 2444
69. Meyer TJ, (1983) Progr Inorg Chem 30: 389
70. Day A, Sanders N (1967) J Chem Soc A 1536
71. Phifer CC, McMillin DR (1986) Inorg Chem 25: 1329
72 Buckner MT, McMillin DR (1978) J Chem Soc Chem Commun 759
73. Blaskie MW, McMillin DR (1980) Inorg Chem 19: 3519
74. Kirchhoff JR, Gamache RE Jr, Blaskie MW, Del Paggio AA, Lengel RK, McMillin DR (1983) Inorg Chem 22: 2380
75. Parker WL, Crosby GA (1989) J Phys Chem 93: 5692
76. Everly RM, McMillin DR (1991) J Phys Chem 95: 9071
77. Wehry EL, Sundararajan (1972) J Chem Soc Chem Commun 1135
78. McMillin DR, Kirchhoff JR, Goodwin KV (1985) Coord Chem Rev 64: 83
79. Stacy EM, McMillinDR (1990) Inorg Chem 29: 393
80. Palmer CEA, McMillin DR, Kirmaier C, Holten D (1987) Inorg Chem 26: 3167
81. Crane DR, Ford PC (1991) J Am Chem Soc 113: 8510
82. Everly RM, McMillin DR (1989) Photochem Photobiol 50: 711
83. Goodwin KV, McMillin DR (1987) Inorg Chem 26: 875
84. Ichinaga AK, Kirchhoff JR, McMillin DR, Dietrich-Buchecker CO, Marnot PA, Sauvage JP (1987) Inorg Chem 26: 4290
85. Karlsson K, Moucheron C, Kirsch-De Mesmaeker A (1994) New J Chem 18: 721
86. Dietrich-Buchecker CO, Marnot PA, Sauvage JP, Kirchhoff JR, McMillin DR (1983) J Chem Soc Chem Commun 513
87. Gushurst AKI, McMillin DR, Dietrich-Buchcker CO, Sauvage JP (1989) Inorg Chem 28: 4070
88. Palmer CEA, McMillin DR (1987) Inorg Chem 26: 3837
89. Sanna G, Pilo MI, Zoroddu MA, Seeber R, Mosca S (1993) Inorg Chim Acta 208: 153
90. Crosby GA, Elfring WH,Jr (1976) J Phys Chem 82: 2206
91. DeArmond MK, Carlin CM (1981) Coord Chem Rev 36: 325
92 Westra J, Glasbeek M (1990) Chem Phys Lett 166(5,6): 535
93. Humbs W, Yersin H (1996) Inorg Chem 35: 2220
94. Nishazawa M, Suzuki TM, Sprouse S, Watts RJ, Ford PC (1984) Inorg Chem 23: 1837
95. Frink ME, Sprouse SD, Goodwin HA, Watts RJ, Ford PC (1988) Inorg Chem 27: 1283
96. Ortmans I, Didier P, Kirsch-De Mesmaeker A (1995) Inorg Chem 34: 3695
97. Didier P, Ortmans I, Kirsch-De Mesmaeker A, Watts RJ (1993) Inorg Chem 32: 5239
98. Pyle AM, Chiang MY, Barton JK (1990) Inorg Chem 29: 4487

99. Frink ME, Sprouse SD, Goodwin HA, Watts RJ, Ford PC (1988) 27: 1283
100. Indelli MT, Carioli A, Scandola F. (1984) J Phys Chem 27: 2685
101. Ballardini R, Varani G, Balzani V (1980) J Am Chem Soc 102: 1719
102. Ohno T (1985) J Phys Chem 89: 5709
103. Indelli M, Ballardini R, Scandola F (1984) J Phys Chem 27: 1283
104. McKenzie ED, Plowman RA (1970) J Inorg Nucl Chem 32: 199
105. Demas JN, Crosby GA (1970) J Am Chem Soc 92: 7262
106 Carstens DHW, Crosby GA (1970) J Mol Spectrosc 34: 113
107. Muin MM, Huang WL (1973) Inorg Chem 12: 1831
108. Skibsted LH (1989) Coord Chem Rev 94: 151
109. Endicott JF, Ramasani T, Tamilarasan R, Lessard RB, Kul Ryu C, Brubaker GR (1987) Coord Chem Rev 77: 1
110. Kew G, DeArmond K, Hanck K (1974) J Phys Chem 78: 727
111. Billadeau MA, Wood KV, Morrison H (1994) Inorg Chem 33: 5780
112. Elving PJ (1976) Bioelectrochem Bioenerg 3: 37
113. Ohno T (1985) Coord Chem Rev 64: 311
114. Lecomte JP, Kirsch-De Mesmaeker A, Orellana G (1994) J Phys Chem 98: 5382
115. De Buyl F, Kirsch-De Mesmaeker A, Tossi A, Kelly JM (1991) J Photochem Photobiol A: Chem 60: 27
116. Chambron JC, Sauvage JP, Amouyal E, Koffi P (1985) New J Chem 9: 527
117. Amouyal E, Homsl A, Chambron JC, Sauvage JP (1990) J Chem Soc Dalton Trans 1841
118. Choi SD, Kim MS, Kim SK, Lincoln P, Tuite E, Norden B (1997) Biochemistry 36: 214
119. Jenkins Y, Friedman AE, Turro NJ, Barton JK (1992) Biochemistry 31: 10,809
120. Turro C, Bossmann SH, Jenkins Y, Barton JK, Turro NJ (1995) J Am Chem Soc 117: 9026
121. Lincoln P, Broo A, Norden B (1996) J Am Chem Soc 118: 2644
122. Dupureur CM, Barton JK (1994) J Am Chem Soc 116: 10,286
123. Dupureur CM, Barton JK (1997) Inorg Chem 36: 33
124. Tuite E, Lincoln P, Norden B (1997) J Am Chem Soc 119: 239
125. Holmlin RE, Barton JK (1995) Inorg Chem 34: 7
126. Steenken S (1989) Chem Rev 89: 503
127. Candeias LP, Steenken S (1989) J Am Chem Soc 111: 1094
128. Kittler L, Löber G, Gollmick FA, Berg H (1980) J Electroanal Chem 503
129. Brabec V, Dryhurst G (1978) J Electroanal Chem 89: 161
130. Lecomte JP, Kirsch-De Mesmaeker A, Feeney MM, Kelly JM (1995) Inorg Chem 34: 6481
131. Jovanovic SV, Simic MG (1986) J Phys Chem 90: 974
132. Jovanovic SV, Simic MG (1989) Biochim Biophys Acta 1008: 39
133. Lecomte JP, Kirsch-De Mesmaeker A, Kelly JM (1994) Bull Soc Chim Belg 103: 193
134. Smith SR, Neyhart GA, Kalsbeck WA, Thorp HH (1994) New J Chem 18: 397
135. Tamilarasan R, Mc Millin DR (1990) Inorg Chem 29: 2798
136. Sigman DS, Spassky A (1989) DNAse activity of 1,10-phenanthroline-copper ion. In: Eckstein F, Lilley DMJ (eds) Nucleic acids molecular biology, vol 3. Springer, Berlin Heidelberg New York, p 13
137. Goyne TE, Sigman DS (1987) J Am Chem Soc 109: 2846
138. Thederahn TB, Kuwabara MD, Larsen TA, Sigman DS (1989) J Am Chem Soc 111: 4941
139. Sigman DS, Graham DR, D'Aurora V, Stern AM (1979) J Biol Chem 254: 12,269
140. Que BG, Downey KM, So AG (1980) Biochemistry 19: 5987
141. Reich KA, Marshall LE, Graham DK, Sigman DS (1981) J Am Chem Soc 103: 3582
142. Veal JM, Rill RL (1988) Biochemistry 27: 1822
143. Sigman DS (1986) Acc Chem Res 19: 180
144. Veal JM, Rill RL (1989) Biochemistry 28: 3243
145. Sigman DS, Landgraf R, Perrin DM, Pearson L (1996) Nucleic acid chemistry of the cuprous complexes of 1,10-phenanthroline and derivatives. In: Sigel A, Sigel H (eds) Metal ions in biological systems, vol 33. Dekker, New York, p 485
146. Perrin DM, Mazumder A, Sigman DS (1996) Prog Nucleic Acid Res Mol Biol 52: 123

147. McMillin DR, Hudson BP, Liu F, Sou J, Bereger DJ, Meadows KA (1993) Adv Chem Ser 238: 211
148. Liu F, Meadows KA, McMillin DR (1993) J Am Chem Soc 115: 6699
149. McMillin DR, Liu F, Meadows KA, Aldridge TK, Hudson BP (1994) Coord Chem Rev 132: 105
150. Graham DR, Sigman DS (1984) Inorg Chem 23: 4188
151. Tamilarasan R, McMillin DR, Liu F (1989) In: Tullius TD (ed) Metal-DNA chemistry. ACS Symposium Series 402, Washington DC, p 48
152. Everly RM, Ziessel R, Suffert J, McMillin DR (1991) Inorg Chem 30: 559
153. Tamilarasan R, Ropartz S, McMillin DR (1988) Inorg Chem 27: 4082
154. Liu F, Meadows KA, McMillin DR (1993) J Am Chem Soc 115: 6699
155. Hogan M, Dattagupta N, Crothers DM (1979) Biochemistry 18: 280
156. Fritzsche H, Triebel H, Chaires JB, Dattagupta N, Crothers DM (1982) Biochemistry 21: 3940
157. Wirth M, Buchardt O, Koch T, Nielsen PE, Norden B (1988) J Am Chem Soc 110: 932
158. Barton JK (1989) Pure & Appl Chem 61: 563
159. Pyle AM, Chiang MY, Barton JK (1990) Inorg Chem 29: 4487
160. Kumar CV, Barton JK, Turro NJ (1985) J Am Chem Soc 107: 5518
161. Goldstein BM, Barton JK, Berman HM (1986) Inorg Chem 25: 842
162. Kirshenbaum MR, Tribolet R, Barton JK (1988) Nucleic Acids Res 16: 7943
163. Sitlani A, Long EC, Pyle AM, Barton JK (1992) J Am Chem Soc 114: 2303
164. Sitlani A, Barton JK (1994) Biochemistry 33: 12,100
165. David SS, Barton JK (1993) J Am Chem Soc 115: 2984
166. Collins JG, Shields TP, Barton JK (1994) J Am Chem Soc 116: 9840
167. Mahnken RE, Billabeau MA, Nikonowicz EP, Morrison H (1992) J Am Chem Soc 114: 9253
168. Kochevar IE, Dunn DA (1990) Photosensitized reactions of DNA: cleavage and addition. In: Morrison H (ed) Bioinorganic photochemistry, vol 1, New York, p 273
169. Cadet J, Vigny P (1990) The photochemistry of nucleic acids. In: Morrison H (ed) Bioinorganic photochemistry, vol 1, New York, p 1
170. Gut IG, Wood PD, Redmond RW (1996) J Am Chem Soc 118: 2366
171. Morin B, Cadet J (1994) Photochem Photobiol 60: 102
172. Bienvenu C, Wagner JR, Cadet J (1996) J Am Chem Soc 118: 11,406
173. Tuite EM, Kelly JM (1993) J Photochem Photobiol B 21: 103
174. Steenken S, Jovanovic SV (1997) J Am Chem Soc 119: 618
175. Melvin T, Botchway SW, Parker AW, O'Neill P (1996) J Am Chem Soc 118: 10,031
176. Pratviel G, Bernadou J, Meunier B (1995) Angew Chem 34: 746
177. Collins JG, Shields TP, Barton JK (1994) J Am Chem Soc 116: 9840
178. Pyle AM, Long EC, Barton JK (1989) J Am Chem Soc 111: 4521
179. Pyle AM, Morii T, Barton JK (1990) J Am Chem Soc 112: 9432
180. Campisi D, Morii T, Barton JK (1994) Biochemistry 33: 4130
181. Chow CS, Behlen LS, Uhlenbeck OC, Barton JK (1992) Biochemistry 31: 972
182. Sitlani A, Dupureur CM, Barton JK (1992) J Am Chem Soc 115: 12,589
183. Terbrueggen RH, Barton JK (1995) Biochemistry 34: 8227
184. Krotz AM, Kuo LY, Shields TP, Barton JK (1993) J Am Chem Soc 115: 3877
185. Shields TP, Barton JK (1995) Biochemistry 34: 15,037
186. Krotz AM, Hudson, BP, Barton JK (1993) J Am Chem Soc 115: 12,577
187. Sardesai NY, Zimmerman K, Barton JK (1994) J Am Chem Soc 116: 7502
188. Sardesai NY, Lin SC, Zimmerman K, Barton JK (1994) Bioconjugate Chem 6: 302
189. Barton JK, Raphael AL (1985) Proc Natl Acad Sci 82: 6460
190. Müller BC, Raphael AL, Barton JK (1987) Proc Natl Acad Sci 84: 1764
191. Kirschenbaum MR, Tribolet R, Barton JK (1988) Nucl Acids Res 16: 7943
192. Kelly JM, Tossi AB, McConnell DJ, OhUigin C (1985) Nucl Acids Res 13: 6017
193. Tossi AB, Kelly JM, (1989) Photochem Photobiol 49: 545
194. Fleisher MB, Waterman KC, Turro NJ, Barton JK (1986) Inorg Chem 25: 3551

195. Aboul-Enein A, Schulte-Frohlinde D (1988) Photochem Photobiol 48: 27
196. Sentagne C, Chambron J-C, Sauvage J-P, Paillous N (1994) J Photochem Photobiol B Biol 26: 165
197. Schoch TK, Hubbard JL, Zoch CR, Yi G-B, Sorlie M (1996) Inorg Chem 35: 4383
198. Arounaguini S, Maiya BG (1996) Inorg Chem 35: 4267
199. Kelly JM, McConnell DJ, OhUigin C, Tossi AB, Kirsch-De Mesmaeker A, Masschelein A, Nasielski J (1987) J Chem Soc Chem Comm 1821
200. Kelly JM, Feeney MM, Tossi AB, Lecomte J-P, Kirsch-De Mesmaeker A (1990) Anti-Cancer Drug Design 5: 69
201. Feeney MM, Kelly JM, Kirsch-De Mesmaeker A, Lecomte J-P, Tossi AB (1994) J Photochem Photobiol B Biol 23: 69
202. Lecomte JP, Kirsch-de Mesmaeker A, Kelly JM, Tossi AB, Görner H (1992) Photochem Photobiol 55: 681
203. Görner H, Stradowski C, Schulte-Frohlinde D (1988) Photochem Photobiol 47: 15
204. Tossi AB, Görner H, Aboul-Enein A, Schulte-Frohlinde D (1989) Free Radical Res Commun 6: 171
205. Tossi AB, Görner H, Schulte-Frohlinde D (1989) Photochem Photobiol 50: 585
206. Mei H-Y, Barton JK (1988) Proc Natl Acad Sci 85: 1339
207. Kelly JM, Tossi AB, McConnell DJ, OhUigin C, Hélène C, LeDoan T (1989) Free radicals, metal ions and biopolymers. Richelieu
208. Floyd RA, West MS, Eneff KL, Schneider JE (1989) Arch Biochem Biophys 273: 106
209. Kasai H, Yamaizumi Z, Berger M, Cadet J (1992) J Am Chem Soc 114: 9692
210. Cadet J, Berger M, Buchko GW, Joshi PC, Raoul S, Ravanat J-L (1994) J Am Chem Soc 116: 7403
211. Cullis PM, Malone ME, Merson-Davies LA (1996) J Am Chem Soc 118: 2775
212. Ravanat J-L, Berger M, Benard F, Langlois R, Ouellet R, van Lier JE, Cadet J (1992) Photochem Photobiol 55: 809
213. Jacquet L, Kelly JM, Kirsch-De Mesmaeker A (1995) J Chem SocChem Commun 913
214. Jacquet L, Davies RJH, Kirsch-De Mesmaeker A, Kelly JM (1997) (submitted for publication) J Am Chem Soc 119: 11763
215. Billadeau MA, Morrison H (1996) Photolytic covalent binding of metal complexes to DNA. In: Sigel A, Sigel H (eds) Metal ions in biological systems, vol 33. Dekker, New York, p 269
216. Harmon HL, Morrison H (1995) Inorg Chem 34: 4937
217. Billadeau MA, Morrison H (1995) J Inorg Biochem 57: 249
218. Hartwig JF, Pil PM, Lippard SJ (1992) J Am Chem Soc 114: 8292
219. Fkyerat A, Demeunynck M, Constant JF, Michon P, Lhomme J (1993) J Am Chem Soc 115: 9952
220. Lecomte JP, Kirsch-De Mesmaeker A, Demeunynck M, Lhomme J (1993) J Chem Soc Faraday Trans 89: 3261
221. Del Guerzo A, Kirsch-De Mesmaeker A, Demeunynck M, Lhomme J, (1997) J Phys Chem B 101: 7012
222. Kumar CV, Ascuncion EH (1993) J Am Chem Soc 115: 8547
223. Wilson WD, Wang YH, Kusuma S, Chandrasekaran S, Yang YH, Boykin DW (1985) J Am Chem Soc 107: 4989
224. Yao S, Bair KW, Cory M, Wilson WD (submitted for publication)
225. Mutaza Z, Zipp AP, Worl LA, Graff D, Jones WE Jr., Bates WD, Meyer TJ (1991) J Am Chem Soc 113: 5113
226. MacQueen DB, Eyler JR, Schanze KS (1992) J Am Chem Soc 114: 1897
227. Lucia LA, Burton RD, Schanze KS (1993) Inorg Chem 208: 103
228. Worl LA, Duesing R, Chen P, Della Ciana L, Meyer TJ (1991) J Chem Soc Dalton Trans 849
229. Werner U, Staerk H, (1993) J Phys Chem 97: 9274
230. Closs GL, Johnson MD, Miller JR, Piotrwiak P (1989) J Am Chem Soc 111: 3751
231. O'Reilly F, Kelly JM, Kirsch-De Mesmaeker A (1996) J Chem Soc Chem Commun 1013

232. Van Gijte O (1997) PhD thesis, Université Libre de Bruxelles
233. Kirsch-De Mesmaeker A, Ortmans I, Van Gijte O, Bannwarth W (1994) Photoreactions of polyazaaromatic mono- and bimetallic Ru(II) complexes on targeted DNA sites. 30th International Conference on Coordination Chemistry, Kyoto, p 68
234. Sigman DS, Bruice TW, Mazumder A, Sutton CL (1993) Acc Chem Res 26: 98
235. Ortmans I, Content S, Kirsch-De Mesmaeker A, Bannwarth W, Constant J.F, Defrancq E, Lhomme J (submitted)
236. Kirsch-De Mesmaeker A, Ortmans I, Bannwarth W, Hiroshima (1994) International Symposium on Molecular Recognition Involving Metal Complexes, 21–23 July, p 12
237. Murphy CJ, Arkin MR, Jenkins Y, Ghatlia ND, Bossmann SH, Turro NJ, Barton JK (1993) Science 262: 1025

Author Index Volumes 1–92